COMPARATIVE MORPHOLOGY OF RECENT CRUSTACEA

COMPARATIVE MORPHOLOGY OF RECENT CRUSTACEA

PATSY A. McLAUGHLIN

COURTESY PROFESSOR, FLORIDA INTERNATIONAL UNIVERSITY, 1977–
ADJUNCT PROFESSOR, FLORIDA ATLANTIC UNIVERSITY, 1979–

W. H. FREEMAN AND COMPANY *San Francisco*

Sponsoring Editor: Arthur C. Bartlett
Project Editor: Patricia Brewer
Copyeditor: Sean Cotter
Designers: Perry Smith and Sharon Smith
Production Coordinator: William Murdock
Illustration Coordinator: Cheryl Nufer
Compositor: Graphic Typesetting Service
Printer and Binder: The Maple-Vail Book Manufacturing Group

Library of Congress Cataloging in Publication Data

McLaughlin, Patsy A
Comparative morphology of Recent Crustacea.

Bibliography: p.
Includes index.
1. Crustacea—Anatomy. I. Title.
QL445.M25 595′.3′044 79–26066
ISBN 0–7167–1121–4

Printed in the United States of America

9 8 7 6 5 4 3 2 1

Dedicated to
Professor Dr. Lipke B. Holthuis
and the memory of
Dr. Waldo L. Schmitt (1887 – 1977)

Two outstanding carcinologists

CONTENTS

INTRODUCTION 1

Basic Characters of the Superclass Crustacea 2
Classification of the Crustacea 3

CLASS CEPHALOCARIDA 4

CLASS BRANCHIOPODA 8

Order Notostraca 8
Order Anostraca 11
Order Conchostraca 16
Order Cladocera 19

CLASS OSTRACODA 23

CLASS MYSTACOCARIDA 28

CLASS COPEPODA 32

Free-Living Copepoda 33
Order Calanoida 33
Order Harpacticoida 33
Order Cyclopoida 33
Order Misophrioida 34
Parasitic Copepoda 37
Order Notodelphyoida 37
Order Monstrilloida 37
Order Siphonostomatoida 38
Order Poecilostomatoida 38

CLASS BRANCHIURA 41

CLASS CIRRIPEDIA 44

Order Acrothoracica 44
Order Thoracica 46
Suborder Lepadomorpha 47
Suborder Verrucomorpha 51
Suborder Balanomorpha 52
Order Ascothoracica 54
Order Rhizocephala 56

CLASS MALACOSTRACA 59

Subclass Phyllocarida 59
Order Leptostraca 59
Subclass Hoplocarida 63
Order Stomatopoda 63
Subclass Eumalacostraca 70
Superorder Syncarida 70
Order Anaspidacea 70
Order Stygocaridacea 74
Order Bathynellacea 75
Superorder Peracarida 79
Order Mysidacea 79
Order Thermosbaenacea 84
Order Spelaeogriphacea 87
Order Cumacea 88
Order Tanaidacea 91
Order Isopoda 95
Free-Living Isopoda 95
Suborder Valvifera 95
Suborder Anthuridea 96
Suborder Flabellifera 96
Suborder Microcerberidea 97
Suborder Asellota 98
Suborder Phreatoicidea 98
Suborder Gnathiidea 99
Suborder Oniscoidea 100
Parasitic Isopoda 106
Suborder Epicaridea 106
Order Amphipoda 109
Gammaridea and Hyperiidea 109
Suborder Gammaridea 109
Suborder Hyperiidea 110
Ingolfiellidea and Caprellidea 114
Suborder Ingolfiellidea 114
Suborder Caprellidea 114
Superorder Eucarida 117
Order Euphausiacea 117
Order Amphionidacea 123
Order Decapoda 126
Suborder Dendrobranchiata 127
Suborder Pleocyemata 132
Infraorder Stenopodidea 132
Infraorder Caridea 133
Infraorder Astacidea 137
Infraorder Austroastacidea 142
Infraorder Palinura 143
Infraorder Anomura 144
Infraorder Brachyura 151

GENERAL CRUSTACEAN REFERENCES 159

GLOSSARY OF MORPHOLOGICAL TERMS 161

INDEX 175

PREFACE

Since Calman's monograph on crustacean morphology was published shortly after the turn of the century, our knowledge of crustacean biology has increased manyfold. Recent interest in the roles played by crustaceans in the environment has taken investigations of this highly diverse group of invertebrates out of the exclusive realms of basic science and fisheries and made them topics of applied research as well. As a result, the literature pertaining to crustaceans is not only prolific, but also widely scattered and often highly specialized and technical.

This book summarizes much of the more recently acquired basic data and the current interpretations of existing information. Its format has been developed with two purposes in mind. The first is the presentation of the material in the form of a text in crustacean biology, complete with step-by-step directions for the study of specific groups. The second is to provide a concise reference to the major morphological and anatomical characters of the higher crustacean taxa for both specialists and nonspecialists.

Many evolutionary and phylogenetic carcinological questions have not yet been resolved. Consequently not all carcinologists will agree with the classification and terminology employed in this text. With regard to the classification, I have adopted a conservative approach, preferring to await further evidence before taking the major step of recognizing the Crustacea as a distinct phylum. With regard to terminology, although I recognize and appreciate the usefulness of specialized terms in particular situations, I believe that the use of uniform terms for homologous structures reduces confusion and makes the subject matter more comprehensible to the general reader.

The field of carcinology is a vast and exciting one in which our present level of scientific knowledge is still relatively limited. I hope that this book provides a basic knowledge of the Crustacea upon which future carcinologists will build.

October 1979 Patsy A. McLaughlin

ACKNOWLEDGMENTS

I am deeply indebted to Drs. L. G. Abele, Florida State University; J. S. Garth, Allan Hancock Foundation, University of Southern California; and D. P. Henry and P. L. Illg, University of Washington, for data and specimens contributed to this study. Much valuable information has been gained by the careful critical reviews of a number of the world's finest carcinologists: Drs. J. S. Garth and R. Brusca and Ms. J. Haig, University of Southern California; Drs. T. E. Bowman, F. A. Chace, Jr., R. F. Cressey, H. H. Hobbs, Jr., B. Kensley, L. S. Kornicker, and R. B. Manning, National Museum of Natural History, Smithsonian Institution; Drs. D. P. Henry and P. L. Illg, University of Washington; Dr. A. J. Provenzano, Jr., Old Dominion University; and Dr. L. B. Holthuis, Rijksmuseum van Natuurlijke Historie, Leiden, The Netherlands; however, final responsibility for inaccuracies and omissions must rest with the author. I also wish to gratefully acknowledge the most valuable assistance of Mr. J. Garcia-Gomez, University of Miami, in acquiring much of the necessary literature and of Mrs. S. F. Treat, Florida International University, for her diligent proofreading of the manuscript. Special thanks also are due my husband, Mac, for his assistance, patience, and encouragement.

The permission granted by the following institutions, journals, publishers, and authors to adapt illustrations for which they hold the copyright is gratefully acknowledged: Albert Bonniers Forlag, Stockholm, representing *Acta Zoologica;* Akademiya Nauk SSSR, representing *Paleontologischeskii Zhurnal;* A. Asher and Company, Amsterdam; Bergen Museum; Biological Society of Washington; British Museum (Natural History); *Bulletin of Marine Science;* Lund University, representing *Lunds Universitets Arsskrift;* Cambridge University Press, London, representing the *Journal of the Marine Biological Association, United Kingdom;* National Institute of Oceanography, Wormley, representing the *Discovery Reports;* Clarendon Press, Oxford, representing the *Quarterly Journal of Microscopical Science;* E. J. Brill, representing *Crustaceana;* Geological Society of America; Hermann, Paris; *Vie et Milieu;* John Wiley and Sons; Royal Netherlands Academy of Sciences at Amsterdam; the University of Kansas Press; Veb Gustav Fisher Verlag; and Dr. Robert Hessler, Scripps Institution of Oceanography.

P.A.M.

COMPARATIVE MORPHOLOGY OF RECENT CRUSTACEA

INTRODUCTION

The superclass Crustacea is one of the largest, most diverse, and most successful groups of invertebrates. The commercial importance of numerous species of shrimps, crabs, and lobsters has long been recognized. Limnologists have been aware of the contributions made by the numerous small crustacean groups to lake and stream ecosystems. Similarly, fishery biologists have learned that the variety of planktonic crustaceans play a significant role in the food chains of fishes. More recently, benthic biologists and ecologists have discovered that a broad spectrum of benthic crustaceans serve vital roles in trophic levels and energy cycles within communities.

The literature on some of the taxa, such as the copepods and decapods, is extremely extensive, highly technical and specialized, and not readily adapted for general usage. In contrast, literature on some of the less familiar or lesser known groups is extremely meager. A few historic monographs have dealt with the morphology of the superclass; for example, H. Milne Edwards (1834–1840), Stebbing (1893), Calman (1909). More recently Moore (1969), in a synopsis of the Crustacea, has reviewed both fossil and Recent taxa; the morphology is complete in many groups, but in others is much less extensive. The second edition of Kaestner's (1969) volume on the Crustacea, translated by Levi and Levi (Kaestner 1970), is a more even treatment of the major taxa; however, although the newer taxa, such as the Cephalocarida and Mystacocarida, are included, the morphological sections add little to the information presented by earliest carcinologists, and most of the illustrations have been taken directly from these earlier publications.

The current emphasis on environmental assessment of human activities has placed an increasing burden on the biologist to identify and evaluate the faunal and floral segments of the environment properly. As a practicing carcinologist and systematist and one frequently engaged in environmental assessment projects, I have become acutely aware of the need for a single reference that would present the basic data of the various major crustacean taxa in a concise and systematic manner. This need, accompanied by the attempt to teach a course in crustacean biology, has convinced me that such a reference work would have great practical application if expanded to a laboratory teaching manual as well. Thus this book has evolved. While one of its purposes is to bring to students of crustacean biology a series of exercises that will permit them to familiarize themselves with the basic morphology of the diverse groups, another is to provide an easy reference to major structures for specialists.

Every attempt has been made to combine a uniform description of morphological structures with the specialized interpretations of carcinologists working with the various taxa. This has not always been possible, and when a choice has been required, it has been made with a preference to the basic morphological patterns. For instance, in copepods and several other major taxa, specialists have tended to disregard thoracic somites that have been incorporated into the cephalon through fusion. Where 2 thoracic somites are fused to the cephalon, the adjacent somite is morphologically the 3rd thoracic somite; however, many specialists would call it the 1st. Each of the thoracic somites (thoracomeres) basically is provided with a pair of appendages, referred to herein as *thoracopods*. In the more primitive crustacean classes there is little or no difficulty in referring to all thoracic appendages as thoracopods. However, in the Malacostraca they are most frequently referred to as pereopods and are numbered according to their relation to the cephalon and its fused thoracic somites (i.e., 1–3). If, for example, the taxon is characterized as having 8 thoracomeres with thoracopods, but the 1st thoracomere is fused to the cephalon and its pair of appendages modified as maxillipeds, the successive 7 pairs of appendages would be referred to as pereopods 1–7. In this text the term *pereopod* has been used primarily for the Eucarida. When the first 3 pairs of thoracic appendages are modified as maxillipeds, the remaining 5 pairs are referred to as pereopods. Another discrepancy in terminology may be found in the designation of cephalic appendages. In several of the major taxa, the cephalic appendages posterior to the mandibles are called either the *maxillulae* and *maxillae* or the *1st* and *2nd maxillae*. The former usage has been adopted throughout, even when it is not customarily used by practicing specialists. Similarly the terms *antennules* and *antennae* are used in preference to *1st* and *2nd antennae*.

In this text, the basic morphology of each major subdivision has been reviewed, in most instances through personal observations, examinations, and dissections. The data have been compiled and presented in systematic tabular form and in general descriptions and instructions for study. Most instructions deliberately have been generalized to permit their adaptation to a variety of representatives of the group. In some instances the illustrations of the external morphology represent specific animals, and in others a composite form has been drawn. Most major morphological structures have been illustrated as well as described. In addition, the major organ systems for at least one generalized representative of the more common or larger taxa have been described and diagrammatically illustrated. A cursory review of the larval stages of the various crustacean groups has been presented to give the reader a brief synopsis of the great diversity in larval forms occurring in the superclass.

References to some of the more classical and recent literature have been compiled for each of the principal subdivisions within the superclass. A review of the more general crustacean literature concludes the text; a glossary of technical terms is appended.

BASIC CHARACTERS OF THE SUPERCLASS CRUSTACEA

1. *Exoskeleton:* Of chitin; often thin, almost membranous and transparent, or thick, impregnated with calcium carbonate.
2. *Somite:* Mesoblastic division of body with exoskeleton usually differentiated into dorsal tergite and ventral sternite and with pair of appendages. Major divisions of body include cephalon with 5 usually fused somites and presegmental region (acron) bearing eyes; thorax with variable number of somites in primitive taxa, generally 8 in higher taxa, some or all sometimes partially or entirely fused with cephalon (cephalothorax); abdomen with variable number in primitive taxa, 6 or rarely 7 in higher taxa, excluding telson.
3. *Carapace:* Reduplication of integument of posterior dorsal border of head extending posteriorly over body and/or anteriorly over cephalon, sometimes fused with tergite(s) of one or more thoracic somites, or more extensively enclosing part or all of body (uni- or bivalved).
4. *Cephalic appendages:* Paired antennules (1st antennae); antennae (2nd antennae); mandibles; maxillulae (1st maxillae); maxillae (2nd maxillae). Cephalon sometimes bearing one or more of following additional structures: (1) rostrum; (2) nauplius or compound eyes; (3) labrum and labium; (4) epistome; (5) paired maxillipeds (modified thoracic appendages on somites usually fused with cephalon).
5. *Thoracic appendages:* Paired thoracopods, typically biramous, with outer exopod and inner endopod arising from common basal protopod, sometimes also with inner lobes (endites) or outer lobes (exites) or branchial structures (epipods). Protopod with coxa and basis, rarely also with precoxa; exopod sometimes reduced, endopod frequently segmented. See Figure 1.
6. *Abdominal appendages:* Paired biramous appendages unique to Malacostraca; first 5 pairs referred to as pleopods, 6th pair as uropods; 1 or 2 pairs often modified as gonopods or copulatory structures, particularly in males; occasionally secondarily reduced or absent.

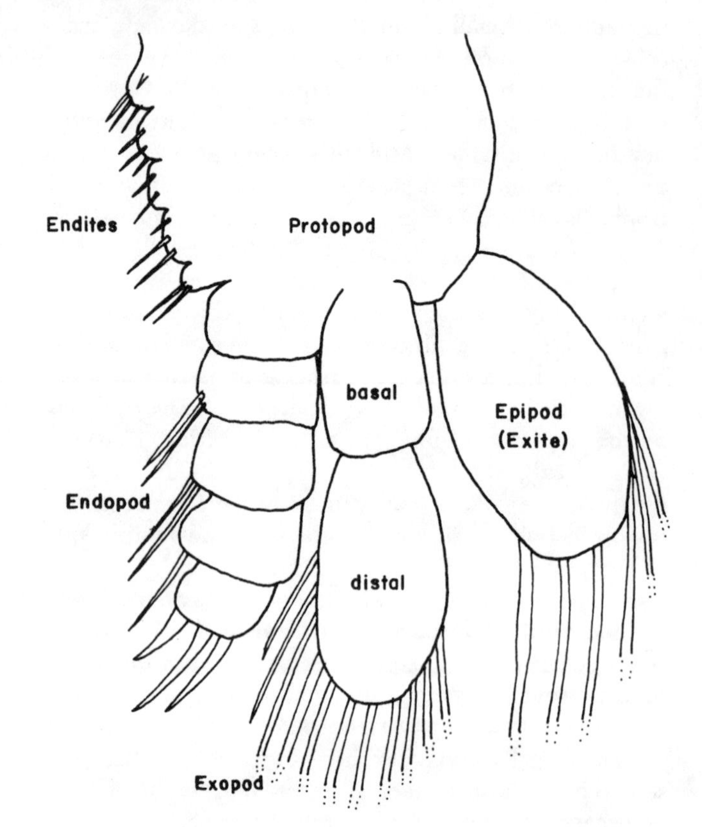

Figure 1 Diagrammatic basic crustacean appendage (thoracopod).

7. *Telson:* Terminal lobe of body, with or without appendages (caudal rami, or taken collectively, a caudal furca) in some of the more primitive taxa*; appendages absent in most of the Malacostraca. Telson not considered a true somite.

*Bowman (1971) has proposed that the "classical" interpretation of a telson is incorrect in that it does not take into consideration the fact that two presumably nonhomologous conditions exist within such a definition, that is, (1) a telson with caudal rami and a terminal anus; and (2) a telson without caudal rami and with a ventrally opening anus. Schminke (1976) has provided evidence purported to refute Bowman's proposal. In this author's opinion, the question of the true nature of the telson remains unresolved. In the text the classical definition of a telson, with or without caudal rami, has been used; however, in those taxa where Bowman disagrees, his interpretation is also indicated.

CLASSIFICATION OF THE CRUSTACEA

- Superclass Crustacea
  - Class Cephalocarida
  - Class Branchiopoda
    - Subclass Calmanostraca
      - Order Notostraca
      - Order Kazacharthra*
    - Subclass Sarsostraca
      - Order Anostraca
      - Order Lipostraca*
      - Order Enantiopoda*
    - Subclass Diplostraca
      - Order Conchostraca
      - Order Cladocera
  - Class Ostracoda
    - Subclass Archaeocopa*
    - Subclass Leperditicopa*
    - Subclass Palaeocopa*
    - Subclass Podocopa
    - Subclass Myodocopa
  - Class Mystacocarida
  - Class Copepoda
    - Order Calanoida
    - Order Harpacticoida
    - Order Cyclopoida
    - Order Misophrioida
    - Order Notodelphyoida
    - Order Monstrilloida
    - Order Siphonostomatoida
    - Order Poecilostomatoida
  - Class Branchiura
  - Class Cirripedia
    - Order Acrothoracica
    - Order Thoracica
    - Order Ascothoracica
    - Order Rhizocephala
  - Class Malacostraca
    - Subclass Phyllocarida
      - Order Leptostraca
      - Order Canadaspidida*
      - Order Hymenostraca*
      - Order Hoplostraca*
      - Order Archaeostraca*
    - Subclass Hoplocarida
      - Order Stomatopoda
      - Order Aeschronectida*
      - Order Palaeostomatopoda*
    - Subclass Eumalacostraca
      - Superorder Syncarida
        - Order Anaspidacea
        - Order Palaeocaridacea*
        - Order Stygocaridacea
        - Order Bathynellacea
      - Superorder Peracarida
        - Order Mysidacea
        - Order Thermosbaenacea
        - Order Spelaeogriphacea
        - Order Cumacea
        - Order Tanaidacea
        - Order Isopoda
        - Order Amphipoda
      - Superorder Eucarida
        - Order Euphausiacea
        - Order Amphionidacea
        - Order Decapoda

*Known only from the fossil record.

CLASS CEPHALOCARIDA Sanders, 1955

Recent species	Nine, in four genera: *Hutchinsoniella, Lightiella, Sandersiella,* and *Chiltoniella.*
Size range	2.0–3.6 mm, excluding caudal rami.
Carapace	Absent; cephalic shield present.
Eyes	Absent.
Antennules	Uniramous; used in locomotion.
Antennae	Biramous, large, and powerful; used in locomotion and food gathering.
Mandibles	Without palp.
Maxillulae	Biramous; gnathobase present.
Maxillae	Biramous; protopod with several endites (gnathobase), endopod, exopod, and pseudepipod.
Maxillipeds	None.
Thoracic appendages	Seven or 8 pairs of biramous thoracopods, generally similar in structure; 8th occasionally without endopod and gnathobase.
Abdominal appendages	None.
Telson	With caudal rami.
Tagmata	Head, thorax, and abdomen.
Somites	Head with 5; thorax with 8; abdomen with 11, excluding telson.
Sexual characters	Comon genital opening on posterior surface of protopod of each 6th thoracopod *(Hutchinsoniella).*
Sexes	Hermaphroditic.
Larval development	Anamorphic; single or pair of embryos during each breeding season; hatch as metanauplii.
Fossil record	Recent.
Feeding types	Nonselective deposit feeders.
Habitat	Species usually occur in areas where surface sediments are covered by layer of flocculent material; however, specimens recently have been collected from bare sand. *Lightiella incisa* and *L. floridana* usually are associated with *Thalassia* beds; other species occur in soft, often highly organic mud or muddy sand, from intertidal depths to approximately 1560 m. *Chiltoniella elongata* has been collected from soft mud having a slight odor of hydrogen sulfide.
Distribution	East coast of North and South America, Gulf of Mexico, and Caribbean; San Francisco Bay, California; West Africa; Peru; New Zealand; New Caledonia; Japan.

The instructions for the study of cephalocaridan morphology are based primarily on specimens of *Lightiella;* however, pertinent differences between this genus and the other three in the class will be pointed out. The paucity of study material, as well as the small size of cephalocarids, makes individual student dissections in the classroom virtually impossible. Therefore, the instructions given are for studying whole-mounted specimens and slide preparations of mouthparts and appendages. Observe the general body plan from both dorsal and ventral views before examining the individual appendages (see Figure 2).

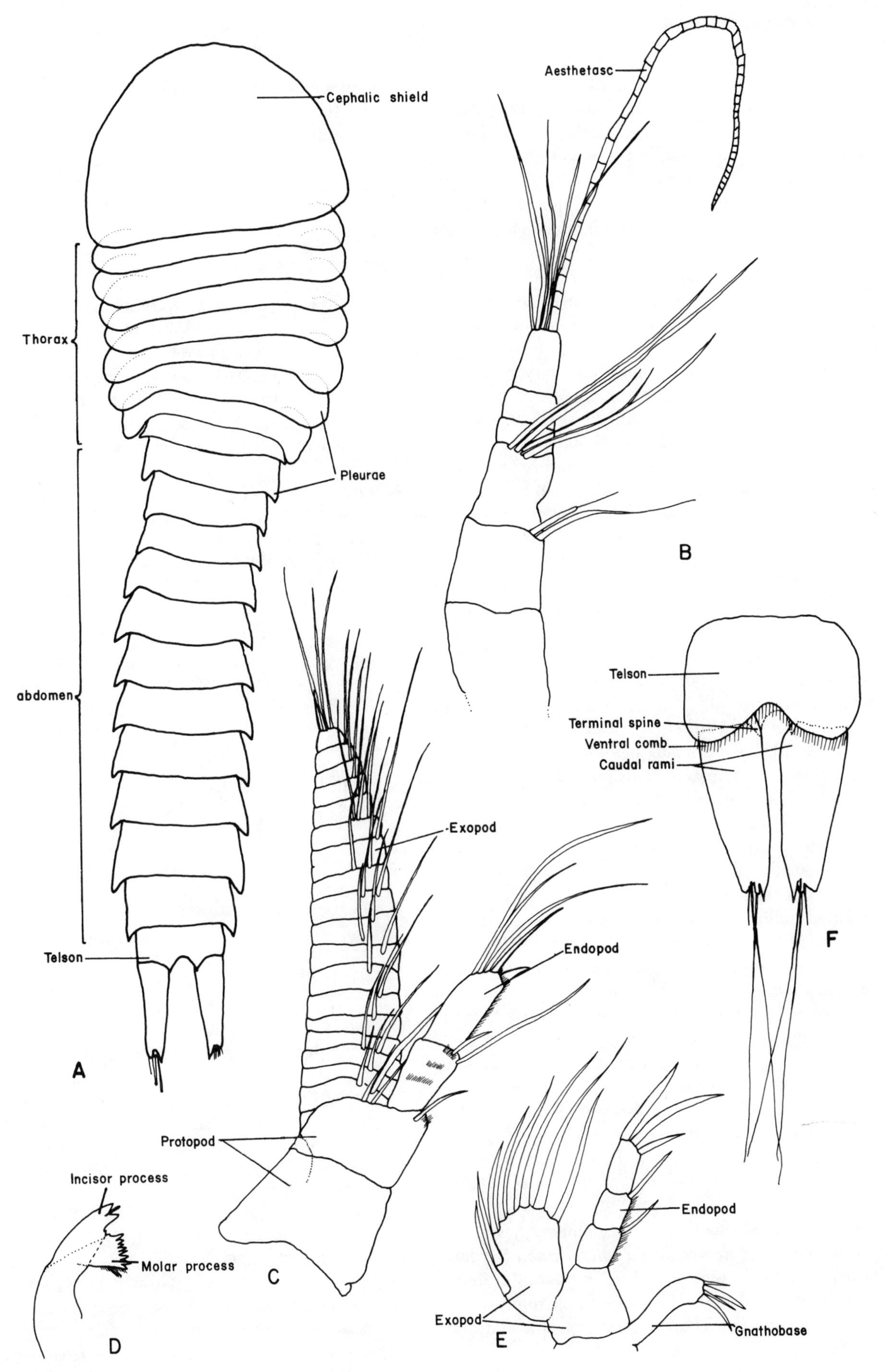

Figure 2 Cephalocarida: A. Whole animal (dorsal view); B. Antennule; C. Antenna; D. Mandible; E. Maxillule; F. Telson and caudal rami (ventral view) [from McLaughlin, 1976].

In dorsal view, an unsegmented cephalic shield formed by the fusion of the cephalic tergites partially overlaps the 1st thoracic somite. Each of the 8 thoracic somites is expanded laterally by pleurae that may be acute or rounded; the pleurae of the 8th somite are reduced much more in *Lightiella* than in the other genera. The 11 abdominal somites also bear lateral pleurae, again less well developed in species of *Lightiella*. The abdomen terminates in a telson and short *(Lightiella)* or long (other genera) caudal rami with terminal setae. [Bowman (1971) refers to the telson as an anal somite and the caudal rami as uropods.]

In ventral view, the antennules and very prominent biramous antennae are located on the inner surface of the cephalic shield anteriorly. In the midline, the bulbous labrum virtually obscures the mandibles and maxillulae. The maxillae, which lie in close approximation to the 1st thoracopods, are differentiated from them by their anterior position and typically fewer spines and setae. Note that there are only 7 pairs of thoracopods in *Lightiella*, and that the 7th pair has a reduced number of setae and endites. Eight pairs of thoracopods are present in the other three genera. In *Sandersiella* the 6th pair is somewhat modified, presumably for reproductive purposes; the 8th pair in *Chiltoniella* lacks endopods. Observe that the telson bears a comb and a pair of posteromedial spines dorsally (spines lacking in *Lightiella serendipita* Jones). In *Hutchinsoniella*, *Sandersiella*, and *Chiltoniella*, the penultimate abdominal somite as well as the telson carries a comb of bristles or setae.

Examine slide preparations of the cephalic and thoracic appendages. The uniramous antennule is 6-segmented and terminates in a sensory seta (aesthetasc), usually of considerable length. The biramous antenna consists of a 2-segmented protopod, 2-segmented endopod, and multisegmented exopod. The 2nd segment of the protopod has a naked *(Chiltoniella)* or setose *(Hutchinsoniella)* knob that is absent in the other two genera. The mandible lacks a palp and usually consists of incisor and molar processes, variously armed. The maxillule (1st maxilla) is biramous and has a well-developed gnathobase. The maxilla (2nd maxilla) and the thoracopods each consist of a protopod bearing a series of endites (usually 4–6), a 5- or 6-segmented endopod, a 2-segmented exopod, and a pseudepipod* (see Figure 3). The posteriormost thoracopods frequently have a reduced number of endites and often a 4-segmented endopod.

Stages in cephalocarid larval development have been reported only for *Hutchinsoniella macracantha* Sanders, *Lightiella incisa* Gooding, and *L. serendipita* Jones. *Lightiella* species hatch with more trunk somites than does *Hutchinsoniella*. There is pronounced anamorphic development of limbs and somites, and the transition from the metanaupliar to juvenile series occurs at the same functional level in all three taxa. The larval limbs are of the same morphology, particularly the cephalic appendages.

*An epipod, but because of its more distal point of origin on the protopod, typically referred to as a pseudepipod.

References

Cals, P., and C. Delamare Deboutteville, 1970. Une nouvelle espèce de Crustacé Céphalocaride de l'Hémisphère austral. *C. R. Acad. Sci.*, *270:* 2444–2447.

Gooding, R. U., 1963. *Lightiella incisa* sp. nov. (Cephalocarida) from the West Indies. *Crustaceana*, *5:* 293–314.

Hessler, A. Y., R. R. Hessler, and H. L. Sanders, 1970. Reproductive system of *Hutchinsoniella macracantha*. *Science*, *168:* 1464.

Hessler, R. R., 1964. The cephalocarids: Comparative skeletomusculature. *Mem. Conn. Acad. Arts Sci.*, *16:* 1–97.

Hessler, R. R., and H. L. Sanders, 1964. The discovery of Cephalocarida at a depth of 300 meters. *Crustaceana*, *7:* 77–78.

———, 1973. Two new species of *Sandersiella* (Cephalocarida), including one from the deep sea. *Crustaceana*, *24:* 181–196.

Jones, M. L., 1961. *Lightiella serendipita* gen. nov., sp. nov., a cephalocarid from San Francisco Bay, California. *Crustaceana*, *3:* 31–46.

Knox, G. A., and G. D. Fenwick, 1977. *Chiltoniella elongata* n. gen. et sp. (Crustacea: Cephalocarida) from New Zealand. *J. Roy. Soc. N.Z.*, *7:* 425–432.

McLaughlin, P. A., 1976. A new species of *Lightiella* (Crustacea: Cephalocarida) from the west coast of Florida. *Bull. Mar. Sci.*, *26:* 593–599.

Saloman, C. H., 1978. Occurrence of *Lightiella floridana* (Crustacea: Cephalocarida) from the west coast of Florida. *Bull. Mar. Sci.*, *28:* 210–212.

Sanders, H. L., 1955. The Cephalocarida, a new subclass of Crustacea from Long Island Sound. *Proc. Natl. Acad. Sci.*, *41:* 61–66.

———, 1957. The Cephalocarida and crustacean phylogeny. *Syst. Zool.*, *6:* 112–129.

———, 1963a. Significance of the Cephalocarida. In H. B. Whittington and W. D. I. Rolfe (eds.), *Phylogeny and evolution of Crustacea*, pp. 163–176. Spec. Publ. Mus. Comp. Zool. Harvard. Cambridge, Mass: Harvard University Press.

———, 1963b. The Cephalocarida: Functional morphology, larval development, comparative external anatomy. *Mem. Conn. Acad. Arts Sci.*, *15:* 1–80.

Sanders, H. L., and R. R. Hessler, 1964. The larval development of *Lightiella incisa* Gooding (Cephalocarida). *Crustaceana*, *7:* 81–97.

Seth, J. D., and W. A. Van Engle, 1969. Record of *Hutchinsoniella macracantha* Sanders, 1955 (Cephalocarida) in Virginia. *Crustaceana*, *16:* 107–108.

Shiino, S. M., 1965. *Sandersiella acuminata* gen. et sp. nov., a cephalocarid from Japanese waters. *Crustaceana*, *9:* 181–191.

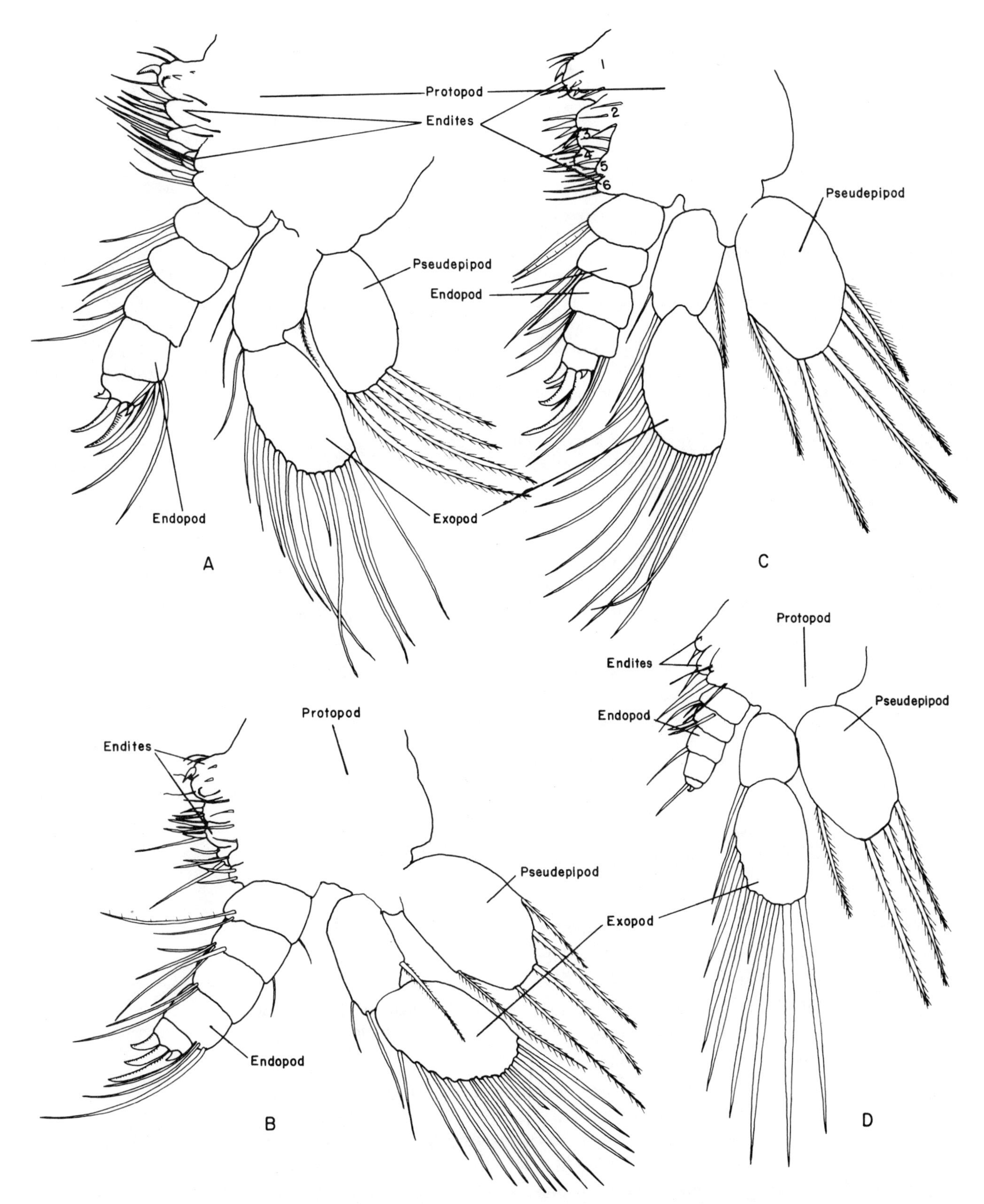

Figure 3 Cephalocarida: A. Maxilla; B. 1st thoracopod; C. 3rd thoracopod; D. 7th thoracopod [from McLaughlin, 1976].

Wakabara, Y., 1970. *Hutchinsoniella macracantha* Sanders, 1955 (Cephalocarida) from Brazil. *Crustaceana, 19:* 102–103.

Wakabara, Y., and S. M. Mizoguchi, 1976. Record of *Sandersiella bathyalis* Hessler and Sanders, 1973 (Cephalocarida) from Brazil. *Crustaceana, 30:* 220–221.

CLASS BRANCHIOPODA Latreille, 1817

The Branchiopoda includes the three subclasses Calmanostraca Tasch, 1969, Sarsostraca Tasch, 1969, and Diplostraca Gerstaecker, 1866. The Calmanostraca and Sarsostraca are represented by the Recent orders Notostraca and Anostraca respectively. The Diplostraca includes the orders Conchostraca and Cladocera. The major characters of the class, although quite variable, do distinguish it from the other crustacean classes. These characters include: (1) The presence, usually, of a carapace, in the form of a dorsal shield or bivalve shell (except in Anostraca). (2) Trunk with variable number of somites; posterior region without appendages and usually terminating in paired caudal rami.* (3) Antennules usually unsegmented. (4) Mandibles usually without palp. (5) Maxillae usually reduced, vestigial, or absent; without maxillipeds. (6) Thoracopods usually of uniform shape, generally foliaceous. (7) Eyes usually present, paired. The four orders are described and discussed individually.

*Bowman (1971) does not always agree with this interpretation.

ORDER NOTOSTRACA Sars, 1867

Recent species	Nine to twelve, in two genera, *Triops* (= *Apus*) and *Lepidurus*.
Size range	10–40 mm.
Carapace	Present, in form of dorsal shield, with cervical groove.
Eyes	Paired, sessile, compound, and nauplius eye.
Antennules	Uniramous, slender; sometimes obscurely segmented.
Antennae	Reduced, vestigial, or absent.
Mandibles	Without palp.
Maxillulae	Uniramous, small, simple; usually with spines on inner margin.
Maxillae	Varying from well developed (*Lepidurus* spp.) to reduced or absent (*Triops* spp.).
Maxillipeds	None.
Thoracic appendages	Eleven pairs of phyllopodlike thoracopods.
Abdominal appendages	Highly variable in number, often nearly twice number of body "rings."
Telson	Heavily chitinized; with segmented caudal rami; frequently with supra-anal plate.
Tagmata	Head, thorax, and abdomen.
Somites	Head with 5 somites; thorax with 11 "rings"; abdomen with variable number of "rings"; sometimes with half rings and spirals. Body rings are not homologous with somites of other crustaceans; each ring may bear more than 1 pair of appendages.
Sexual characters	Gonopores on 11th thoracic "ring"; mature female with ovisac; no marked sexual dimorphism.
Sexes	Separate or functional hermaphrodites.
Larval development	Primarily anamorphic; hatch as metanauplii.
Fossil record	Upper Carboniferous.

Feeding types — Primarily detritus feeders; sometimes bodies of dead or living larger animals, e.g., oligochaetes, mollusks, tadpoles, and frog eggs, actively gnawed.

Habitat — Temporary fresh, brackish, or saline ponds; shallow lakes, peat bogs, and moors. Desiccation required in some species to activate hatching mechanisms with return of hydration.

Distribution — Generally worldwide; in the United States not known east of Mississippi River.

The instructions for observations and dissections of notostracan specimens are based on species of *Lepidurus;* the relatively few major morphological differences between *Triops* and this genus will be pointed out. Begin your examination by observing the structures on the dorsal surface of the intact animal (see Figure 4). The head and most of the body rings of the thorax are covered by a broad, somewhat horseshoe-shaped dorsal shield or carapace. Anteriorly, in the midline, the pair of closely set, sessile compound eyes are immediately apparent beneath a capsulelike swelling of the shield. Anterior to, and slightly beneath, the compound eyes is a small simple nauplius or median eye. Directly posterior to the compound eyes, in the midline, observe the nuchal or dorsal organ (circular in *Lepidurus* and somewhat triangular in *Triops).* This structure, whose function is unknown, apparently is not homologous with the nuchal or "neck" organ of anostracans. Posterior to the dorsal organ are 2 transverse grooves: the anterior mandibular and the posterior cervical; the latter delineates the head from the thorax. A low median keel extends posteriorly from the cervical groove. Laterally, on either side of the keel, the maxillary glands can be distinguished between the layers of the carapace. Distinguish the abdomen and observe the number of abdominal body rings visible in dorsal view and the few to numerous small spines on each ring. Are any half rings or spiral rings present? The telson may or may not extend between the pair of caudal rami as a supra-anal plate. In species of *Lepidurus* this plate is well developed; it is absent from all but one species of *Triops,* and in this species it is only rudimentary. The caudal rami are segmented and often armed with minute spinules and moderately long setae. [Bowman (1971) considers that *Lepidurus* species have a telson and that most species of *Triops* have an anal somite.]

Examine the animal in ventral view. The broad, flattened, underturned anterior portion of the dorsal shield, often referred to as the subfrontal plate, medially approximates the basal portion of the very prominent labrum. Below the subfrontal plate and laterad of the labrum, identify the rather small antennules and even smaller antennae. In some species the antennae may be vestigial or entirely wanting. Observe the relationship of the mouthparts to the labrum, phyllopods, and food groove.

The first thoracic appendages are greatly modified; the subsequent phyllopods generally are all similar in structure but are progressively reduced in size. Count the number of thoracic and abdominal appendages. How does this number compare with the number of body rings? Remove and examine a 1st thoracic appendage and one of the more typical thoracic phyllopods. An example of the latter has been illustrated and labeled in Figure 4D. The 6th endite has been called an endopod by Linder (1952). Identify the corresponding structures on the 1st thoracopod. Examine the 11th pair of thoracopods. In mature females ovisacs are attached to this pair of appendages. The presence of ovisacs is the only reliable external means of sex determination in the notostracans.

Before removing the mouthparts from one side of the animal, examine the maxilla carefully to find the duct leading to it from the maxillary gland. You should also be able to identify the external opening of the gland on the maxilla. Having identified these structures, remove, examine, and sketch the maxilla and other mouthparts. In some species of *Triops* the maxillae are reduced to small lobelike structures that receive the duct from the maxillary gland. The maxillulae are small, simple, single-lobed structures. Lying between the maxillulae and the mandibles are another pair of lobelike structures that variously have been interpreted as additional lobes of the maxillulae or paragnaths. Most recent evidence suggests that the latter interpretation is correct. The massive mandibles have broadly toothed triturating surfaces.

Notostracan larvae typically hatch as metanauplii, with the first 3 pairs of appendages (antennules, antennae, and mandibles) well developed and some segmentation apparent. Segments and appendages are added progressively through a series of instars.

References

Akita, M., 1976. Classification of Japanese tadpole shrimps. *Dobutsugaku Dasshi, Zool. Mag.,* 85: 237–247.

Barnard, K. H., 1929. Contributions to the crustacean fauna of South Africa. No. 10. A revision of the South African Phyllopoda. *Ann. S. Afr. Mus.,* 29: 181–272.

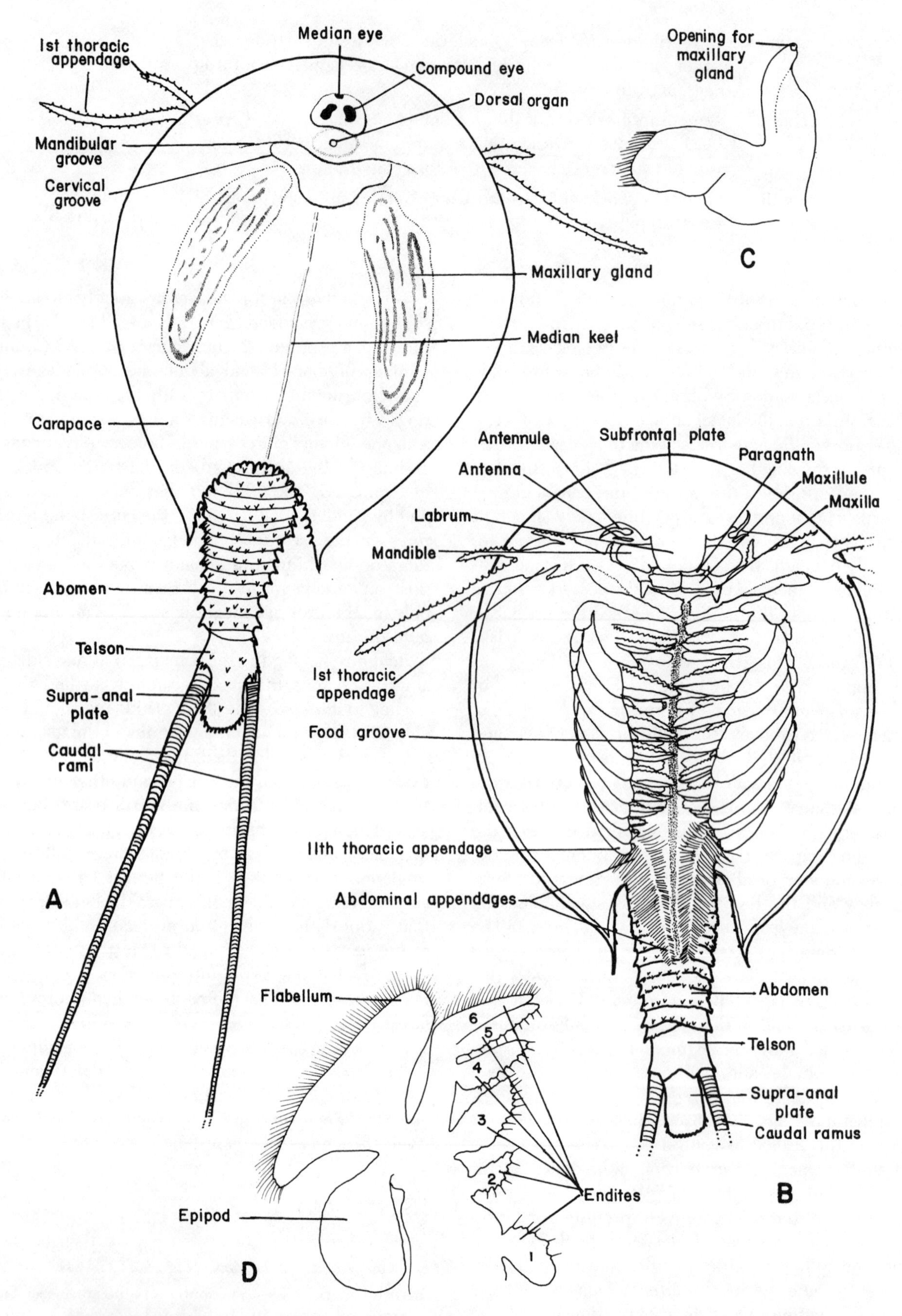

Figure 4 Notostraca: A. Whole animal (dorsal view); B. Whole animal (ventral view); C. Maxilla; D. Typical thoracopod.

Elofsson, R., 1966. The nauplius eye and frontal organs of the non-Malacostraca (Crustacea). *Sarsia,* no. 25: 1–128.

Johansen, F., 1922. Euphyllopod Crustacea of the American Arctic. *Rept. Canadian Arctic Expedition, 1913–1918, 7* (G): 3–34.

Linder, F., 1947. Abnormal body-rings in Branchiopoda Notostraca. *Zool. Bidr. Uppsala, 25:* 378–385.

———, 1952. Contributions to the morphology and taxonomy of the Branchiopoda Notostraca with special reference to North American species. *Proc. U.S. Natl. Mus., 102:* 1–69.

———, 1959. Notostraca. In W. T. Edmondson (ed.), *Freshwater biology,* 2nd ed., pp. 572–576. New York and London: John Wiley and Sons.

Longhurst, A. R., 1954. Reproduction in Notostraca (Crustacea). *Nature,* London, *173:* 781–782.

———, 1955a. Evolution in the Notostraca. *Evolution, 9:* 84–86.

———, 1955b. A review of the Notostraca. *Brit. Mus. (Nat. Hist.) Zool., Bull., 3:* 1–54.

Takahashi, F., 1977. Pioneer life of the tadpole shrimps, *Triops* spp. (Notostraca: Triopsidae). *Appl. Entomol. Zool., 12:* 104–117.

ORDER ANOSTRACA Sars, 1867

Recent species	Approximately 175, in seven families.
Size range	Up to 1 cm in length.
Carapace	Absent.
Eyes	Paired, stalked compound; median nauplius eye often present.
Antennules	Uniramous; short, thin, frequently unsegmented.
Antennae	Uniramous; usually reduced in females; stout and usually 2-segmented, often with accessory appendages in males; basal segments sometimes coalesced forming clypeus or frontal plate.
Mandibles	Without palp, except in *Polyartemia.*
Maxillulae	Uniramous; usually small, simple.
Maxillae	Usually reduced or vestigial.
Maxillipeds	None.
Thoracic appendages	Usually 11, occasionally 17–19, pairs of phyllopod-type, biramous thoracopods; generally similar in structure; posterior pair(s) sometimes reduced.
Abdominal appendages	None.
Telson	With long, setose caudal rami.
Tagmata	Head, thorax, and abdomen.
Somites	Head with 5; thorax with 11–19; abdomen with 8, excluding telson.
Sexual characters	Gonopores of both sexes on coalesced 1st and 2nd abdominal somites; females with ovisac produced, males with paired penes. Antennae of males usually well developed and modified for clasping females during copulation.
Sexes	Separate; however, parthenogenesis may occur.
Larval development	Primarily anamorphic; hatch as metanauplii, or nauplii.
Fossil record	Lower Devonian.
Feeding types	Filter feeders, primarily on plankton, occasionally predatory.
Habitat	Small, temporary, alkaline or freshwater pools; occasionally also in very saline lakes.
Distribution	Worldwide.

The instructions for the study of anostracan morphology are based on a species composite. Not all characters discussed necessarily will be found in a single taxon. Anostracans in general exhibit marked sexual dimorphism, so be sure to examine specimens of both sexes. Observe the overall body structure and segmentation before beginning the dissection (see Figure 5).

The long, clearly segmented body is divided into a head, thorax (bearing phyllopod thoracopods), and abdomen. The first 2 abdominal somites are coalesced and bear reproductive appendages. The abdomen terminates in a telson bearing a pair of caudal rami. [Once again, Bowman (1971) considers that the anostracan body terminates in an anal somite and that its appendages are uropods.]

The head, which is slightly swollen, is clearly distinguishable from the thorax. Laterally the head bears a pair of prominent, stalked compound eyes; medially and distally a small sessile nauplius eye frequently can be observed. Middorsally a nuchal ("neck") organ, presumably a sensory structure, usually is apparent through the integument. The segmentation of the head is indicated primarily by its appendages; however, the presence of a mandibular groove gives the superficial appearance of segmentation. The 1st pair of cephalic appendages, the antennules, are uniramous, thin, generally unsegmented (although frequently there may be a superficial suggestion of segmentation), and often bear terminal setae. The 2nd pair of appendages, the uniramous antennae, usually are reduced, often significantly so, in females. In males they exhibit considerable diversity in form but with few exceptions (Polyartemiidae) consist of 2 segments. The basal segments are short, cylindrical, and frequently have sensory hairs and outgrowths, the antennal appendages. In some taxa the basal segments fuse medially to form a platelike structure, the frontal appendage or clypeus. In addition to the prominent, posteriorly directed labrum that covers the mouth ventrally, the mouthparts include a pair of massive mandibles, a pair of moderately small and simple maxillulae, and a pair of even more reduced maxillae. Only the mandibles can be observed easily without dissection.

The thorax consists of 11 somites except in species of *Polyartemia* (19) and *Polyartemiella* (17). Each somite carries a pair of phyllopod appendages; however, in the female of a few species, the posteriormost pairs are reduced or absent. The somites are concave ventrally between the bases of the thoracopods, forming a longitudinal food groove.

The abdomen consists of 8 somites, excluding the telson; however, in a few taxa the division between the 8th somite and the telson is indistinct. The 2 anterior abdominal somites, which are partially or completely coalesced, bear reproductive appendages: in the male a pair of penes; in the female a single ovisac. The successive pretelsonic somites lack appendages. The unsegmented caudal rami frequently are fringed with long plumose setae. In a few species the telson is expanded to form a structure reminiscent of the supra-anal plate present in some notostracans.

Having observed the general morphology of both males and females, remove, examine, and sketch one or more of each type of appendage. The antennules, located in close proximity to the stalked eyes, are removed most easily from the dorsal side. Suction discs, hooks, and other modifications for grasping the females may be observed on the antennae of males. Examine these appendages carefully, also noting any frontal appendages, sensory setae, or other copulatory adaptations. Is a clypeus present? With the antenna removed from one side, the large, bulbous labrum can be observed. It need not be removed but may have to be pulled anteriorly to facilitate dissection of the other mouthparts. To either side of the labrum, and partially covered by it, are the prominent mandibles. With the exception of *Polyartemia*, the mandibles are without palps. Remove one of the mandibles and examine the triturating surface. Although anostracans in general are reported to lack paragnaths, the structures between the mandibles and maxillulae described by Lynch (1937) for *Branchinecta gigas* Lynch might well be interpreted as paragnaths. Similar structures are present in several other taxa as well, and apparently do not represent endites of the maxillulae. The latter appendages are relatively simple in form and usually bear a series of filtering spines or setae. The maxillae are either vestigial or reduced to a small simple lobe situated near the base of each 1st phyllopod.

Remove several of the phyllopods and identify the various elements. Are there structural differences between anterior and posterior pairs? How do anostracan thoracopods compare with the thoracopods of cephalocarids?

Examine the coalesced 1st and 2nd abdominal somites and the associated appendages in both sexes. The paired oviducts of the female are united and expanded to form the unpaired ovisac. The penes are eversible, thus they may appear broken or damaged. When fully extended, each apical segment usually is elongate, cylindrical, and frequently armed with serrations, denticulations, or spines. From each penis the looped vas deferens usually can be traced to the basal segment of the appendage. Characters afforded by the male, such as antennae, penes, and seminal vesicles, are some of the primary characters used in identification of taxa.

Using cleared specimens, examine the major internal organ systems (see Figure 6). Dorsally throughout the length of the body is the heart, with 14–18 pairs of segmental ostia. Blood enters the heart through the ostia from the underlying pericardial sinus and flows anteriorly. There are no arteries in anostracans; blood exits from the heart into the hemocoele through an anterior cardiac opening. The pericardial sinus is separated from

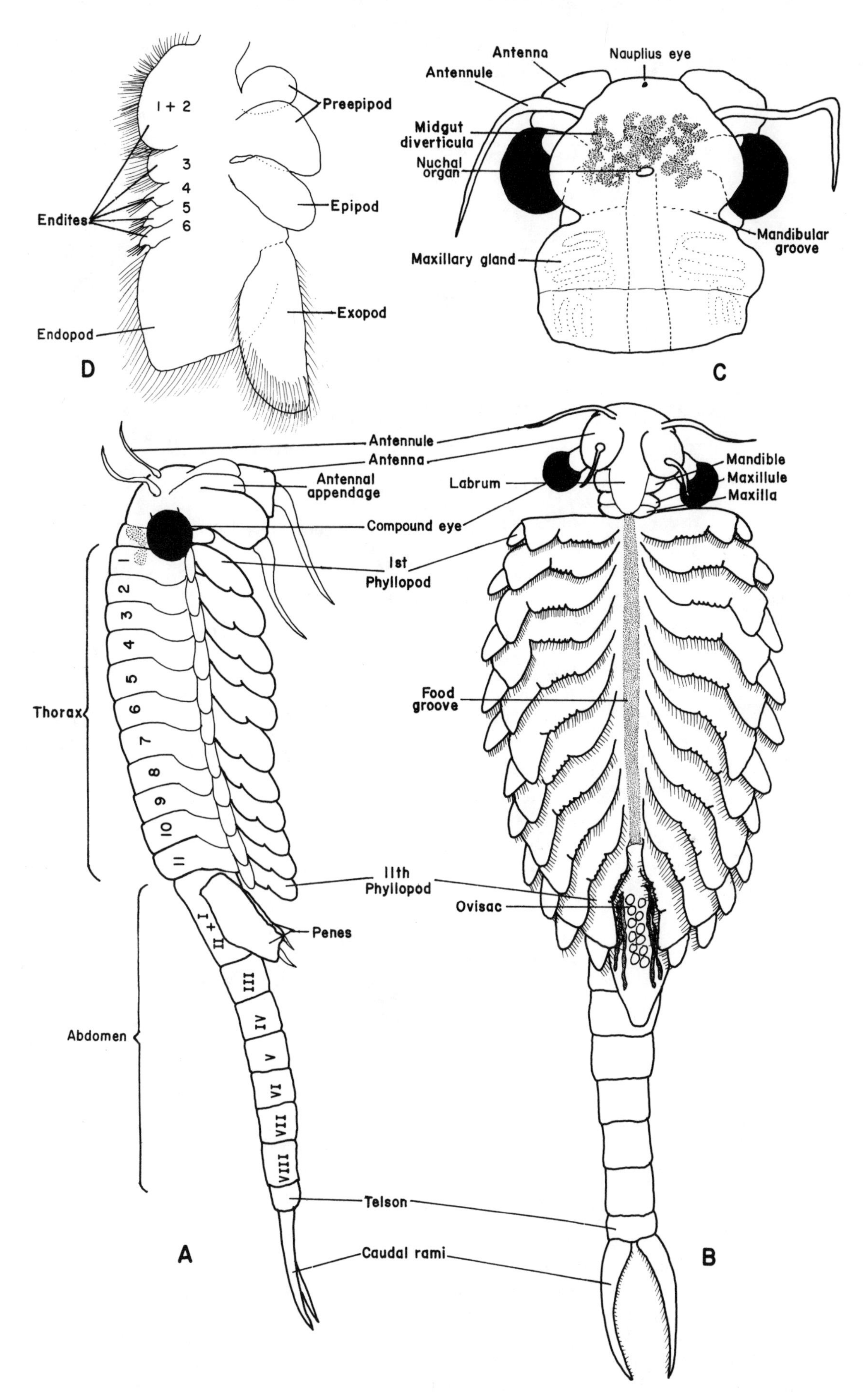

Figure 5 Anostraca: A. Male (lateral view); B. Female (ventral view); C. Cephalic region and 1st thoracic somite (dorsal view); D. Typical thoracopod.

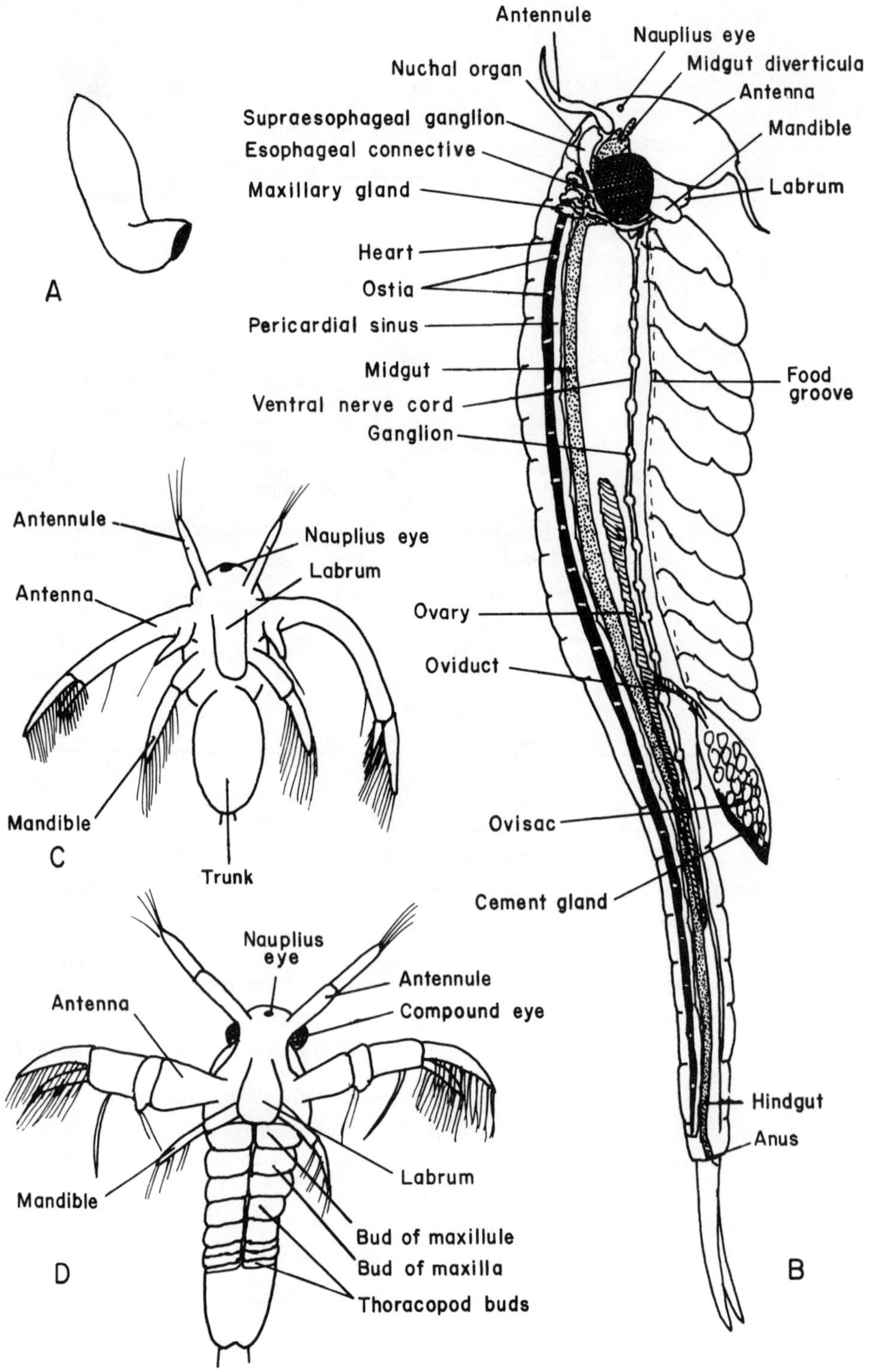

Figure 6 Anostraca: A. Left mandible; B. Diagrammatic anostracan in lateral view, illustrating major organ systems; C. Nauplius; D. Metanauplius [C and D after Snodgrass, 1956].

the hemocoele by a pericardial septum, but a series of openings provide for the return of blood to the sinus.

Directly beneath the heart and pericardial sinus, identify the midgut that extends almost the entire length of the body. Posteriorly the very short hindgut terminates in an anal opening located between the bases of the caudal rami. Food passes from the mouth through the short, vertically directed esophagus to the midgut; 2 cecalike diverticula, located in the head, join the midgut in the postesophageal region. Laterally and posteriorly the maxillary glands, which extend into the 1st thoracic somite, can be distinguished. The antennal glands, which in most crustaceans serve an excretory function, are present in larval anostracans but soon degenerate and are replaced in the adult by the maxillary glands. The ducts of these glands open between the maxillulae and the 1st pair of thoracopods at the level of the vestigial or reduced maxillae.

Located between the lobes of the midgut diverticula, several sensory structures should be identified. In the dorsal midline observe the previously mentioned nuchal organ and more anterior nauplius eye consisting of 2 lateral and 1 ventral cups. In close proximity to the latter are the X-organ and paired ventral frontal organs; however, these structures will not be apparent without special staining. The latter structure is similar to but apparently not a homologue of the ventral frontal organ of the Malacostraca. Posteroventral to the nuchal organ is the supraesophageal ganglion and the esophageal connectives, which encircle the esophagus. As these nerve elements may be obscured by the midgut diverticula or the compound eyes, it is advisable to open the chitinous exoskeleton of the head and carefully remove or push aside the ceca. Trace the esophageal connectives to their junction with the subesophageal ganglia and ventral nerve cord. Ganglia are present in each somite through the 2nd postgenital somite.

The testes usually extend from the 3rd or 4th abdominal somite anteriorly into the 1st genital somite. Near the point of fusion of the 2 genital somites the paired vas deferens leave the testes and are directed ventrally and posteriorly to the penes. A large distention of the vas deferens near its junction with the penis serves as a seminal vesicle. The ovaries usually originate in the 4th or 5th abdominal somite and extend anteriorly to the 8th, or more often, 7th thoracic somite. The oviducts arise near the middle of the fused genital somite, or near the boundary between the 2 genital somites in instances where fusion has not occurred, and lead into the ovisac. The latter frequently will have a pair of cement glands dorsal and proximal to the egg mass.

Development in anostracans generally is anamorphic, with hatching occurring as either the nauplius or metanauplius. The naupliar stage is characterized by the presence of 3 pairs of appendages (antennules, antennae, and mandibles) and an unsegmented trunk. At the metanaupliar stage, not only are the 3 primary appendages present, but buds of the maxillulae, maxillae, and several thoracopods as well. Changes beyond the 1st metanaupliar stage are gradual and often between 14 and 17 instars are passed through before adulthood is reached.

References

Barnard, K. H., 1924. Contributions to a knowledge of the fauna of South-West Africa. II. Crustacea Entomostraca, Phyllopoda. *Ann. S. Afr. Mus.*, *20:* 213–230.

———, 1929. Contributions to the crustacean fauna of South Africa. No. 10. A revision of South African Branchiopoda (Phyllopoda). *Ann. S. Afr. Mus.*, *29:* 181–272.

Bond, R. M. 1934. Report on phyllopod Crustacea (Anostraca, Notostraca, Conchostraca) including a revision of the Anostraca of the Indian Empire. [Yale North India Expedition] *Mem. Conn. Acad. Arts Sci.*, *10:* 26–62.

Cannon, H. G., and F. M. C. Leak, 1933. VIII. On the feeding mechanism of the Branchiopoda. *Phil. Trans. Roy. Soc. London*, (B) *222:* 267–352.

Creaser, E. P., 1929. The Phyllopoda of Michigan. *Papers Mich. Acad. Sci.*, *11:* 381–388.

Daborn, G. R., 1977. On the distribution and biology of an arctic fairy shrimp *Artemiopsis stefanssoni* Johansen, 1921 (Crustacea: Anostraca). *Can. J. Zool.*, *55:* 280–287.

Daday, E., 1910. Monographie systématique des Phyllopodes Anostracés. *Ann. Sci. Nat.*, (9) *11:* 91–489.

Dexter, R. W., 1953. Studies on North American fairy shrimps with the description of two new species. *Am. Midland Natur.*, *49:* 751–777.

———, 1959. Anostraca. In W. T. Edmondson (ed.), *Freshwater biology,* 2nd ed., pp. 558–571. New York and London: John Wiley and Sons.

Elofsson, R., 1966. The nauplius eye and frontal organs of the non-Malacostraca (Crustacea). *Sarsia,* No. 25: 1–128.

Fryer, G., 1966. *Branchinecta gigas* Lynch, a nonfilter feeding raptatory anostracan, with notes on the feeding habits of certain other anostracans. *Proc. Linn. Soc. London, 177:* 19–34.

Heath, H., 1924. External development of certain phyllopods. *J. Morph.*, *38:* 453–483.

Johansen, F., 1922. Euphyllopoda Crustacea of the American Arctic. *Rept. Canadian Arctic Expedition, 1913–18,* 7 (G): 1–34.

Linder, F., 1941. Contributions to the morphology and the taxonomy of the Branchiopoda Anostraca. *Zool. Bidr. Uppsala, 20:* 101–302.

Lynch, J. E., 1937. A giant new species of fairy shrimp of the genus *Branchinecta* from the state of Washington. *Proc. U.S. Natl. Mus.*, *84:* 555–562.

———, 1960. The fairy shrimp *Branchinecta campestris* from the northwestern United States (Crustacea: Phyllopoda). *Proc. U.S. Natl. Mus.*, *112:* 549–561.

———, 1964. Packard's and Pearse's species of *Branchinecta:* Analysis of a nomenclatural involvement. *Am. Midland Natur.*, *71:* 466–488.

Packard, A. S., 1883. A monograph of the phyllopod Crustacea of North America with remarks on the order Phyllocarida. *Twelfth Ann. Rept. U.S. Geol. Surv. Terr., 1878,* Sect. 2: 295–592.

Pearse, A. S., 1918. The fairy shrimps (Phyllopoda). In H. B. Ward and G. C. Whipple (eds.), *Fresh-water biology,* 1st ed., pp. 661–665. New York: John Wiley and Sons.

Shantz, H. L., 1905. Notes on the North American species of *Branchinecta* and their habitats. *Biol. Bull.*, *9:* 249–259.

Verrill, A. E., 1870. Observations on phyllopod Crustacea of the family Branchipidae with descriptions of some new genera and species from America. *Proc. Am. Assoc. Adv. Sci.*, 18th Meeting, 1869, pp. 230–247.

ORDER CONCHOSTRACA Sars, 1867

Recent species	Approximately 180 in six families.
Size range	Approximately 5–17 mm.
Carapace	Bivalve, laterally compressed, with or without growth lines; completely enclosing head, body, and appendages.
Eyes	Sessile, compound, dorsal, and usually closely set; and single nauplius eye.
Antennules	Uniramous, short; unsegmented or with 2 segments, usually with series of dorsal sensory papillae.
Antennae	Biramous; each ramus variously segmented according to taxa.
Mandibles	Without palp.
Maxillulae	Small, simple single lobe.
Maxillae	Vestigial or absent.
Maxillipeds	None.
Thoracic appendages	Thorax and abdomen not delimited; 10–32 pairs of trunk appendages.
Abdominal appendages	As above; generally phyllopod in structure, progressively decreasing in size posteriorly.
Telson	Broad, truncate, with pair of clawlike caudal rami or cercopods.
Tagmata	Head and trunk (postcephalic body).
Somites	Head with 5, trunk with 10–32 somites, excluding telson.
Sexual characters	Gonopores of both sexes on 11th trunk somite. Males with modified 1st and/or 2nd postcephalic appendages; carapace often exhibiting sexual dimorphism.
Sexes	Generally separate, but with some parthenogenetic reproduction.
Larval development	Primarily anamorphic; hatch as nauplii or metanauplii.
Fossil record	Lower Devonian.
Feeding types	Filter feeders on detritus and plankton.
Habitat	Free swimming in littoral regions of lakes, ponds, and temporary freshwater pools. Most seem to prefer warmer water than anostracans.
Distribution	Broadly distributed through temperate regions; however, certain species may be restricted to type localities or have very restricted distributional patterns.

The instructions for observations and dissections of conchostracans are based on specimens of *Caenestheriella*. Differences between this genus and the others in the family Cyzicidae are in the strength of the occipital point, the shape of the rostrum, and the number of antennal articles. Major taxonomic differences between the other conchostracan families will be pointed out when such differences are pertinent to the study of the general morphology of the group.

The conchostracan carapace is a bivalve shell, hinged in the midline, that encloses the head, trunk, and appendages (see Figure 7). A distinct umbonal region usually can be distinguished. Except in the Lynceidae, the shell is marked with a varying number of concentric lines, referred to as growth lines. As the carapace is not shed with the animal's exoskeleton during molting, the growth lines may reflect successive molts. The carapace is attached to the body anteriorly by an attachment ligament in the region of the umbones and by a pair of adductor muscles. By cutting the adductor muscle and ligament on one side, one valve of the carapace can be removed to expose the animal.

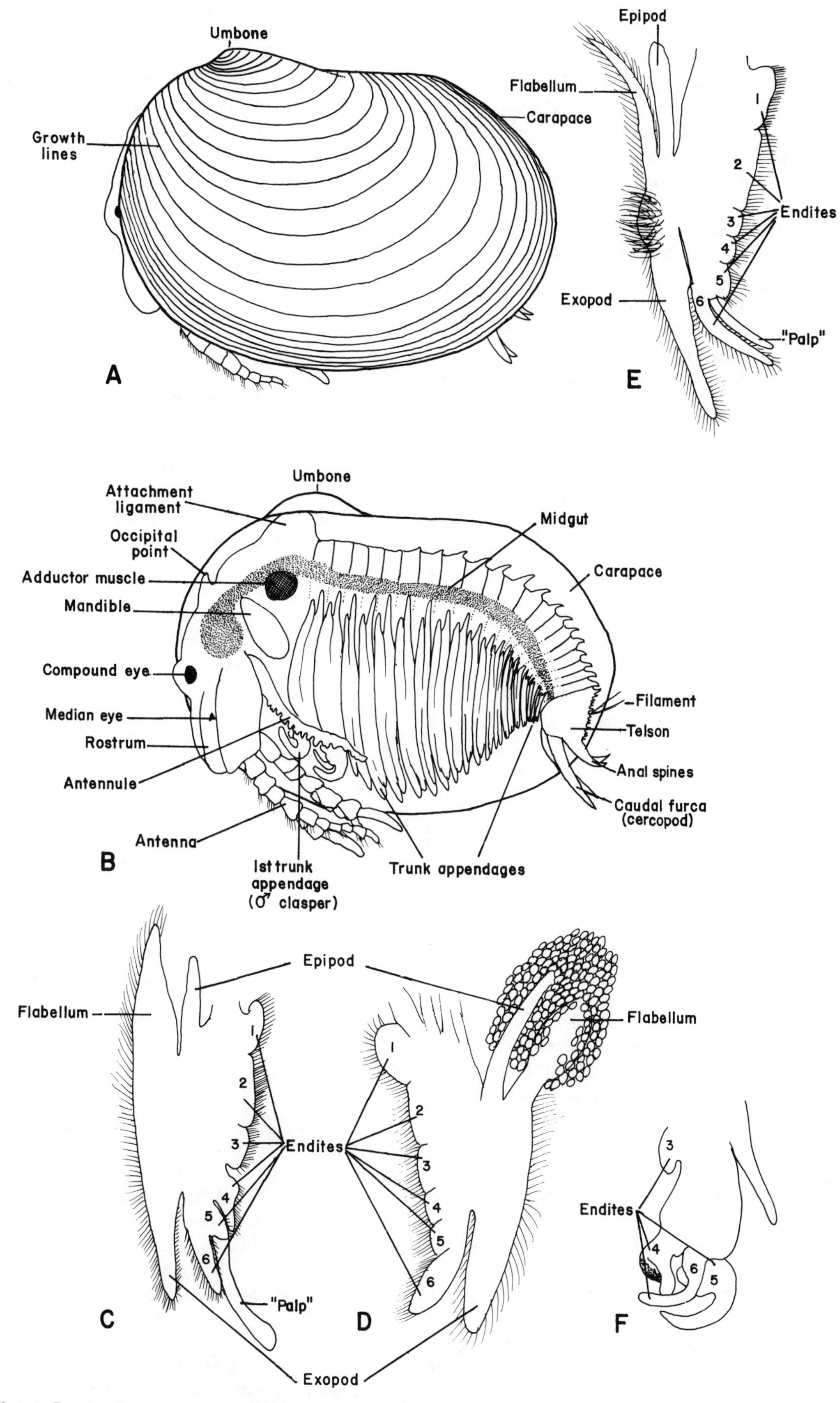

Figure 7 Conchostraca: A. Whole animal (lateral view); B. Animal with left valve removed; C. 10th thoracopod of male; D. 10th thoracopod of female; E. 1st thoracopod of female; F. 1st thoracopod of male.

The conchostracan body is divided into 2 parts, the head (cephalon) and trunk (postcephalic body). Several structures of the head will be immediately apparent. The sessile compound eyes usually are closely set but in a few taxa may be fused to form a single large eye. The median or nauplius eye is small and frequently triangular in shape. One of the most conspicuous structures of the head is the usually well-developed rostrum, which in many species is pointed, spatulate, and notched and may have a terminal spine. The occipital point and notch are not well developed in some species of *Caenestheriella* but in other taxa often are very prominent. In the family Limnadiidae a pyriform frontal organ projects from the middorsal surface of the head. This structure appears to be associated with the nauplius eye (cf. Elofsson, 1966). Note the moderately short uniramous antennules, the very well-developed biramous antennae, the prominent mandibles, and the small maxillulae. The abdomen is not distinguishable; the postcephalic body consists of 10 to 32 somites, excluding the telson. Each somite carries a pair of appendages, usually phyllopod in structure, but in males the 1st or the 1st and 2nd pairs are modified as claspers to grasp the female during copulation. In many species the dorsal surfaces of the posteriormost somites are drawn out into a series of spinelike projections. The telson is a broad structure terminating in a pair of caudal rami (cercopods or furcal claws). [Bowman (1971) considers only the triangular terminal lobe of the ultimate segment the telson; the remainder is an anal somite, and its appendages are uropods.] The morphological dorsal surface of the telson appears to be the posterior margin because of the flexure of the trunk. Usually it is armed with a series of spines; the most prominent posterior ones are referred to by some authors as "anal spines." A biramous filament also is present.

Begin your detailed examination by first removing the antenna and antennule from one side of the body. The biramous antenna has a stout protopod and moderately long, segmental rami. The antennule is uniramous, usually unsegmented, but may have 2 segments, with a series of dorsal sensory papillae. The chitinous mandibles are massive; the toothed margins usually are hidden by the labrum. The very small, single-lobed maxillulae carry only a few terminal setae. The maxillae are vestigial or absent.

Sexual dimorphism is most clearly expressed in the trunk appendages, so be sure to examine these appendages in both sexes. The 1st, sometimes the 1st and 2nd, pair in the male usually is modified for grasping. Compare one of these appendages with one of the more typical phyllopod appendages that follow. Identify the endites, epipods, and exopods of each. In the female the 1st and 2nd pairs of phyllopods are unmodified; however, modifications for egg carrying occur on the 9th to 11th pairs. The eggs are cemented in raftlike fashion to the slender and elongate flabellum and epipod. How do the trunk appendages of conchostracans compare with the thoracopods of the anostracans and notostracans?

The conchostracan nauplius is atypical in that the antennules are reduced or vestigial. The shell typically is produced by a pair of integumental folds in the metanauplius; however, the nauplii of some have a broad dorsal shield resembling that of the Notostraca. The later larval stages frequently resemble the adult structure of the Cladocera.

References

Baird, W., 1849. Monograph of the family Limnadiidae, a family of entomostracous Crustacea. *Proc. Zool. Soc. London*, Pt. 17: 84–90.

Barnard, K. H., 1929. A revision of the South African Branchiopoda (Phyllopoda). *Ann. S. Afr. Mus.*, *20:* 181–272.

Brehm, V., 1933. Phyllopoden. Mitteilungen von der Wallacea-Expedition Woltereck., 5. *Zool. Anz.*, *104:* 31–40.

Daday, E., 1915. Monographie systématique des phyllopodes conchostracés. *Ann. Sci. Nat. Zool.*, (9) *20:* 39–330.

———, 1927. Monographie systématique des phyllopodes conchostracés. *Ann. Sci. Nat. Zool.*, (10) *10:* 1–112.

Elofsson, R., 1966. The nauplius eye and frontal organs of the non-Malacostraca. *Sarsia*, no. 25: 1–128.

Johansen, F., 1922. Euphyllopod Crustacea of the American Arctic. *Rept. Canadian Arctic Expedition, 1913–18*, 7 (G): 1–34.

Linder, F., 1945. Affinities within the Branchiopoda, with notes on some dubious fossils. *Ark. Zool.*, *37* (A): 1–28.

Mattox, N. T., 1957. A new estheriid conchostracan with a review of other North American forms. *Am. Midland Natur.*, *58:* 367–377.

———, 1959. Conchostraca. In W. T. Edmondson (ed.), *Fresh-water biology*, 2nd ed., pp. 577–586. New York and London: John Wiley and Sons.

Novojilov, N., 1958. Recueil d'articles sur les Phyllopodes Conchostracés. *Ann. Serv. Inform. Geól. B.R.G.G.M.*, *26:* 1–135.

Packard, A. S., 1875. Synopsis of the fresh-water phyllopod Crustacea of North America. *Ann. Rept. U.S. Geol. Geogr. Surv.*, *1873:* 613–622.

———, 1883. A monograph of the phyllopod Crustacea of North America. *12th Ann. Rept. U.S. Geol. Geogr. Surv.*, *1878* (sec. 1): 295–592.

Sars, G. O., 1896. Phyllocarida og Phyllopoda. Fauna Norvegiae, *Vidensk. Selsk. Forhandl. Kristiana*, *1:* 1–140.

Smirnov, S., 1936. Zweiter Beitrag zur Phyllopodenfauna Transkaukasiens. *Zool. Anz.*, *113:* 311–320.

———, 1949. New species of the genus *Eulimnadia* Packard from Uzbekistan. *C. R. Acad. Sci. Moscow* (n.s.) *67:* 1159–1162.

Šrámek-Hušek, R., M. Straškraba, and J. Brtek, 1962. *Lupernonozci (Branchiopoda).* Fauna CSSR, *16:* 1–470.

Stingelin, Th., 1908. *Phyllopodes.* Catalogue des Invertébrés de la Suisse, *2:* xi + 156 pp. Genève.

Straškraba, M., 1965. Taxonomic studies on Czechoslovak Conchostraca, 1. Family Limnadiidae. *Crustaceana, 9:* 263–273.

Uéno, M., 1927. The freshwater Branchiopoda of Japan, 1. *Mem. Coll. Sci. Kyoto Imp. Univ.*, (B) *2:* 259–311.

———, 1967. Two new species of Conchostraca (Branchiopoda) from Nepal and Iran. *Crustaceana, 13:* 249–256.

Wootton, D. M., and N. T. Mattox, 1958. Notes on the distribution of conchostracans in California. *Bull. S. Cal. Acad. Sci., 57:* 122–124.

Zwolski, W., 1959. Materialy do znajmosci liscionogow wláscivych (Euphyllopoda) Polski. *Ann. Univ. Mariae-Curie Sklodowska*, (C) *11:* 1–23.

ORDER CLADOCERA Latreille, 1829

Recent species	Approximately 420 in eleven families.
Size range	0.2–18 mm.
Carapace	Univalve, laterally compressed, and folded over; usually enclosing trunk and appendages; occasionally reduced.
Eyes	Sessile compound, coalesced, medial; occasionally absent; often with small nauplius eye; occasionally only with latter.
Antennules	Uniramous, usually short, slender, with tuft of terminal sensory setae; occasionally modified in males.
Antennae	Biramous, strong; principal appendages of locomotion.
Mandibles	Without palp.
Maxillulae	Small, simple.
Maxillae	Occasionally rudimentary or vestigial; usually absent.
Maxillipeds	None.
Thoracic appendages	Thorax and abdomen not readily delimited; 4–6 pairs of biramous trunk appendages, usually foliaceous; in some predatory species modified for grasping prey.
Abdominal appendages.	See above.
Telson	Usually not delimited; typically with clawlike caudal furca; occasionally with long caudal process or appendix.
Tagmata	Head and trunk; posterior portion of trunk recurved (postabdomen).
Somites	Variable, not easily distinguished.
Sexual characters	Genital ducts opening laterally on "thoracic" portion of trunk. Antennules usually more developed in males, occasionally modified for clasping female; rostrum sometimes exhibiting sexual dimorphism.
Sexes	Sometimes separate; parthenogenetic reproduction most common.
Larval development	Primarily epimorphic; in *Leptodora*, released from brood pouch as metanauplii.
Fossil record	Oligocene to Recent.
Feeding types	Primarily filter feeders; occasionally particulate feeders or predators.
Habitat	Generally restricted to fresh or brackish waters, occasionally marine. Representatives occasionally also found in acid bogs or mud and saline lakes.
Specialization	Cyclomorphosis, i.e., seasonal changes in morphology that drastically alter individual appearances.
Distribution	Worldwide.

The instructions for observations and dissections of the Cladocera are based on species of *Daphnia*. Cladocerans do exhibit a considerable amount of interspecific morphological variation; however, with the exceptions of members of the families Leptodoridae and Polyphemidae, they generally follow the patterns described for species of *Daphnia*. Your attention will be directed to any major exceptions. If living specimens are not available for observation, some cleared or stained specimens also should be available, as the opaqueness of most preserved material makes study of the internal anatomy difficult. Some staining techniques, such as chlorazol black in lactophenol, frequently will distort the carapace.

Parthenogenesis is the most common form of reproduction among cladocerans, therefore samples may contain few if any males. Should males be present, they usually can be recognized by their smaller size, the longer and occasionally modified antennules, the somewhat modified postabdomen, and the 1st trunk appendage modified as a hook or claw for grasping the female.

Observe the general body features before attempting a dissection of the appendages and mouthparts (see Figure 8). The head is exposed; the body usually is enclosed in a bivalvelike carapace. Actually the carapace is a single valve arising from an integumental fold behind the head; it extends posteriorly and laterally to enclose the trunk and appendages. It may be extended posteriorly as a prominent carapace or shell spine. In a few taxa, the carapace is greatly reduced and serves only as a brood pouch. The junction of the head and body frequently is delimited by a depression, the cervical notch or sinus. Anteriorly the head usually is produced into a beaklike rostrum. The head region between the rostrum and the compound eye is referred to as the vertex. In addition to the large sessile compound eye (the result of the fusion of the typically paired eyes), many cladocerans also retain the small median or nauplius eye located in the midline between the mouth and the compound eye. Elofsson (1966) has reported that the frontal organ present in some taxa is probably an additional light-sensitive structure.

The appendages of the head consist of paired antennules, antennae, mandibles, and maxillulae. The maxillae are absent except in a few taxa where they are rudimentary. The antennules usually are quite short, situated in close proximity to the rostrum, and provided with sensory setae. The antennae are the very well-developed biramous appendages arising laterally from the head; they are the primary locomotory appendages in cladocerans.

Cut the carapace along the dorsal midline to expose the body and appendages on one side or remove it entirely. In most cladocerans the trunk segmentation is not apparent, but the posterior part of the abdomen is ventrally flexed. This region, referred to as the postabdomen, abreptor, or pygidium, usually terminates in a pair of caudal rami (caudal furca); occasionally a long caudal process or appendix may be present. On the morphological dorsal surface of the postabdomen, which appears to be in a posterior position as a result of the abdominal flexure, are a series of small anal spines and a pair of abdominal setae. Frequently in females the brood pouch with developing embryos will be observed, and occasionally the developing ephippium and ephippial eggs will be present. Posterior to the brood pouch are a pair of abdominal processes. These are thought to aid in closing off the brood chamber. Depending upon the particular taxon under study, 4 to 6 pairs of trunk appendages will be present. Before beginning the dissection of the mouthparts and trunk appendages, observe the relationships of these appendages.

For ease of dissection it is recommended that you remove the antenna first. Then carefully remove the trunk appendages from one side of the body. In *Daphnia* and related taxa there are 5 pairs. These appendages usually are flattened and foliaceous; they are used for filtering food particles. In the Polyphemidae and Leptodoridae these appendages usually are prehensile. Examine each appendage and identify the structural elements. How do these appendages, particularly the 3rd and 4th pairs in *Daphnia* and related genera, compare with and/or differ from those of other Branchiopoda?

With the anterior trunk appendages and antenna removed, the mouthparts will be much more accessible. The prominent labrum lies in the midline and covers the median portions of the well-developed mandibles. Remove 1 of the mandibles and observe the toothed or ridged grinding surface. A pair of very small, simple maxillulae are located between the mandibles and the median, conical paragnath (lower lip).

Examine a cleared specimen and identify the components of the major organ systems. Dorsally the subovate heart is located anterior to the brood pouch. It may appear reddish in color as hemoglobin frequently is present. There are no arteries or well-developed sinuses in *Daphnia* and related genera, but an anterior aorta has been reported in *Leptodora*. In *Daphnia* blood flow is channeled through the hemocoele and into the appendages by thin membranes. Blood enters the heart through a single pair of ostia.

Food is passed along the food groove to the mouth and then through the short esophagus to the elongate, straight or coiled midgut. In *Daphnia* and related genera a pair of ceca, situated posteriorly in the head, join the midgut just posterior to its junction with the esophagus. In some genera these ceca are absent, and in others a 3rd cecum is present posteromedially. This cecum is thought to be derived from the hindgut. The latter is short and terminates in an anal opening at the base of the caudal furca.

In addition to the large compound eye, 1 to 4 small nauplius eyes may be present. These are innervated by nerves from the supraesophageal ganglion, which lies just anterior to the midgut-esophagus junction and

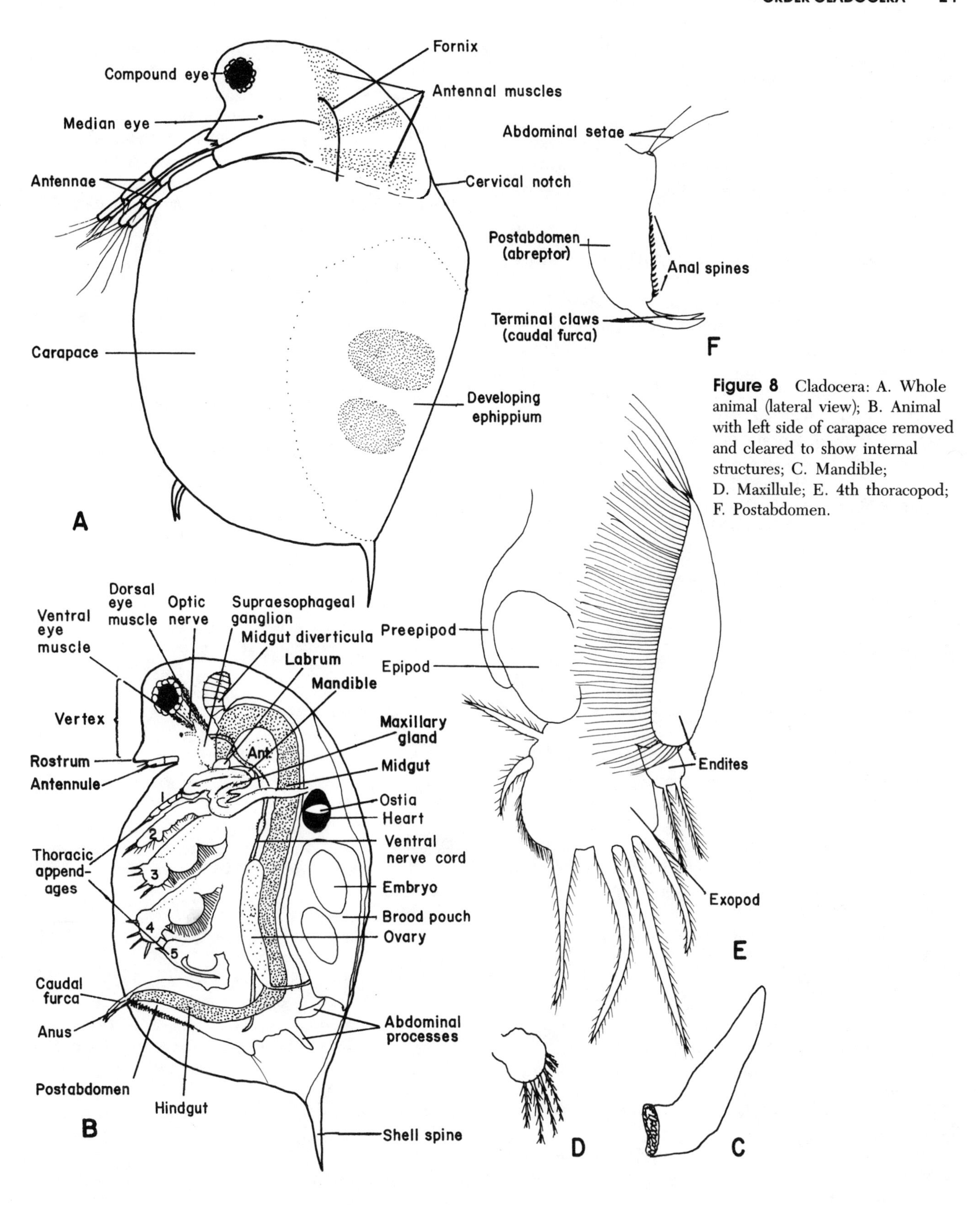

Figure 8 Cladocera: A. Whole animal (lateral view); B. Animal with left side of carapace removed and cleared to show internal structures; C. Mandible; D. Maxillule; E. 4th thoracopod; F. Postabdomen.

below the midgut diverticula. Trace the large optic nerve from the ganglion to the compound eye. The eye is moved by pairs of dorsal and ventral optic muscles. The pair of esophageal connectives that encircle the esophagus, as well as the antennular and antennal nerves, may be difficult to identify unless the specimen has been stained with one of the vital stains. The ganglia of the ventral nerve cord usually can be distinguished without special staining. In *Daphnia* 3 ganglia are present in the anterior part of the trunk.

Identify the paired ovaries lying behind the midgut in the posterior half of the trunk. A slender duct leads from each ovary to the brood pouch. In males the paired testes are in a similar position. The vas deferens are directed ventrally to gonopores on the furca, or occasionally to penes when present.

Only in *Leptodora* is a metanauplius hatched. In all other cladocerans complete development takes place within the egg. Under routine conditions one to several parthenogenetic generations may be produced without any males at all. This does not, however, mean that all parthenogenetic offspring will be identical. At least in some species of *Daphnia* there is a type of meiosis that occurs within the nuclear membrane and there may be some gene exchange before bivalents conjugate during anaphase. Under certain circumstances, presumably ones of stress, both males and haploid eggs that require fertilization are produced. These are ephippial eggs, rich in yolk, and capable of withstanding drying or freezing. They are enclosed in strongly sclerotized egg cases, or ephippia, which, after being discharged into the water, begin to float and can be transported by wind, birds, and other animals.

References

Berg, K., 1929. A faunistic and biological study of Danish Cladocera. *Vidensk. Meddr. Dansk Naturh. Foren.*, *88:* 31–111.

Brooks, J. L., 1957. The systematics of North American *Daphnia. Mem. Conn. Acad. Arts Sci.*, *13:* 1–180.

———, 1959. Cladocera. In W. T. Edmondson (ed.), *Freshwater biology,* 2nd ed., pp. 587–656. New York and London: John Wiley and Sons.

Cannon, H. G., 1922. On the labral glands of a cladoceran *(Simocephalus vetulus)* with a description of its mode of feeding. *Quart. J. Micro. Sci.*, *66:* 213–234.

———, 1933. On the feeding mechanism of the Branchiopoda. *Phil. Trans. Roy. Soc. Lond.*, (B) *222:* 267–352.

Elofsson, R., 1966. The nauplius eye and frontal organs of the non-Malacostraca (Crustacea). *Sarsia*, no. 25: 1–128.

Eriksson, S., 1934. Studien über die Fangapparate der Branchiopoden. *Zool. Bidr. Uppsala*, *15:* 24–287.

Flössner, D., 1962. Zur Cladocerenfauna des Stechlinsee-Gebietes. I. Über Morphologie und Variabilität einiger Formen und über Funde seltener Arten. *Limnologica* (Berl.), *1:* 217–229.

———, 1964. Zur Cladocerenfauna des Stechlinsee-Gebietes II. Ökologische Untersuchungen über die litoralen Arten. *Limnologica* (Berl.), *2:* 35–103.

Frey, D. G., 1959. The taxonomic and phylogenetic significance of the head pores of the Chydoridae (Cladocera). *Int. Revue Ges. Hydrobiol. Hydrogr.*, *44:* 27–50.

———, 1969. Phylogenetic relationships in the family Chydoridae (Cladocera). *Mar. Biol. Assoc. India, Symp.*, Pt. 1: 29–37.

Fryer, G., 1963. The functional morphology and feeding mechanism of the chydorid cladoceran *Eurycercus lamellatus* (O. F. Müller). *Trans. Roy. Soc. Edinb.*, *65:* 335–381.

———, 1968. Evolution and adaptive radiation in the Chydoridae (Crustacea: Cladocera): A study in comparative functional morphology and ecology. *Phil. Trans. Roy. Soc. Lond.*, (B) *254:* 221–385.

———, 1972. Observations on the ephippia of certain macrothridicid cladocerans. *Zool. J. Linnean Soc.*, *51:* 79–96.

———, 1974. Evolution and adaptive radiation in the Macrothricidae (Crustacea: Cladocera): A study in comparative functional morphology and ecology. *Trans. Roy. Soc. Lond.*, (B) *269:* 137–274.

Goulden, C. E., 1968. The systematics and evolution of the Moinidae. *Trans. Amer. Phil. Soc. Philadelphia*, n.s., *58:* 1–101.

Lilljeborg, W., 1900. *Cladocera Sueciae.* Nova Acta. Regiae Soc. Sci. Uppsal., (3) *19:* 1–101.

Longhurst, A. R., and D. L. R. Seibert, 1972. Oceanic distribution of *Evadne* in the eastern Pacific (Cladocera). *Crustaceana*, *22:* 239–248.

Mordukhai-Boltovskoi, Ph.D., 1967a. On the taxonomy of the genus *Cornigerius* (Cladocera, Polyphemidae). *Crustaceana*, *12:* 74–86.

———, 1967b. On the males and gamogenetic females of the Caspian Polyphemidae (Cladocera). *Crustaceana*, *12:* 113–123.

Poulsen, E. M., 1940. The zoology of East Greenland. Freshwater Entomostraca. *Meddel. Gronland*, *121:* 1–73.

Richard, J., 1894. Revision des Cladocères. Part I. Sididae. *Ann. Sci. Nat. Zool.*, *7:* 279–389.

———, 1896. Revision des Cladocères. Part II. Daphnidae. *Ann. Sci. Nat. Zool.*, *8:* 187–363.

Röben, P., 1974. Wasserflöhe (Cladocera) und Ruderfluss-krebse (Copepoda) der Isel Tenerife unter Berücksichtigung irher Verbreitung in Nord-und West-afrika. *Zool. Anz.*, *193:* 110–126.

Scourfield, D. J., and J. P. Harding, 1966. *A key to the British freshwater Cladocera.* Freshwater Biol. Assoc., Sci. Publ., no. 5 (3rd ed.): 1–55.

Smirnov, N. N., 1976. Macrothricidae i Moinidae fauny mira. *Fauna CCCP Rakoobraznye*, *1:* 1–236.

Šrámek-Hušek, R., 1962. Die Mitteleuropäische Cladoceren- und Copepodenge-meinschaften und deren Verbreitung in den Gewässern der CSSR. Sbornik Vysoké Skoly chemtechnol. Praze. *Technol. Vody*, *6:* 99–133.

Šrámek-Hušek, R., M. Straškraba, and J. Brtek, 1962. Lupenonozci-Branchiopoda. *Fauna CSSR. Svazek*, *16:* 1–470.

Tsi-Chung, Ch., and L. S. Clemente, 1954. The classification and distribution of freshwater cladocerans around Manila. *Nat. Appl. Sci. Bull.*, *14:* 85–150.

Wagler, E., 1927. Branchiopoda, Phyllopoda. In W. Kukenthal and T. Krumbach (eds.), *Handbuch der Zoologie*, *3* (1): 309–398. Berlin: Der Gruyter.

CLASS OSTRACODA Latreille, 1806

Recent species	Approximately 2000.
Size range	0.4–30 mm.
Carapace	Bivalve, laterally compressed; completely enclosing head, body, and most appendages.
Eyes	Usually present; single nauplius eye, sometimes also sessile compound eyes.
Antennules	Uniramous; endopod with 5–8 segments (including 2 of protopod); sometimes sexually dimorphic.
Antennae	Biramous; sometimes sexually dimorphic.
Mandibles	With endopodal palp, exopod, and gnathobase.
Maxillulae	Well developed, variable (usually referred to as maxilla).
Maxillae	None (1st thoracopod sometimes referred to as 2nd maxilla).
Maxillipeds	None (1st thoracopod sometimes referred to as maxilliped).
Thoracic appendages	One to 3 pairs of thoracopods, usually biramous; variously developed and specialized.
Abdominal appendages	None.
Telson	Not delimited; with caudal rami, variable in structure and form.
Tagmata	None delineated.
Somites	Segmentation indicated by paired appendages.
Sexual characters	Genital ducts opening anterior to caudal furca; usually on genital lobe; female often with sperm receptacle; male frequently with penes. Appendages and carapace sometimes sexually dimorphic.
Sexes	Separate; sometimes parthenogenetic reproduction.
Larval development	Anamorphic; eggs brooded beneath carapace or cemented to plants or substrate; 6 naupliar substages (nauplius atypical, with bivalve carapace).
Fossil record	Cambrian to Recent; approximately 12,000 fossil species described.
Feeding types	Diverse; filter feeders (plankton and detritus), predators, scavengers, and herbivores.
Habitat	Benthic and planktonic; in virtually all aquatic habitats, including sulfur springs, stagnant ponds, swamps, streams, brackish lagoons, estuaries, tide pools, salt marshes, epicontinental seas, and ocean basins. Two species are reported to be terrestrial in moss; one group is commensal with freshwater isopods, crayfish, and crabs.
Distribution	Worldwide.

The instructions for the examination of ostracod external morphology are based on a composite taken from examples of both Recent subclasses, the Myodocopa and Podocopa. Specific differences that may be encountered by students having representatives of only one of the subclasses available for study will be pointed out. Although our present interest is with Recent crustaceans, fossil ostracods have played an extremely important role in ostracod taxonomy, phylogeny, and morphology. Consequently, ostracod specialists have developed a terminology and vocabulary that is not necessarily applicable to other crustacean groups. This is particularly true with reference to the ostracod shell, and shell structure is given a great deal of attention by specialists of this class. An in-depth study of various types of ostracod carapace sculpture is beyond the scope of this text, so only the more generalized aspects of the shell morphology will be described.

The carapace is truly bivalved, being articulated dorsally along a hinge (see Figure 9). The 2 valves, which may be equal, subequal, or unequal in size, enclose the entire body of the animal. Generally the carapace is composed of a hard calcium carbonate layer and a soft epidermis, although some taxa possess very thin carapaces. The calcium carbonate layer is coated with a layer of chitin and the epidermis is enclosed in chitin. The calcium layer may be subdivided into an outer lamella and a duplicature. The duplicature (calcified inner lamella) extends along the free margin of the valve and is fused to the outer lamella. In contrast to conchostracans, ostracods shed their hard carapaces in a series of molts. Environmental conditions may contribute to shell thickness and to bizarre ornamentation developed after a molt. The shape and the outline of the carapace are of diagnostic significance. When the valves of the carapace are opened, a free edge and a contract margin can be determined. The latter is the narrow portion of each valve that comes in contact with its opposing valve upon closure. The contact margin can be simple or complex. In its simplest form, a single ridge, called the selvage, extends along the free margin. When the valves are closed, the selvage fits into the selvage groove of the opposite valve. When a duplicature is present, the contact margin is complex. Two other ridges may be present: a small ridge, the list, proximal to the selvage; and distal to the selvage a ridge (flange) that is part of the outer lamella. When both a selvage and a flange occur, they are separated by a flange groove. The lateral surfaces of the valves may be smooth or ornamented by granules, pustules, striae, costae, pits, spines, or reticula. The surface may also be pierced by pores or have lobes, sulci, furrows or sutures.

Remove one valve of the carapace and observe the animal while it is still attached to the other valve. Notice the general body form and the orientation of the appendages. The ostracod body is short, laterally compressed, and shows no trace of segmentation. Recent subclasses of ostracods may be separated, in part, by the development of the antennae. In the Myodocopa the antennae are strongly developed swimming appendages; in the Podocopa they are developed for walking. The latter lack a rostrum and rostral incisure that usually are present in the Myodocopa. Before removing the animal from its other valve, determine if sessile eyes, a nauplius eye, or a lens or pair of lenses are present dorsal, and usually posterior, to the base of the antenna. A frontal organ also may be observed, usually in close proximity to the eyes.

Remove the animal from the other valve but do not discard the valve. Remove and examine the appendages of one side of the body. After the antennule and antenna have been removed, the thoracic appendages should be removed before attempting to dissect the mouthparts. Depending upon the taxon, 1 to 3 pairs of thoracopods are present. The protopod has 1, 2, or less frequently, 3 segments (coxa, basis, and possibly precoxa). They usually have an endopod and exopod and sometimes additional structures such as epipods and endites. Identify the structural components of the thoracopods of your specimen(s). The thoracopods usually are rather elongate in podocopid ostracods; in many myodocopids the 3rd pair of thoracopods are markedly altered in structure; they are multiarticulate and directed dorsally and posteriorly.

The mouthparts consist only of the mandibles and maxillulae. Remove and examine these appendages. The mouth is formed by an upper lip (labrum) and a lower lip (labium or hypostome). In most ostracods the latter is preceded by a pair of rake-shaped chitinous structures embedded in the tissue.

A definitive abdomen is lacking in ostracods; the body terminates in a caudal furca. [Bowman (1971) interprets the posteriorly produced terminal end of the abdomen in the podocopid genus *Loxoconcha* as a telson and the rounded lobes ventral to the anus and bearing setae possibly as nonarticulated uropods. He also suggests that the myodocopid furca may be a deeply incised telson.] Very frequently a genital lobe can be distinguished anteroventrally of the caudal furca. In some taxa, a very large and complex penial structure is developed in males.

Returning to the shell, examine the interior and identify the muscle scars and hinge structure. The patterns of these scars and the details of the hinge are of taxonomic importance.

The internal organ systems are quite variable in ostracods. The description presented is based on specimens of the myodocopid family Cypridinidae (see Figure 10). Major differences between cypridinids and other myodocopids, as well as differences between the Myodocopa and Podocopa, will be pointed out. Begin by removing one half of the carapace of a cleared specimen and observing the general musculature of the body and appendages. To examine the major internal organ systems, carefully remove the cephalic and trunk appendages from the exposed side of the body.

A heart is present only in myodocopids. In many taxa it is positioned just dorsal to the mandible; in the species illustrated it is located anterodorsally near the hinge. A single pair of ostia is present in the ostracod heart; blood flows from the heart either through an anterior opening or through a definitive aortic artery. Identifiable arteries are absent in podocopids and in many

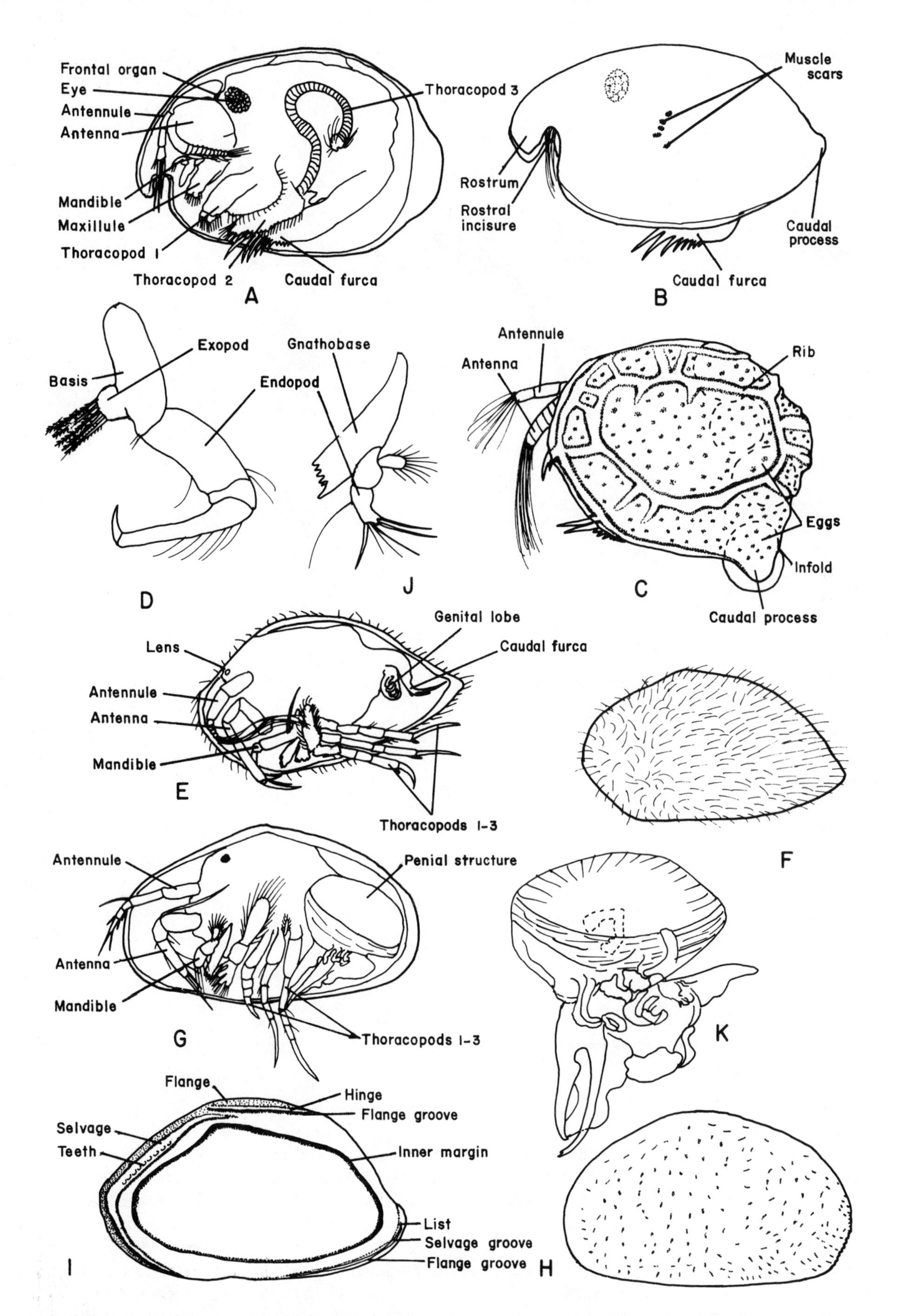

Figure 9 Ostracoda: A–C, Myodocopa; D–K, Podocopa. A. Lateral view of animal with left valve removed; B. Whole animal (lateral view); C. Whole animal with sculptured carapace (lateral view); D. 3rd thoracopod; E. Lateral view of animal with left valve removed; F. Left valve (external view); G. Lateral view of animal with left valve removed; H. Left valve (external view); I. Right valve (internal view); J. Mandible; K. Penial structure enlarged.

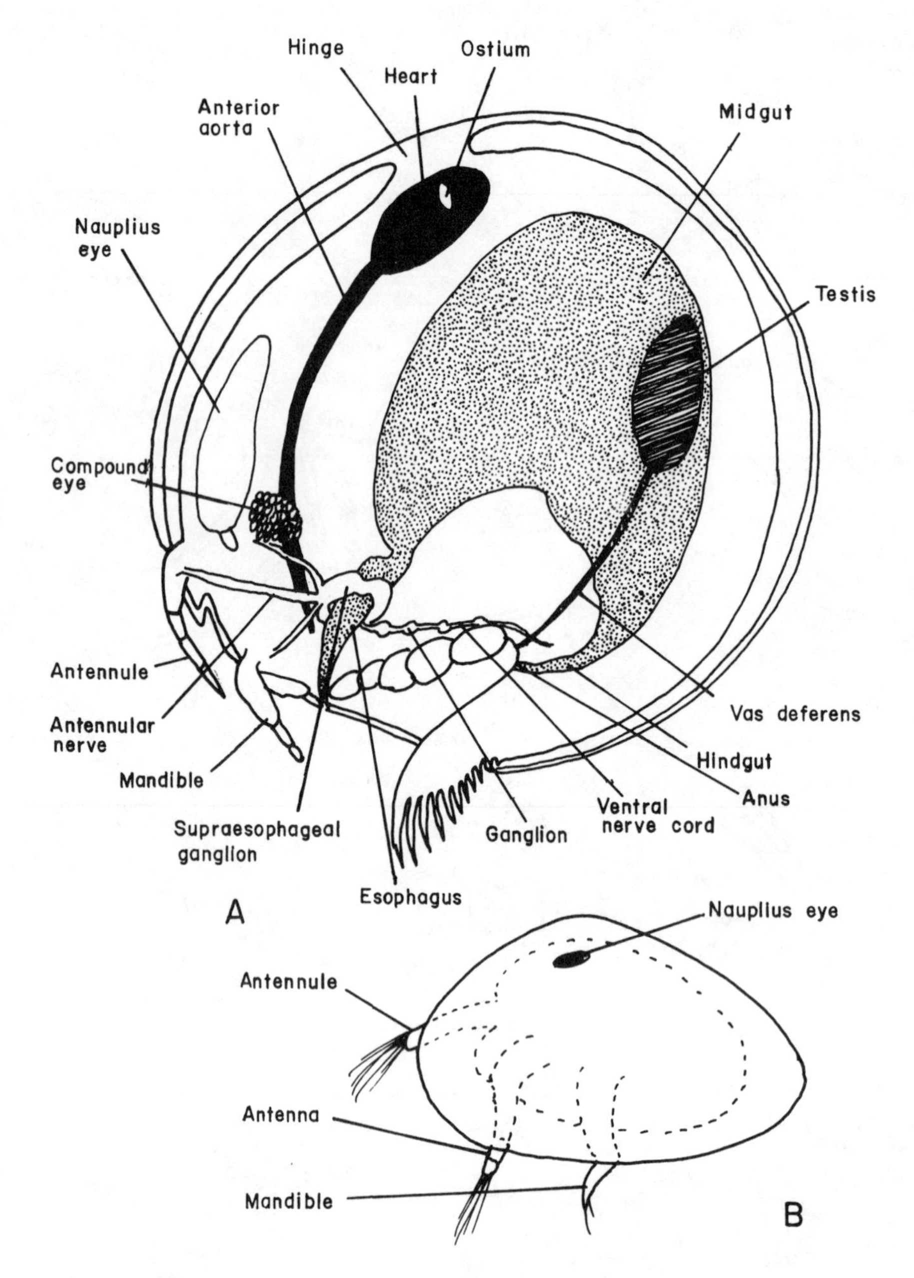

Figure 10 A. Diagrammatic myodocopid ostracod with musculature removed to show major organ systems; B. Nauplius.

myodocopids; blood circulation is through blood spaces or a network of blood channels.

Food is passed from the mouth through the muscular esophagus to the midgut or stomach. In myodocopids such as the species illustrated in Figure 10, the midgut is very large and occupies much of the body space; the hindgut is short and opens to the exterior through a ventrally positioned anus. In podocopids the midgut is small and usually subdivided by a constriction; the hindgut is enlarged and the anus is dorsal to the furca. A single large pair or numerous small midgut ceca are present in many ostracods, but are frequently absent in cypridinids. In some the gut appears to be surrounded by parenchymal tissue. Excretory functions are carried out by antennal glands in some ostracods and by maxillary glands in others.

Observe the prominent "nerve ring" consisting of the supraesophageal ganglion and esophageal connectives encircling the esophagus. Trace the relatively large nerves leading from the nerve ring to the antennules, antennae, and mouthparts. From the nerve ring, follow the parallel branches of the ventral nerve cord posteriorly through the body, identifying the ganglia of the trunk appendages, which often lie in close proximity to the bases of the appendages. Frequently the ganglion of the last pair of thoracopods will be very small.

In females the ovaries usually are located in the posterior part of the body, although in the Cyprididae they are in the hypodermis. If possible, trace the oviducts from the ovaries to their openings, usually on a genital lobe anterior to the caudal furca. Are seminal receptacles present on your specimen(s)? The male reproductive system basically consists of paired testes, vas deferens, and usually penes; however, in many taxa ejaculatory ducts or a complex penial structure are present. In a male specimen locate the testes (simple in myodocopids, usually 4-lobed in podocopids) and trace one of the vas deferens. Are penes, penial apparatus, or ejaculatory ducts present?

Larval development in ostracods differs from that in other Crustacea in that the newly hatched nauplius possesses a bivalve shell. As with typical nauplii, the appendages of the 1st naupliar substage consist of the paired antennules, antennae, and mandibles. Usually the buds of a new pair of appendages appear after each molt. The number of larval instars varies among the taxa.

References

Angel, M. V., 1977. Studies on Atlantic halocyprid ostracods: vertical distributions of the species in the top 1000 m in the vicinity of 44°N, 13°W. *J. Mar. Biol. Assoc. U.K.*, *57:* 239–251.

Benson, R. H., 1975. Morphologic stability in Ostracoda. *Bull. Amer. Paleontol.*, *65:* 13–45.

Cannon, H. G., 1931. On the anatomy of a marine ostracod, *Cypridina (Doloria) levis* Skogsberg. *Discovery Rept.*, *2:* 435–482.

———, 1940. On the anatomy of *Gigantocypris mulleri*. *Discovery Rept.*, *19:* 185–244.

Danielopol, D. L., 1972. Supplementary data on the morphology of *Neoesidea* and remarks on the systematic position of the family Bairdiiae (Ostracoda: Podocopida). *Proc. Biol. Soc. Wash.*, *85:* 39–48.

———, 1977. Recherches sur les Ostracodes Entocytheridae. Données sur *Sphaeromicola cegennica juberthiei* nov. ssp. et *Sphaeromicola cirolanae* Rioja. *Int. J. Speleol.*, *9:* 21–41.

Deevey, G. B., 1974. Pelagic ostracods collected on Hudson 70 between the equator and 55°S in the Atlantic. *Proc. Biol. Soc. Wash.*, *87:* 351–380.

———, 1975. Two new species of *Bathyconchoecia* (Myodocopa Halocyprididae) from the Caribbean Sea. *Proc. Biol. Soc. Wash.*, *88:* 141–158.

Delamare Deboutteville, C., 1960. *Biologie des eaux souterraines littorales et continentales.* 740 pp. Paris: Hermann.

Hart, C. W., and D. Hart, 1969. Evolutionary trends in the Ostracoda family Entocytheridae, with notes on the distributional patterns in the southern Appalachians. In P. C. Holt (ed.), The distributional history of the biota of the southern Appalachians Pt. I Invertebrates, *Virginia Polytechnic Inst. Res. Div. Monogr. 1:* 179–190.

———, 1975. Ostracoda I. Podocopa I Fam. Entocytheridae. *Crustaceorum Catalogus*, pt. 4, pp. 1–64.

Hart, D. G., and C. W. Hart, 1974. The ostracod Family Entocytheridae. *Monogr. Acad. Nat. Sci. Philadelphia*, no. 18, 239 pp.

Hazel, J. E., 1967. Classification and distribution of the Recent Hemicytheridae and Trachyleberididae (Ostracoda) off northeastern North America. *Geol. Survey Prof. Pap. 564*, 49 pp.

Herbst, H. V. von, 1965. Zwei bermerkenswerte Ostracoda (Crustacea) aus dem Rheinland. *Gewässer und Abwässer*, *39/40:* 32–40.

Hobbs, H. H., Jr., and D. J. Peters, 1977. The entocytherid ostracods of North Carolina. *Smiths. Contr. Zool.*, no. 247, 73 pp.

Howe, H. V., R. V. Kesling, and H. W. Scott, 1961. Morphology of living ostracods. In R. C. Moore (ed.), *Treatise on invertebrate paleontology*, Pt. Q Arthropoda 3, Crustacea Ostracoda, pp. Q3–Q17. Lawrence, Kansas: Geol. Soc. America and Univ. Kansas Press.

Kornicker, L. S., 1967. The myodocopid ostracod families Philomedidae and Pseudophilomedidae (new family). *Proc. U.S. Natl. Mus.*, *121:* 1–35.

———, 1976a. Myodocopid Ostracoda from southern Africa. *Smiths. Contr. Zool.*, no. 214, 39 pp.

———, 1976b. Benthic marine Cypridinacea from Hawaii (Ostracoda). *Smiths. Contr. Zool.*, no. 231, 24 pp.

———, 1977. West African myodocopid Ostracoda (Cypridinidae, Philomedidae). *Smiths. Contr. Zool.*, no. 241, 100 pp.

Kornicker, L. S., and M. V. Angel, 1975. Morphology and ontogeny of *Bathyconchoecia septemspinosa* Angel, 1970 (Ostracoda: Halocyprididae). *Smiths. Contr. Zool.*, no. 195, 21 pp.

Kornicker, L. S., and M. Bowen, 1976. *Sarsiella ozotothrix*, a new species of marine Ostracoda (Myodocopina) from the Atlantic and Gulf Coast of North America. *Proc. Biol. Soc. Wash.*, *88:* 497–502.

Kornicker, L. S., and F. E. Caraion, 1974. West African myodocopid Ostracoda (Cylindorleberididae). *Smiths. Contr. Zool.*, no. 179, 76 pp.

Kornicker, L. S., and F. P. C. M. van Morkhoven, 1976. *Metapolycope*, a new genus of bathyl Ostracoda from the Atlantic (suborder Cladocopina). *Smiths. Contr. Zool.*, no. 225, 29 pp.

Maddocks, R. F., 1977. Anatomy of *Australoecia* (Pontocyprididae, Ostracoda). *Micropaleon.*, *23:* 206–214.

McKenzie, K. G., 1977. Gonaducecytheridae, a new family of cytheracean Ostracoda, and its phylogenetic significance. *Proc. Biol. Soc. Wash.*, *90:* 263–273.

Poulsen, E. M., 1977. Zoogeographical remarks on marine pelagic Ostracoda. *Dana Rept.*, no. 87, 34 pp.

Skogsberg, T., 1920. Studies on marine Ostracoda. *I. Zool. Bidr. Uppsala* (supplement), pp. 1–784.

Tressler, W. L., 1949. Marine Ostracoda from Tortugas, Florida. *J. Wash. Acad. Sci.*, *39:* 334–342.

CLASS MYSTACOCARIDA Pennak and Zinn, 1943

Recent species	Nine plus three subspecies in one genus, *Derocheilocaris*.
Size range	0.525–1.01 mm; stage 1 larvae reported greater than 2.0 mm (Hessler, 1969)
Carapace	Absent.
Eyes	Ocelli, generally 4; each typically with lens, usually without pigment.
Antennules	Uniramous, sometimes with terminal aesthetasc; primarily sensory in function, also used in locomotion.
Antennae	Biramous; used in locomotion and feeding.
Mandibles	Biramous; used in locomotion and feeding.
Maxillulae	Uniramous; used in locomotion and feeding.
Maxillae	Uniramous, without exopod; gnathobase usually well developed.
Maxillipeds	One pair; usually biramous; without exopod in 1 species.
Thoracic appendages	Four pairs of small, uniramous, unsegmented thoracopods; function as yet undetermined.
Abdominal appendages	None.
Telson	With clawlike caudal furca; used in grooming and as lever in movement.
Tagmata	Head, thorax, and abdomen; or preferably, head, maxilliped somite, and thoracoabdomen.
Somites	Head with 5; maxilliped somite (1); thoracoabdomen with 9, excluding telson.
Sexual characters	Gonopore on ventral surface of 3rd thoracoabdominal somite, or reportedly possibly paired on appendages of 3rd somite; 4th, sometimes also 3rd, pair of appendages exhibiting sexual dimorphism.
Sexes	Separate.
Larval development	Anamorphic; hatch as nauplii or metanauplii; 7–10 juvenile substages.
Fossil record	Recent.
Feeding types	Deposit feeders, feeding on detritus, bacterial film, and microorganisms.
Habitat	Interstitial on intertidal and subtidal beaches. Sand type and particle size appear to be important factors.
Distribution	Maine to Miami Beach, Florida; Gulf of Mexico; Mediterranean and Portugal; east and west coasts of Africa; west coast of South America.

Mystacocarids basically are very conservative in their morphology. Individual species can be distinguished by the characters of the antennulary portion of the cephalon, the setation of the appendages, and the configuration and armature of the telsonic combs and caudal furca. The extremely small size, as well as the limited supply of specimens, precludes individual student dissections. Consequently, the instructions are for observations of whole mounts and slide preparations of the appendages.

Observe the general body plan, both dorsally and ventrally, before studying the individual appendages (see Figure 11). In dorsal view, the cephalon will be seen to

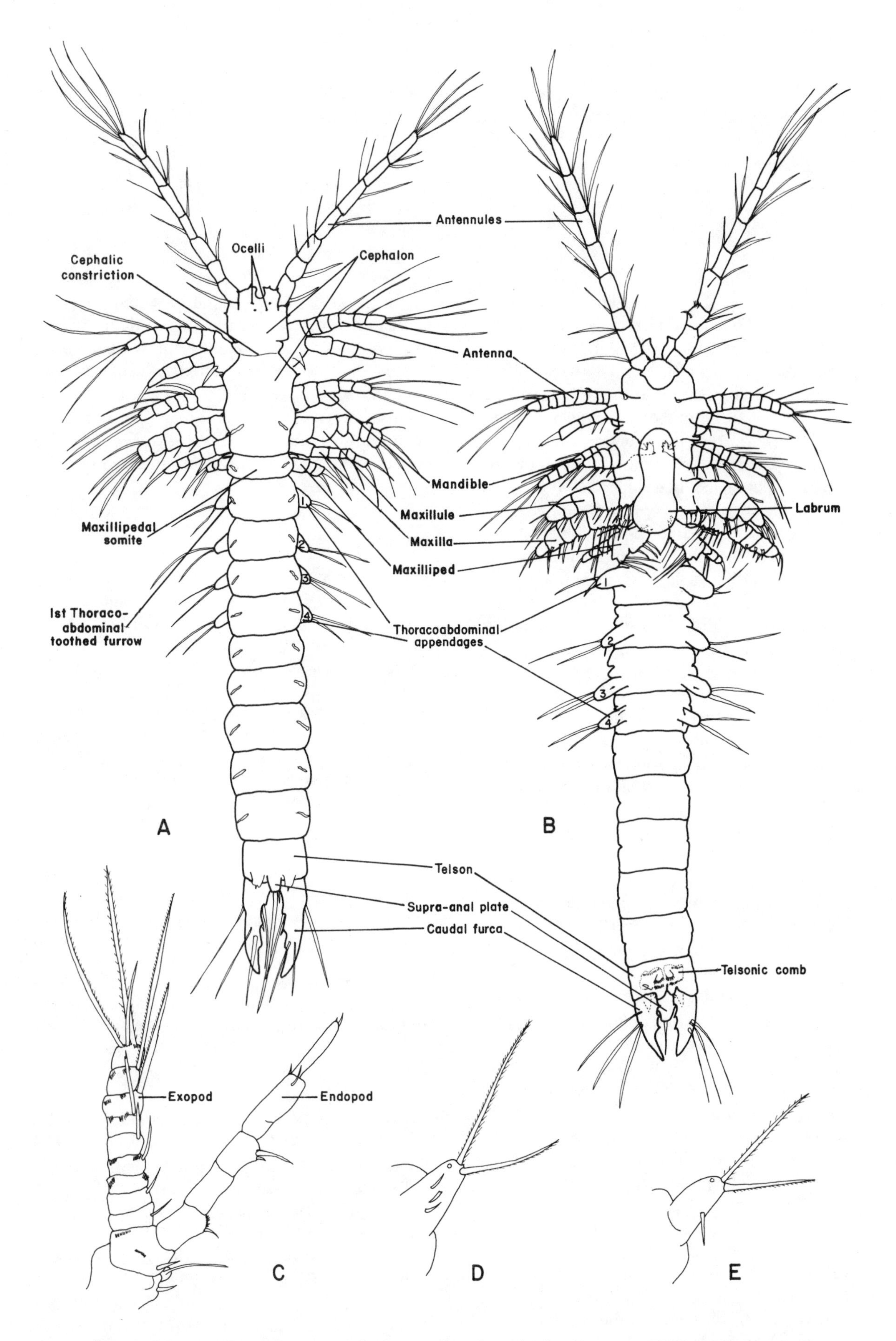

Figure 11 Mystacocarida: A. *Derocheilocaris* (dorsal view) [after Kaestner, 1970]; B. Whole animal (ventral view) [after Hessler and Sanders, 1966, and Hessler, 1972]; C. Antenna; D. Male 4th thoracopod (ventral view); E. Female 4th thoracopod (ventral view) [C–E after Hessler, 1969].

be very elongate, with an anterior or antennulary part delineated by a cephalic constriction. The anterior and lateral margins of the antennulary part are cleft, forming 2 prominent anteromedial lobes and 2 anterolateral lobes respectively. Usually 4 simple ocelli are present. In a few species neither lens nor pigment spots can be detected. Laterally on the posterior part of the cephalon, the 1st pair of toothed furrows can be observed. The structure and extent of the development of these furrows and the succeeding pairs may be diagnostic among the taxa. The function of the toothed furrows is not known.

Posterior to the cephalon are 10 body somites, excluding the telson and caudal furca. [Bowman (1971) considers the supra-anal plate to be the telson, and the caudal rami probably to be uropods; however, he subsequently has proposed that the thoracoabdomen consists of 11 somites and that the telson is absent (personal communication).] There has been additional disagreement among zoologists as to the proper subdivision of the somites. Some have included the maxilliped somite with the cephalon because the maxilliped functions as a cephalic appendage. Other zoologists have proposed that the maxilliped somite be considered the 1st thoracomere of a 5-somite thorax. A third alternative interpretation, and one that appears to have received support from the majority of mystacocarid specialists, is that the maxilliped somite, which is fused neither to the cephalon nor to the thorax, be considered as a distinct unit. The remaining 9 trunk somites then would be collectively referred to as thoracoabdominal somites, the first 4 of which bear small appendages. This latter interpretation has been adopted herein. A pair of lateral toothed furrows is present on each of the thoracoabdominal somites. The telson, dorsally, carries a median supra-anal process, flanked by a pair of spinous processes. The length of the major seta of the supra-anal process often is an important diagnostic character. Dorsal to the insertion of the major seta, 2 raised pores are apparent in some species.

In ventral view, the cephalic appendages and the 4 thoracoabdominal appendages can be distinguished more clearly. The antennules are relatively long; each consists of 8 segments, often also with a terminal aesthetasc. The antennae are relatively short and biramous. The biramous mandibles are similar in structure to the antennae but have gnathobases, which lie beneath the anterior part of the bulbous labrum. The maxillulae and maxillae are uniramous. The maxillipeds, except in one species, are biramous; the protopods often bear 2 to several endites. The thoracoabdominal appendages are small, single-segmented structures with only a few setae. In some species, a slight difference between the sexes can be observed in the 4th and occasionally also in the 3rd pair of appendages.

A genital pore may be observed on the ventral surface of the 3rd thoracoabdominal somite. In some species the possibility of a pair of genital pores at the bases of the appendages of the 3rd somite has been reported. The ventral surface of the telson bears the anus and 2 pairs of ventral combs. The combs, which are variously developed, are one of the more distinct diagnostic characters for species determination. The ventral setae of the caudal furca or furcal claws also are important characters in identification.

Having examined the appendages on the intact animal, examine the slide preparations of the individual appendages. It would be well to sketch the cephalic appendages so that you can compare their structure with those of the other relatively primitive and conservative groups that you have studied, as well as with other members of the class.

Larval development in the mystacocarids is poorly documented. One species is known to hatch as a metanauplius, and 10 larval instars have been reported for a second species.

References

Armstrong, J. C., 1949. The systematic position of the crustacean genus *Derocheilocaris* and the status of the subclass Mystacocarida. *Amer. Mus. Nov.*, 1413: 1–6.

Buchholz, H. A., 1953. Die Mystacocarida. Eine neue Crustaceenordnung aus dem Lüchensystem der Meeressande. *Mikrokosmos, 43:* 13–16.

Chappuis, P. A., A. Remane, and C. Delamare Deboutteville, 1951. Découverte, sur les côtes du Roussillon d'un ordre des Crustacés nouveau pour l'ancien monde: les Mystacocarida Pennak et Zinn. *Vie et Milieu, 2:* 129–130.

Dahl, E., 1952a. A new species of the Mystacocarida (Crustacea). *Nature*, London, *170:* 75–76.

———, 1952b. Mystacocarida. Reports of the Lund University Chile Expedition 1948–49, 7. *Lunds Univ. Arssk.*, n.s., (2) *48:* 1–41.

Delamare Deboutteville, C., 1953a. La faune des eaux souterraines littorales des plages de Tunisie. *Vie et Milieu, 4:* 141–170.

———, 1953b. L'ecologie et la répartition du Mystacocaride *Derocheilocaris remanei* Delamare et Chappuis. *Vie et Milieu, 4:* 321–380.

———, 1953c. Description d'un appareil pour la capture de la faune des eaux souterraines littorales sous la mer. *Vie et Milieu, 4:* 411–422.

———, 1953d. Révision des Mystacocarides du genre *Derocheilocaris* Pennak et Zinn. *Vie et Milieu, 4:* 459–469.

———, 1953e. La faune des eaux souterraines littorales d'Algerie. *Vie et Milieu, 4:* 470–503.

———, 1954a. Développement postembryonnaire des Mystacocarides. *Arch Zool. Exp. Gén., 91:* 25–34.

———, 1954b. Premières recherches sur la faune souterraine littorale en Espagne. *Inst. Biol. Aplic.*, Barcelona, *17:* 119–129.

———, 1954c. L'ecologie et la répartition du Mystacocaride *Derocheilocaris remanei* Delamare et Chappuis. *Rassegna Speleog. Italiana, 4:* 119–122.

———, 1954d. L'ecologie du Mystacocaride, *Derocheilocaris remanei* f. *biscaynesis* Del. sur la côte du Golfe de Gascogne. *Vie et Milieu, 5:* 310–329.

———, 1954e. Eaux souterraines littorales de la côte catalane française (Mise au point faunistique). *Vie et Milieu, 5:* 408–452.

———, 1956. L'ecologie et la répartition du Mystacocaride *Derocheilocaris remanei* Del. et Chapp. *Proceed. 14th Intern. Congr. Zool. Copenhague:* 504–505.

———, 1957. Le Mystacocaride *Derocheilocaris biscayensis* sur la côte du Portugal. *Vie et Milieu, 8:* 110.

———, 1960. *Biologie des eaux souterraines littorales et continentales*, 740 pp. Paris: Hermann.

Delamare Deboutteville, C., and P. A. Chappuis, 1951. Présence de l'ordre des Mystacocaria Pennak & Zinn dans la sable des plages du Roussillon: *Derocheilocaris remanei* n. sp. *C. R. Acad. Sci.*, Paris, *233:* 437–439.

———, 1954. Morphologie des Mystacocarides. *Arch Zool. Exp. Gén., 91:* 7–24.

———, 1957. Constitution à l'etude de la faune interstitielle marine des côtes d'Afrique. I. Mystacocarides, Copépodes et Isopodes. *Bull. I.F.A.N. 19*(A): 491–500.

Delamare Deboutteville, C., S. Gerlach, and R. Siewing, 1954. Recherches sur la faune des eaux souterraines littorales du Golfe de Gascogne, littoral des landes. *Vie et Milieu, 5:* 373–407.

Delamare Deboutteville, C., and A. de Barros Machado, 1954. Présence de la sous-classe des Mystacocarides sur les côtes de l'Angola. *Subsidio Estudo Biol. Lunda*, Lisbonne, *23:* 119–124.

Friauf, J. J., and L. Bennett, 1974. *Derocheilocaris hessleri:* a new Mystacocarida from the Gulf of Mexico. *Vie et Milieu, 24:* 487–496.

Hall, J. R., 1972. Aspects of the biology of *Derocheilocaris typica* (Mystacocarida, Crustacea): II. Distribution. *Mar. Biol.*, Berlin, *12:* 42–52.

Hall, J. R., and R. R. Hessler, 1971. Aspects in the population dynamics of *Derocheilocaris typica* (Mystacocarida, Crustacea). *Vie et Milieu, 22:* 305–326.

Hessler, R. R., 1969. A new species of Mystacocarida (Crustacea) from Maine. *Vie et Milieu, 20:* 105–116.

———, 1971. Biology of the Mystacocarida: A prospectus. *Smiths. Cont. Zool.*, no. 76: 87–90.

———, 1972. New species of Mystacocarida from Africa. *Crustaceana, 22:* 259–273.

Hessler, R. R., and H. L. Sanders, 1966. *Derocheilocaris typicus* Pennak & Zinn (Mystacocarida) revisited. *Crustaceana, 11:* 141–155.

Jànsson, B. A., 1966. The ecology of *Derocheilocaris remanei. Vie et Milieu, 17:* 143–186.

McLachlan, A. 1977. The larval development and population dynamics of *Derocheilocaris algoensis* (Crustacea, Mystacocarida). *Zool. Afr. 12*: 1–14.

McLachlan, A., and J. R. Grindley, 1975. A new species of Mystacocarida (Crustacea) from Algoa Bay, South Africa. *Ann. S. Afr. Mus., 66*: 169–175.

Masry, D., and F. D. Por, 1970. A new species and a new subspecies of Mystacocarida (Crustacea) from the Mediterranean shores of Israel. *Israel J. Zool., 19:* 95–103.

Noodt, W., 1954. Crustacea Mystacocarida von Süd-Afrika. *Kieler Meeresforsch., 10:* 243–246.

———, 1961. Nuevo hallazgo de *Derocheilocaris galvarini* Dahl en Chile Central (Crustacea, Mystacocarida). *Inv. Zool.*, Chilenas, *7:* 97–99.

Pennak, R. W., and D. J. Zinn, 1943. Mystacocarida, a new order of Crustacea from intertidal beaches in Massachusetts and Connecticut. *Smiths. Misc. Coll., 103:* 1–11.

Tuzet, O., and A. Fize, 1958. La spermategénèse de *Derocheilocaris remanei* Del. et Ch. (Crust. Mystacocarida). *C. R. Acad. Sci.*, Paris, *156*: 3669–3671.

Zaffagnini, F., 1969. L'appareill reproducteur de *Derocheilocaris remanei* Del. et Chapp. (Crustacea, Mystacocarida). *Cah. Biol. Mar., 10:* 103–107.

CLASS COPEPODA H. Milne Edwards, 1840

Recent species Approximately 7500.

Size range < 0.26– 17 mm in free-living species; up to 320 mm in some parasitic species.

Carapace Absent; cephalic shield may project forward as rostrum.

Eyes Usually nauplius eye; occasionally absent.

Antennules Uniramous, sometimes markedly reduced; in males often well developed and geniculate.

Antennae Uni- or biramous, well developed, or reduced; sometimes prehensile; occasionally absent.

Mandibles Usually well developed, with biramous palp; sometimes reduced, absent, or modified as piercing stylet.

Maxillulae Usually biramous; sometimes prehensile; occasionally reduced or absent.

Maxillae Uniramous; sometimes reduced or absent.

Maxillipeds One or 2 pairs; uniramous, sometimes prehensile; occasionally reduced or absent.

Thoracic appendages Four or 5 pairs; last pair often modified, reduced, or vestigial; occasionally absent.

Abdominal appendages None.

Telson Absent; anal somite with caudal rami.

Tagmata Cephalosome, metasome, and urosome; or prosome (cephalosome + metasome) and urosome.

Somites Head with 5 + 1 or 2 thoracic (maxillipeds) [= cephalosome]; thorax basically with 5 or 4 (varying from 2– 5) [= metasome]; abdomen with 5 or fewer, including genital and anal somites [= urosome].

Sexual characters Gonopore(s) of female on ventral surface of genital somite (1st or fused 1st and 2nd abdominal somite), gonopore(s) of male also on genital somite. Female often with seminal receptacle(s); 5th pereopods of male usually modified as copulatory structures; male antennules sometimes also modified.

Sexes Separate; occasionally parthenogenetic reproduction.

Larval development Metamorphic; eggs carried in ovisacs or shed freely; 6 naupliar substages → 5 copepodid substages → adult.

Fossil record L. Cretaceous to Recent.

Feeding types Calanoids generally filter feeders; some calanoids and many cyclopoids predatory; harpacticoids generally grazers on sand grains. Parasitic orders adapted for piercing and sucking or absorption.

Habitat Generally marine, although many species occur in freshwater; some terrestrial harpacticoids have been reported. Calanoids primarily pelagic, harpacticoids generally benthic or interstitial; approximately 2000 parasitic species with many kinds of hosts.

Distribution Worldwide.

The members of this class have successfully adapted to a variety of habitats. Of the eight orders usually recognized, the members of four are parasitic, at least at some stage in the life cycle; that is, Notodelphyoida, Monstrilloida, Caligoida, and Lernaeopodioda. Some representatives of the Cyclopoida and a few of the Harpacticoida also are parasitic. Members of the Calanoida and Misophrioida all are free-living. Kabata (in press) has proposed changes that will greatly alter the hierarchical classification, particularly of the parasitic groups, and these changes appear to have a sound basis in morphological and phylogenetic fact (T. Bowman, personal communication). In Kabata's classification most parasitic taxa have been removed from the Cyclopoida; the names Poecilostomatoida and Siphonostomatoida have been reintroduced for those taxa having falcate and styletlike mandibles, respectively. The free-living and parasitic types will be discussed separately; the major characters of each taxon are given.

FREE-LIVING COPEPODA

ORDER CALANOIDA Sars, 1903

Antennules	16–26 articles; 1 antennule often modified.
Antennae	Biramous.
Mandibles	Biramous; with gnathobase.
Maxillulae	Biramous.
Maxillae	Uniramous; often modified.
Maxillipeds	Uniramous.
Flexure	Between 6th thoracic somite and abdomen.
Habitat	Free-living, usually planktonic.

ORDER HARPACTICOIDA Sars, 1903

Antennules	Ten or fewer articles; often geniculate in male.
Antennae	Biramous; exopod reduced.
Mandibles	Biramous; usually with gnathobase.
Maxillulae	Usually biramous; broad, platelike.
Maxillae	Uniramous.
Maxillipeds	Uniramous; sometimes subchelate.
Flexure	Between 5th and 6th thoracic somites.
Habitat	Usually free-living; primarily benthic; occasionally parasitic.

ORDER CYCLOPOIDA Sars, 1903

Antennules	10–16 articles; often geniculate in male.
Antennae	Usually uniramous; exopod absent or reduced.
Mandibles	Usually biramous.
Maxillulae	Usually biramous.
Maxillae	Uniramous.
Maxillipeds	Uniramous.
Flexure	Between 5th and 6th thoracic somites (except in Lernaeidae).
Habitat	Usually free-living; often planktonic; frequently in fresh water.

ORDER MISOPHRIOIDA Gurney, 1933

Antennules	Sixteen or fewer but not less than 11 articles; prehensile.
Antennae	Biramous.
Mandibles	Biramous; with gnathobase.
Maxillulae	Biramous.
Maxillae	Uniramous.
Maxillipeds	Uniramous.
Flexure	Between 5th and 6th thoracic somite.
Habitat	Free-living epibenthic.

The Calanoida, Harpacticoida, and Cyclopoida all have a great many marine and freshwater representatives; the Misophrioida is known from very few species. The discussion of copepod morphology that follows is quite general to facilitate its usefulness. Examine the sampling of copepods, which should be a mixture of several taxa. Two characters can be used to distinguish calanoid species from harpacticoid and cyclopoid taxa quickly: the number of antennular articles (15 or more) and the position of the body flexure or articulation (between the 6th thoracic somite and the genital somite). Distinctions between harpacticoid and cyclopoid copepods are not as easily made. In general, the antennular flagella of harpacticoid copepods have fewer articles (5–10) and only a slight constriction, if any, exists between the 5th and 6th thoracic somites (somites of the 4th and 5th pereopods). In cyclopoids, the flagella of the antennules usually have more articles (6–17) and the constriction between the 5th and 6th thoracic somites usually is pronounced. Another distinguishing character is the basal segment of the 5th pereopods. This segment is enlarged on the inner margin in harpacticoids but not in cyclopoids. Determine the order(s) to which the copepods in your sample belong.

Begin your study of copepods by examining the whole animal(s) (see Figure 12). A carapace is absent; however, the cephalic shield may be produced anteriorly to form a small rostrum. Eyes may be present or absent; when present, they frequently lack pigment. The cephalon is fused with the 1st thoracic somite (Calanoida) or with the 1st and 2nd (some Calanoida, Cyclopoida, Harpacticoida, and Misophrioida); the composite is referred to as the cephalosome. The remaining thoracic somites become progressively narrower posteriorly in some calanoids, cyclopoids, and the misophrioids, but remain generally the same width in harpacticoids. As previously indicated, a flexure of the body occurs between the 5th and 6th thoracic somites in all but the Calanoida. The flexure in the latter is between the 6th thoracomere and the genital somite. The free thoracic somites anterior to the flexure collectively are referred to as the metasome, and the somites (thoracic and abdominal) posterior to the flexure as the urosome. The abdomen usually consists of 4 or 5 somites, the 1st of which is fused in females with the last thoracic somite to form a genital somite. The last or anal somite of the abdomen, called a telson by some authors, is not homologous with a true telson. A pair of caudal rami are terminal.

Examine the animal from a lateral or ventral view and identify the appendages before beginning your dissection. In most calanoids the antennules are very elongate; one antennule may be geniculate in males; in some species the maxillae may be extremely enlarged and/or modified. The 1st thoracopods are modified as maxillipeds. Four pairs of pereopods (thoracopods 2–5) are well-developed biramous appendages used in locomotion. The 5th pair (6th thoracopods) may be reduced or strongly modified, and are sometimes absent in females. The characters afforded by this pair in males frequently are of considerable importance in species determination. Remove the pereopods and identify the structural components, such as protopod, endopod, and exopod. Frequently both members of a pair will be removed simultaneously. This is because they are bound together by a basal connection that insures that they will move in unison.

Before removing the mouthparts, observe the position of the maxillipeds in relation to the maxillae. Usually they are attached medial to the maxillae. Remove the mouthparts from one side of the body and examine their structure. Would you expect your study specimens to be filter feeders, grazers, or predators? On what characters have you based your answer?

Using cleared specimens, it is possible to identify most of the major organ systems without extensive dissection (see Figure 13) unless the species under study is one with prominent lateral muscle bands. Carefully remove any cephalic or thoracic appendages that obscure a clear view of the body. If prominent lateral muscle bands are present carefully cut the integument and remove them from one side. In calanoids and misophrioids a heart is present dorsally in the midline of the body. It is a short organ with 1 pair of lateral ostia and 1 ventral ostium in calanoids. Blood is carried anteriorly to the head by a

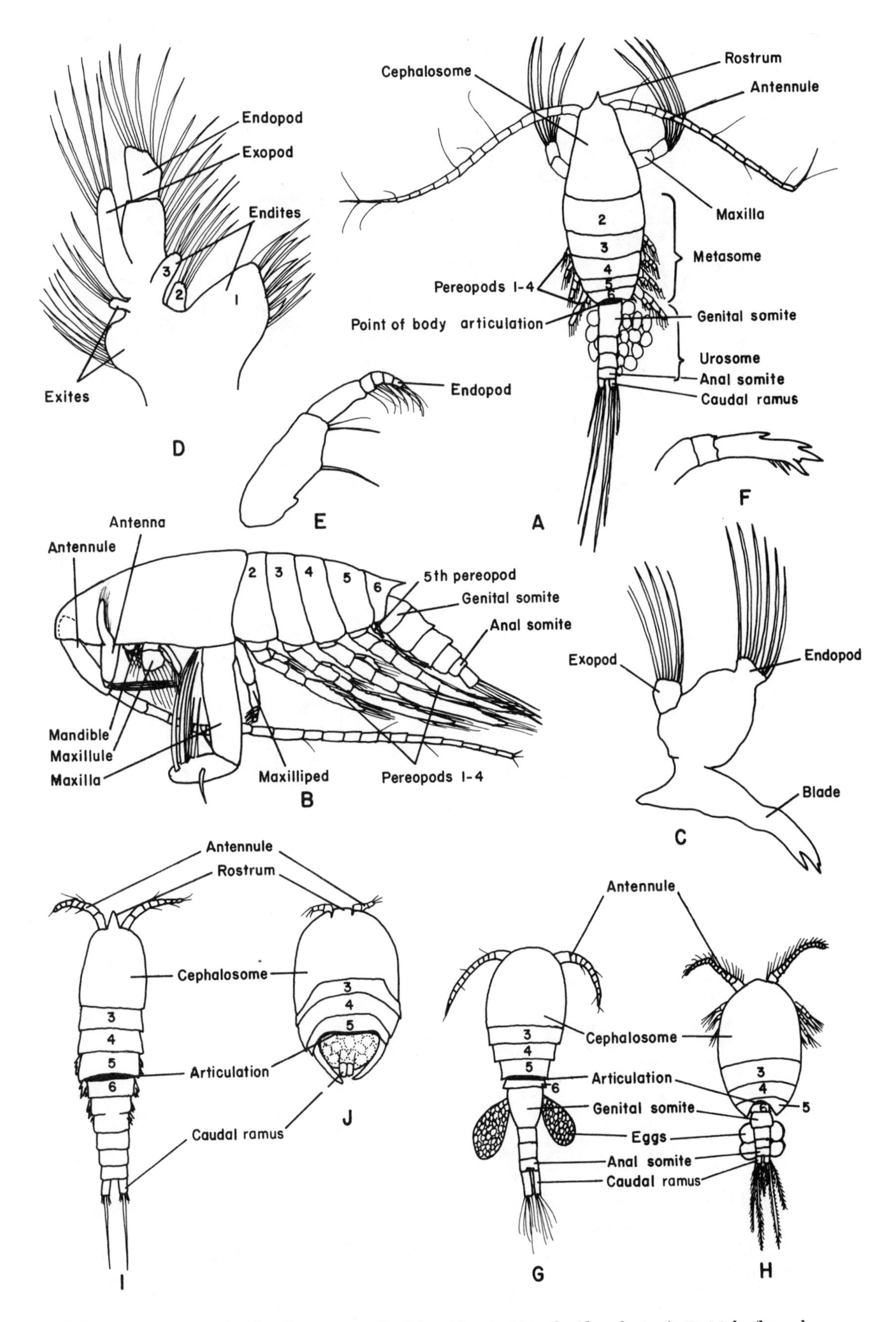

Figure 12 Copepoda, free-living: A—F, Calanoida. A. Female (dorsal view); B. Male (lateral view); C. Mandible of male; D. Maxilla of female; E. Maxilliped of male; F. 5th pereopod of male; G. Female Cyclopoida (dorsal view); H. Female Misophrioida (dorsal view [after Sars, 1903]); I. Male Harpacticoida (dorsal view); J. Female Harpacticoida (dorsal view).

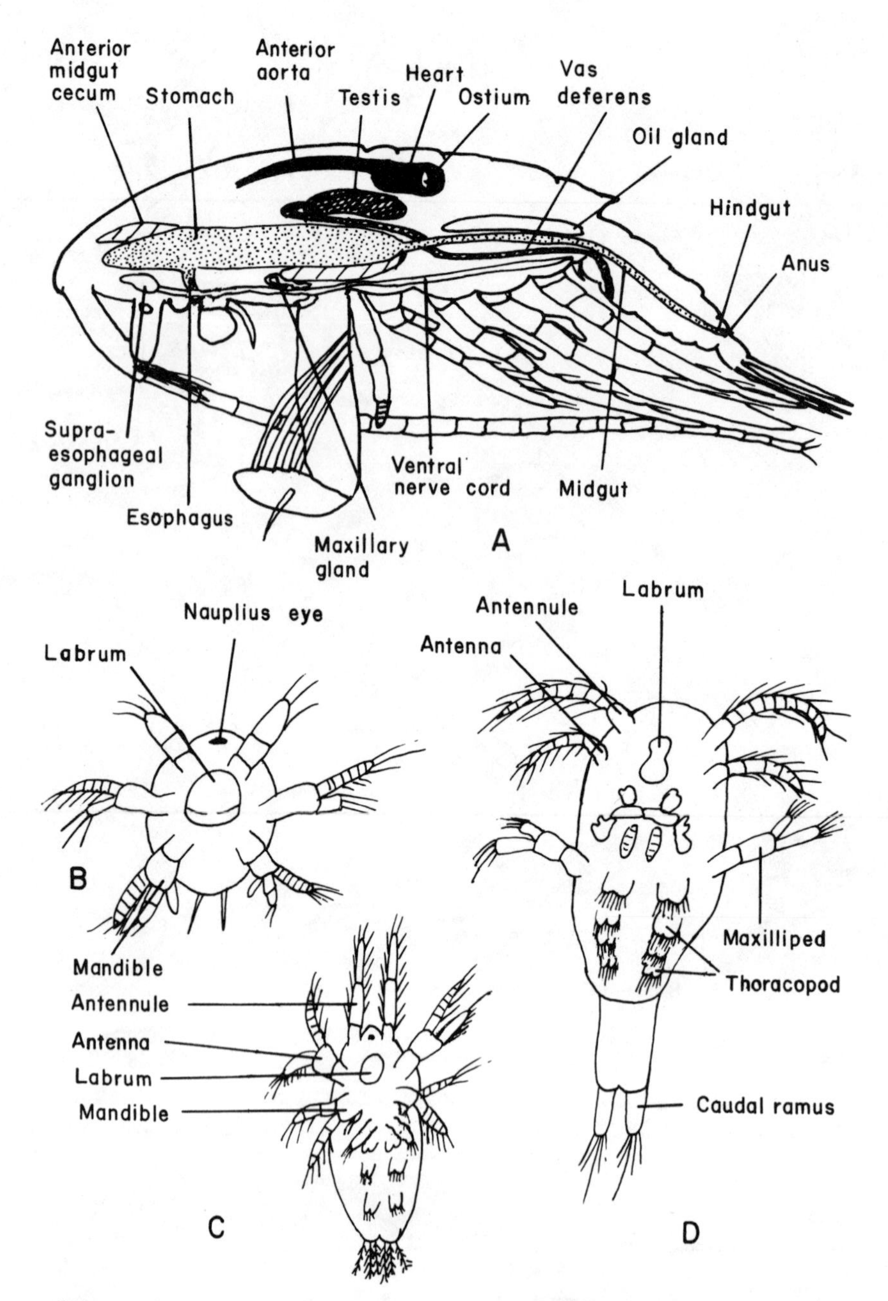

Figure 13 Copepoda, free-living: A. Diagrammatic calanoid copepod with musculature removed to show major internal organ systems; B. Diagrammatic nauplius I; C. Diagrammatic nauplius IV; D. Diagrammatic copepodid I.

short anterior aorta; no other blood vessels are present. Blood circulates to the thoracic appendages through a ventral sinus. In harpacticoids and cyclopoids no heart or arterial system are present; body movements and the pulsating movements of the gut provide the necessary movement of blood in the hemocoel.

The digestive system is simple, consisting of an esophagus, stomach, midgut, hindgut, and usually ceca. The anterior portion of the enlarged stomach, as well as the esophagus, are of ectodermal origin; the remainder of the stomach and the midgut, of endodermal. The short hindgut extends approximately the length of the anal somite and terminates dorsally in the anus. One or a pair of midgut ceca usually arise at the junction of the stomach and midgut. In *Calanus* a prominent oil gland is present above the midgut in the metasome. Excretion is through maxillary glands in adult copepods, but through antennal glands in the nauplius.

In free-living copepods the gonad (ovary or testis) typically is an unpaired organ lying in the midline above the midgut. Anteriorly from the ovary trace the pair of oviducts forward toward the head. In mature females large diverticuli from the oviducts occupy much of the dorsal region of the cephalosome. From the diverticuli follow the oviducts posteroventrally to the genital somite, where they join just before entering the atrium (calanoids) or separately open through a pair of gonopores (cyclopoids). Calanoids have a pair of seminal receptacles that open into the atrium; cyclopoids have a median seminal receptacle which connects with the gonopores via short, often kinked, fertilization ducts. A 3rd opening into this receptacle is a copulatory pore. In the male locate the testis in approximately the same position as the female's ovary was found. One (calanoids) or a pair (cyclopoids) of vas deferens leave the testis anteriorly and proceed posteriorly and then ventrally to the genital somite. Histologically it is possible to distinguish regions of the vas deferens as seminal vesicle, spermatophoric sac, and ejaculatory duct; it may also be possible to observe changes in the spermatophores as they pass from the testis to the genital somite.

Anterior to or anteroventrally of the stomach observe the supraesophageal ganglion. Typically innervation of the sense organs of the head occurs from this ganglion but will be difficult to follow unless the specimen has been specially stained. From the supraesophageal ganglion follow one of the large esophageal connectives around the esophagus to the ventral nerve cord. If you are studying a specimen of *Calanus*, locate the giant nerve fibers that innervate the muscles of the pereopods and metasome. Posteriorly, identify the ganglia of the nerve cord. Ganglia are not present in the abdomen.

Larval development in copepods is metamorphic; young hatch as 1st substage nauplii having 3 pairs of appendages (antennules, antennae, and mandibles) but lacking external segmentation of the body. Additional appendages are added during successive naupliar substages. Six naupliar substages are present in most calanoids; in harpacticoids and cyclopoids 5 commonly occur, occasionally fewer. A metamorphic molt to the copepodid stage follows the last naupliar substage. The body form is more like that of the adult and body segmentation is apparent. Additional metameres and appendages are added during the 4 subsequent copepodid substages.

PARASITIC COPEPODA

ORDER NOTODELPHYOIDA Sars, 1903

Antennules	Often reduced, sometimes modified.
Antennae	Uniramous; 3-segmented or reduced; sometimes modified as claw.
Mandibles	Biramous; with gnathobase; occasionally reduced.
Flexure	Usually between 6th thoracic somite and abdomen.
Habitat	Parasitic or commensal in tunicates; males sometimes free-living.

ORDER MONSTRILLOIDA Sars, 1903

Antennules	Usually well developed.
Antennae	Absent in adults.
Mandibles	Absent in adults.
Flexure	Usually between 5th and 6th thoracic somites.
Habitat	Parasitic as larvae on polychaetes and rarely echinoderms; free-living, presumably nonfeeding as adults.

ORDER SIPHONOSTOMATOIDA Thorell, 1859

Antennules	Usually reduced or minute.
Antennae	Frequently prehensile; sometimes biramous with exopod reduced or vestigial.
Mandibles	Modified as piercing stylets.
Flexure	Between 4th and 5th or 5th and 6th thoracic somites; occasionally lacking.
Habitat	Parasitic on marine and freshwater fishes and invertebrates.

ORDER POECILOSTOMATOIDA Thorell, 1859

Antennules	Well developed or reduced.
Antennae	Uniramous.
Mandibles	Uniramous; falcate.
Flexure	Between 5th and 6th thoracic somites; often lost.
Habitat	Parasitic on fishes and invertebrates, usually marine.

Among the four orders of parasitic copepods, there is a great diversity of body form; in fact, in many instances it is difficult to recognize the parasite as being a copepod. The degree of specialization or the amount of reduction in appendages also varies greatly within individual orders. Examples showing some of this diversity have been illustrated in Figure 14. A caligid representative of the Siphonostomatoida will be described and may serve as a model if demonstration materials are available.

The term cephalothorax frequently is used in referring to the cephalon and fused thoracic somites of most parasitic copepods. The number of thoracic somites that are fused varies with the taxa. In dorsal view, it will be noticed that the cephalothorax is covered by a broad shield; this is not a true carapace. In certain taxa, a pair of lunules is present anteriorly and the 2- or 3-segmented antennules usually can be distinguished. A nauplius eye also may be observed in the midline anteriorly. In some taxa, the shield may be produced posteriorly as a pair of alae, as frequently also occurs in Branchiura. One or more pedigerous segments may be present. Posteriorly the genital somite is enlarged, particularly in sexually mature females, and varies in shape; the posterolateral surfaces may be expanded. Developing embryos are carried in a pair of egg strings. The abdomen may be segmented or not; the former condition is more typical in males. The abdomen terminates in a pair of caudal rami.

In ventral view, the antennules and lunules are more clearly observed. The antenna is reduced; 1 to 3 short segments frequently terminate in hooks or claws. Laterad of the antennae in many taxa is a pair of postantennal processes or "maxillary hooks," which are not, however, components of the maxillulae or maxillae. The mandibles usually are small and obscurely implanted in the mouth cone (tube). The maxillulae typically are small, often nodular or prehensile, but may be reduced or vestigial. The maxillae usually have 1 or 2 segments and often terminate in weakly developed hooks or spines. Each maxilliped frequently terminates in a strong clawlike structure. A pair of post-oral processes and a sternal furca also may be present. The first 7 pairs of thoracic appendages are uni- or biramous and variously developed; the 5th and 6th pairs, on the genital somite, usually are greatly reduced and even may be absent.

References

Borutskii, E. V., 1952. Freshwater Harpacticoida. *Fauna of USSR; Crustacea, 3*, 396 pp. [Translation for the Smithsonian Institution and National Science Foundation, 1964.]

Bowman, T. E., 1971. The distribution of calanoid copepods off the southeastern United States between Cape Hatteras and southern Florida. *Smiths. Contr. Zool.*, no. 96: 1–58.

Bowman, T. E., and L. S. Kornicker, 1967. Two new crustaceans: The parasitic copepod *Sphaeronellopsis monothrix* (Choniostomatidae) and its myodocopid ostracod host *Parasterope pollex* (Cylindroleberidae) from the southern New England coast. *Proc. U.S. Natl. Mus., 123:* 1–28.

Boxshall, G. A., 1977. The planktonic copepods of the northeastern Atlantic Ocean: some taxonomic observations on the Oncaeidae (Cyclopoida). *Bull. Brit. Mus. (Nat. Hist.) Zool., 31:* 101–155.

Bradford, J. M., and J. B. A. Jillett, 1974. A revision of generic definitions in the Calanidae (Copepoda, Calanoida). *Crustaceana, 27:* 5–16.

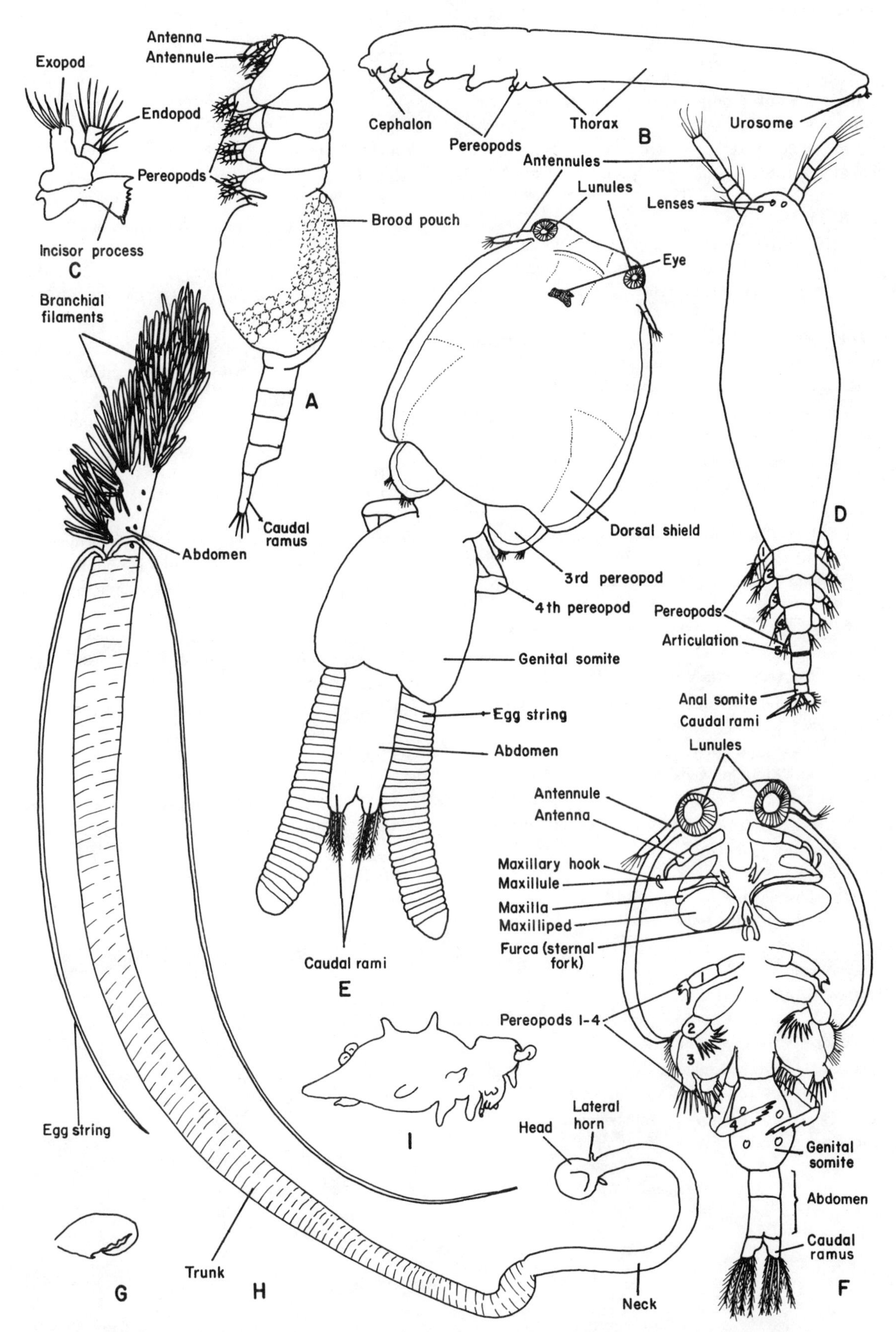

Figure 14 Copepoda, parasitic: A—C, Notodelphoida [after Illg, 1958]. A. Female, *Notodelphys* (lateral view); B. Female, *Scolecodes* (lateral view); C. Mandible, *Notodelphys;* D. Monstrilloida, female (dorsal view) [after Wilson, 1932]; E—H, Siphonostomata. E. Female (dorsal view); F. Male (ventral view); G. Maxilliped of male; H. Female, *Penella;* I. Poecilostomata, female [after Ho, 1972].

Brandorff, G-O., 1976. The geographic distribution of the Diaptomidae in South America (Crustacea, Copepoda). *Rev. Brasil. Biol.*, *36:* 613–627.

Brodsky, K. A., 1950. Veslonogie rachki Calanoida dal'nevostochnikh morei SSSR i poliarnogo basseina. Akad. Nauk, SSSR, Opredeliteli po faune SSSR, *Izd. Zool. Inst. Akad. Nauk SSSR*, *35:* 1–441.

Coull, B. C., 1973. Harpacticoid copepods (Crustacea) of the family Tetragonicipitidae Lang: A review and revision with keys to the genera and species. *Proc. Biol. Soc. Wash.*, *86:* 9–24.

Cressey, R. F., 1967a. Genus *Gloiopotes* and a new species with notes on host specificity and intraspecific variation (Copepoda: Caligoida). *Proc. U.S. Natl. Mus.*, *122:* 1–22.

———, 1967b. *Catrius*, a new genus of caligoid, with a key to the genera of Caliginae. *Proc. U.S. Natl. Mus.*, *123:* 1–8.

Cressey, R. F., and C. Patterson, 1973. Fossil parasitic copepods from a Lower Cretaceous fish. *Science*, *180:* 1283–1285.

Dudley, P. L., 1966. *Development and systematics of some Pacific marine symbiotic copepods. A study of the biology of the Notodelphyidae, associates of ascidians.* 282 pp. Seattle and London: University of Washington Press.

González, J. G., and T. E. Bowman, 1965. Planktonic copepods from Bahía Fosforescente, Puerto Rico, and adjacent waters. *Proc. U.S. Natl. Mus.*, *117:* 241–304.

Gurney, R., 1933. Notes on some Copepoda from Plymouth. *J. Mar. Biol. Assoc.*, U.K., *19:* 299–304.

Hamond, R., 1968. Some marine copepods (Misophrioida, Cyclopoida, and Notodelphyoida) from Norfolk, Great Britain. *Crustaceana*, Suppl. 1: 37–60.

Heegaard, P., 1947. Contribution to the phylogeny of the arthropods: Copepoda. *Spolia Zool. Mus. Nauniensis*, *8:* 1–236.

———, 1948. Discussion of the mouth appendages of copepods. *Ark. Zool.*, *40A:* 1–8.

Ho, J-S., 1972. Four new parasitic copepods of the family Chondracanthidae from California inshore fishes. *Proc. Biol. Soc. Wash.*, *85:* 523–540.

Illg, P. L., 1958. North American copepods of the family Notodelphyidae. *Proc. U.S. Natl. Mus.*, *107:* 463–649.

Isaac, M. J., 1974. Monstrillid copepods in the Zoological museum, Berlin. *Mitt. Zool. Mus. Berlin*, *50:* 131–135.

———, 1975 Copepoda. Suborder: Monstrilloida. *Zooplankt. Cons. Int. Explor. Mer.* No. 144–145, 10 pp.

Jakobi, H., 1976. Über ökologische und biogeographische Trends innerhalb der Harpactioiden (Copepoda-Crustacea). *Int. J. Speleol.*, *8:* 93–106.

Kabata, Z., in press. *British parasitic copepods.*

Lang, K., 1946. A contribution to the question of the mouthparts of the Copepoda. *Ark. Zool. Uppsala*, *38A:* 1–24.

———, 1948a. Copepoda "Notodelphyoida" from the Swedish west coast, with an outline on the systematics of the copepods. *Ark. Zool. Uppsala*, *40A:* 1–36.

———, 1948b. *Monographie der Harpacticiden*, *1*, *2* 1682 pp. Stockholm: Ohlsson.

Lawson, T. J., 1977. Community interactions and zoogeography of the Indian Ocean Candaciidae (Copepoda: Calanoida). *Mar. Biol.*, *43:* 71–92.

Lowe, E., 1936. Anatomy of a marine copepod. *Trans. Roy. Soc. Edinburgh*, *58:* 561–603.

Marshall, S. M., and A. P. Orr, 1955. *The biology of a marine copepod, Calanus finmarchicus.* 188 pp. Edinburgh: Oliver and Boyd.

Moraitou-Apostolopoulou, M., 1974. An ecological approach to the systematic study of planktonic copepods in a polluted area (Saronic Gulf—Greece). *Boll. Pesca. Pisciolt. Idrobiol.*, *29:* 29–47.

Owre, H. B., and M. Foyo, 1967. *Copepods of the Florida current.* Fauna Caribaea, No. 1, Crustacea, Pt. 1: Copepoda, 137 pp. Miami, Fla.: Institute of Marine Science.

Sars, G. O., 1903. *An account of the Crustacea of Norway . . . 5: Copepoda Harpacticoida*, Pts. 1 and 2. Bergen: Bergen Museum.

Stock, J. H., A. G. Humes, and R. U. Gooding, 1963. Copepoda associated with West Indian Invertebrates. Parts 3 and 4. *Stud. Fauna Curaçao Carib. Is.*, *17:* 1–37, 1–74.

Vervoort, W., 1969. Caribbean Bomolochidae (Copepoda: Cyclopoida). *Stud. Fauna Curaçao Carib. Is.*, *28;* 1–125.

Wilson, C. B., 1932. The copepods of the Woods Hole region. *Bull. U.S. Natl. Mus.*, 158: 1–635.

Wilson, M. S., and H. C. Yeatman, 1966. Free-living Copepoda. In W. T. Edmondson (ed.), *Fresh-water biology*, 2nd ed., pp. 735–861. New York and London: John Wiley and Sons.

Yamaguti, S., 1963. *Parasitic copepods and Branchiura of fishes.* 1104 pp. New York and London: John Wiley and Sons.

CLASS BRANCHIURA Thorell, 1864

Recent species	Approximately 130 in four genera: *Argulus*, *Chonopeltis*, *Dolops*, and *Dipteropeltis*.
Size range	0.5–30 mm.
Carapace	Dorsoventrally flattened; usually broad.
Eyes	Sessile, compound, and usually 1–3 median nauplius eyes.
Antennules	Reduced, usually with small hooks or spines; sometimes absent.
Antennae	Uniramous; reduced or absent.
Mandibles	Highly modified; in *Argulus*, at least, usually incorporated in medial proboscis system.
Maxillulae	Uniramous, without endites; highly modified; with distal claws in *Dolops;* other three genera with paired, stalked suction discs.
Maxillae	Usually with hooks or spines.
Maxillipeds	None.
Thoracic appendages	Four pairs of biramous thoracopods; used in swimming; usually sexually dimorphic.
Abdominal appendages	None.
Telson	Absent; anal somite terminating in caudal rami.
Tagmata	Cephalothorax and abdomen; or cephalon, pereon, and abdomen.
Somites	Head with 5 + 1st thoracic; thorax or pereon usually with 3; abdomen unsegmented.
Sexual characters	Gonopores of female at bases of 4th thoracopods (oviducts apparently function alternatively); male with single gonopore medially on ventral surface of last cephalothoracic somite; female with paired spermatheca on abdomen anteroventrally. Thoracic appendages usually sexually dimoprhic.
Sexes	Separate.
Larval development	Anamorphic or epimorphic; eggs deposited on stones or plants; hatch as nauplius (some *Argulus* spp.) or naupliar and metanaupliar stages passed through in egg, hatch as juveniles.
Fossil record	Recent.
Feeding types	Parasitic, adapted for piercing and sucking.
Habitat	All are temporary to permanent ectoparasites of fishes, occurring in marine, brackish, and freshwater. Temporary parasites are good swimmers and apparently are not host specific.
Distribution	*Argulus* —worldwide; *Chonopeltis* —Africa; *Dolops* —primarily South America, but also known from Tasmania and Africa; *Dipteropeltis* — South America.

The instructions for the study of the Branchiura are based on specimens of *Argulus* (see Figure 15). The branchiurans are morphologically adapted for a parasitic existence. The somewhat flattened carapace arises at the base of the cephalic region and extends posteriorly as a pair of lobes or alae, partially covering the thoracic somites. The carapace is fused only with the first thoracic somite. The cephalon is delineated posteriorly by the

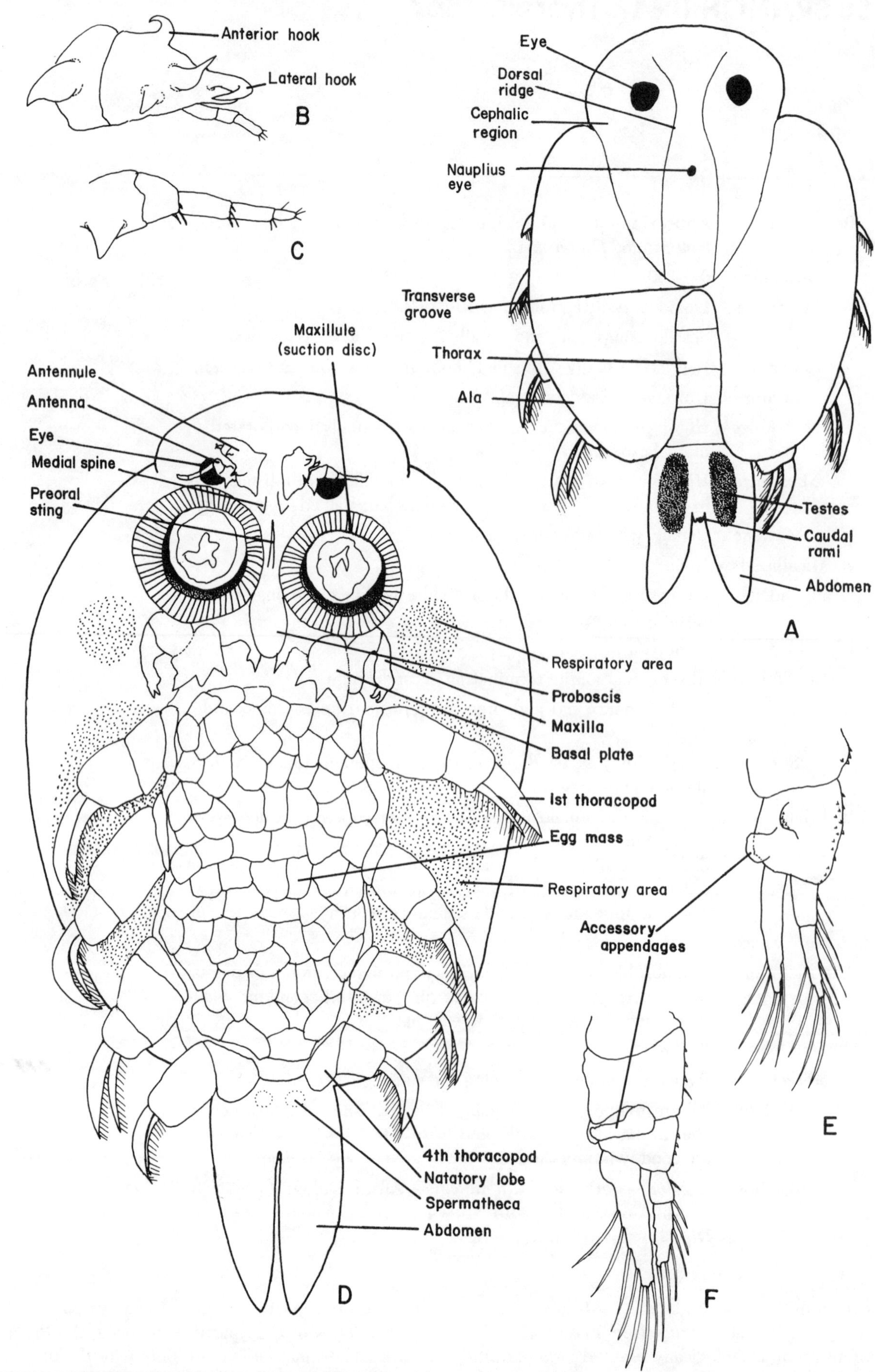

Figure 15 Branchiura: A. Male (dorsal view); B. Antennule; C. Antenna; D. Female (ventral view); E. Male 3rd thoracopod; F. Male 4th thoracopod.

transverse or cervical groove and laterally by ridges. In dorsal view, the prominent paired compound eyes will be clearly visible. Frequently a single small median or nauplius eye also is present. Usually a pair of longitudinal dorsal ridges, situated between the eyes, can be observed. In most specimens, a pair of respiratory areas, lying on the ventral surface of the carapace, may be distinguished from the ventral side. Between the alae the free thoracic somites also can be seen. The abdomen is unsegmented but lobed posteriorly. Distinguish the caudal rami between the lobes. [Bowman (1971) considers that these rami are uropods.]

In ventral view, examine the cephalic and thoracic appendages. The antennules and antennae usually are reduced in size or occasionally absent; in argulids, the basal segments of both generally are broadened and frequently armed with spines or hooks. The most prominent appendages are the maxillulae, which are modified as suction discs. In the midline and somewhat anteriorly, locate the preoral sting, a retractile piercing mechanism with a poison gland at its base. Posterior to the sting is the large proboscis. The mandibles are completely enclosed in the proboscis and thus are not easily observable. The maxillae usually have broadened, toothed basal plates and often 2 pairs of postmaxillary spines.

Remove and examine 1 or more of the thoracopods. The first 2 pairs may be provided with epipods. In the male, the 2nd through 4th thoracopods have accessory appendages that are used in copulation. The 4th thoracopods usually are basally expanded into natatory lobes.

In males the testes usually are very apparent as a pair of elongate bodies lying in the abdomen to each side of the midline. Be sure to examine specimens of both sexes. In females observe the pair of spermatheca located on the ventral surface of the abdomen.

Larval development in branchiurans may be one of two types. In the first, a form of anamorphic development occurs; the young hatch as nauplii, swim by means of the antennae, mandibles, and 1st thoracopods, and attach to a host by means of the distal hooks of the maxillulae. After one molt all 4 pairs of thoracopods are functional. The second type of development is epimorphic; the young hatch with functional appendages and usually are similar to the adult form except in size and sucker development.

The Branchiura were, until recently, classified as a taxon of the Copepoda. What characters can you find that justify the separation of this group of fish parasites into a separate class? In which characters do they agree with the Copepoda?

References

Cressey, R., 1971. Two new argulids (Crustacea: Branchiura) from the eastern United States. *Proc. Biol. Soc. Wash., 84:* 253–258.

Fryer, G., 1961. Larval development in the genus *Chonopeltis* (Crustacea; Branchiura). *Proc. Zool. Soc. London, 137:* 61–69.

Gurney, R., 1948. The British species of fish louse of the genus *Argulus. Proc. Zool. Soc. London, 118:* 553–558.

Ivánfi, E., 1926. Morphologische und biologische Untersuchungen an des Karpfenlaus *(Argulus foliaceus). Arch. Balatonicum, 1:* 145–163.

Markewitsch, A. P., 1931. Parasitsche Copepoden und Branchiuren des Aralsees, nebst systematischen Bermerkungen über die *Ergasilus* Nordmann. *Zool. Anz., 96:* 121–143.

Martin, M. F., 1932. On the morphology and classification of *Argulus. Proc. Zool. Soc. London, 1932:* 771–806.

Meehan, L. O., 1940. A review of the parasitic Crustacea of the genus *Argulus* in the collection of the United States National Museum. *Proc. U.S. Natl. Mus., 88:* 459–522.

Monod, T., 1928. Les Argulidés de musée du Congo. Inventaire systématique, comprenent la description d'*Argulus schoutedeni* nov. sp., et liste génerale critique des branchiures africains, tant marins que dulca quicoles. *Rev. Zool. Bot. Afr., 21:* 1–36.

Thiele, J., 1904a. Diagnosen neuer Arguliden-Arten. *Zool. Anz., 23:* 46–48.

———, 1904b. Beiträge zur Morphologie der Arguliden. *Mitteil. Zool. Mus. Berlin, 2:* 1–51.

Wilson, C. B., 1902. North American parasitic copepods of the family Argulidae, with a bibliography of the group and a systematic review of all known species. *Proc. U.S. Natl. Mus., 25:* 635–742.

———, 1944. Parasitic copepods in the United States National Museum. *Proc. U.S. Natl. Mus., 94:* 529–582.

Yamaguti, S., 1963. *Parasitic Copepoda and Branchiura of fishes.* 1104 pp. New York and London: John Wiley and Sons.

CLASS CIRRIPEDIA Burmeister, 1836

The Cirripedia is a relatively large, highly successful group of sedentary crustaceans whose members are noted for their diversity in form and habitat. And, at least among free-living taxa, they also are the most uncommon-looking crustaceans in the superclass. Until the mid–nineteenth century, when the crustacean affinities of their larvae were discovered, cirripeds were thought to be a type of mollusk. The class is comprised of four orders: Acrothoracica, Thoracica, Ascothoracica, and Rhizocephala. Representatives of the first two orders usually are free-living; those of the latter two are parasitic.

ORDER ACROTHORACICA Gruvel, 1905

Recent species	40–45 in eight genera.
Size range	0.4–20.0 mm.
Carapace	Mantle thin, rarely with 1 pair of calcareous plates.
Eyes	Nauplius eye in nauplius, with compound eyes in cyprid; nauplius eye sometimes retained in adult.
Antennules	Uniramous in cyprid and male; absent in female.
Antennae	Only in nauplius, biramous, occasionally absent.
Mandibles	With palp in adult female; absent in male.
Maxillulae	Present only in female.
Maxillae	Present only in female.
Maxillipeds	May be represented by pair of mouth cirri.
Thoracic appendages	In female, 3–5 pairs of uni- or biramous terminal cirri; absent in male.
Abdominal appendages	None.
Telson	Absent.
Tagmata	Head and thorax in adult female; not differentiated in male.
Somites	Head with 5 + 1 thoracic (= presoma); thorax usually with 6 (7 + 8 fused).
Sexual characters	Openings of oviducts near bases of 1st cirri in females; male with or without penis and with or without specialized attachment structures.
Sexes	Separate; males reduced.
Larval development	Metamorphic; nauplius → cyprid; frequently released as cyprid.
Fossil record	Carboniferous to Recent.
Feeding types	Filter feeders (female); male nonfeeding.
Habitat	Marine; in burrows in living and dead coral and limestone and shell.
Distribution	Worldwide.

The Acrothoracica differ from all other cirripeds in being burrowers. The female is permanently attached to the burrow and usually lacks calcareous plates. The reduced male lacks mouthparts. The Acrothoracica are more closely related to the Thoracica than to the other orders. They differ in several characters: (1) the posterior cirri (sometimes reduced in number) are terminal in position; (2) calcareous plates usually are not developed; (3) the entire exoskeleton, instead of only the inner lining and appendages, is molted. Sexes always are separate. Naupliar substages usually are passed through in the egg. Acrothoracicans always are free-living; their specialized habitats, such as burrows in coral and shell substrates, do not make them easy to study.

The acrothoracican usually is not visible, but its presence is marked by a tapered slit in the substrate. To remove the female acrothoracican from her burrow, one must either chip away the matrix surrounding the burrow or dissolve the matrix in acid. Shell material probably is best decalcified by placing the specimen in dilute hydrochloric acid (1 or 2%), unless the specimens have been collected alive. In the latter case, placement of the specimens in a large volume of Bouins' solution will serve both to preserve the animals and to decalcify the shelly matrix. Even with decalcification, most specimens will not be freed completely and must be extracted gently with a pair of fine forceps. Specimens treated in either of the manners described usually will take on a dark red coloration that is an aid in their recognition.

Any more than a superficial examination of the gross morphology of the female will not be possible under routine classroom conditions (see Figure 16). Males may be attached to the female's body or to the walls of the burrow. The females are small but the males are even smaller, and care should be taken to examine both the burrow and the female for them. Males often are attached to the periphery of the "horny" attachment knob. This region of the mantle is firmly attached to the burrow and apparently is not cast routinely with each molt. Consequently, as the female grows, this area usually expands in a series of concentric rings.

The most obvious external features of the typical acrothoracican are the lips of the mantle aperture. The thickened plates may be homologous or analogous to the occludent margin of the mantle of the thoracicans. These lips frequently are armed with teeth, hooks, or spines. Observation of most taxonomically diagnostic characters requires treatment of the specimens with heated potas-

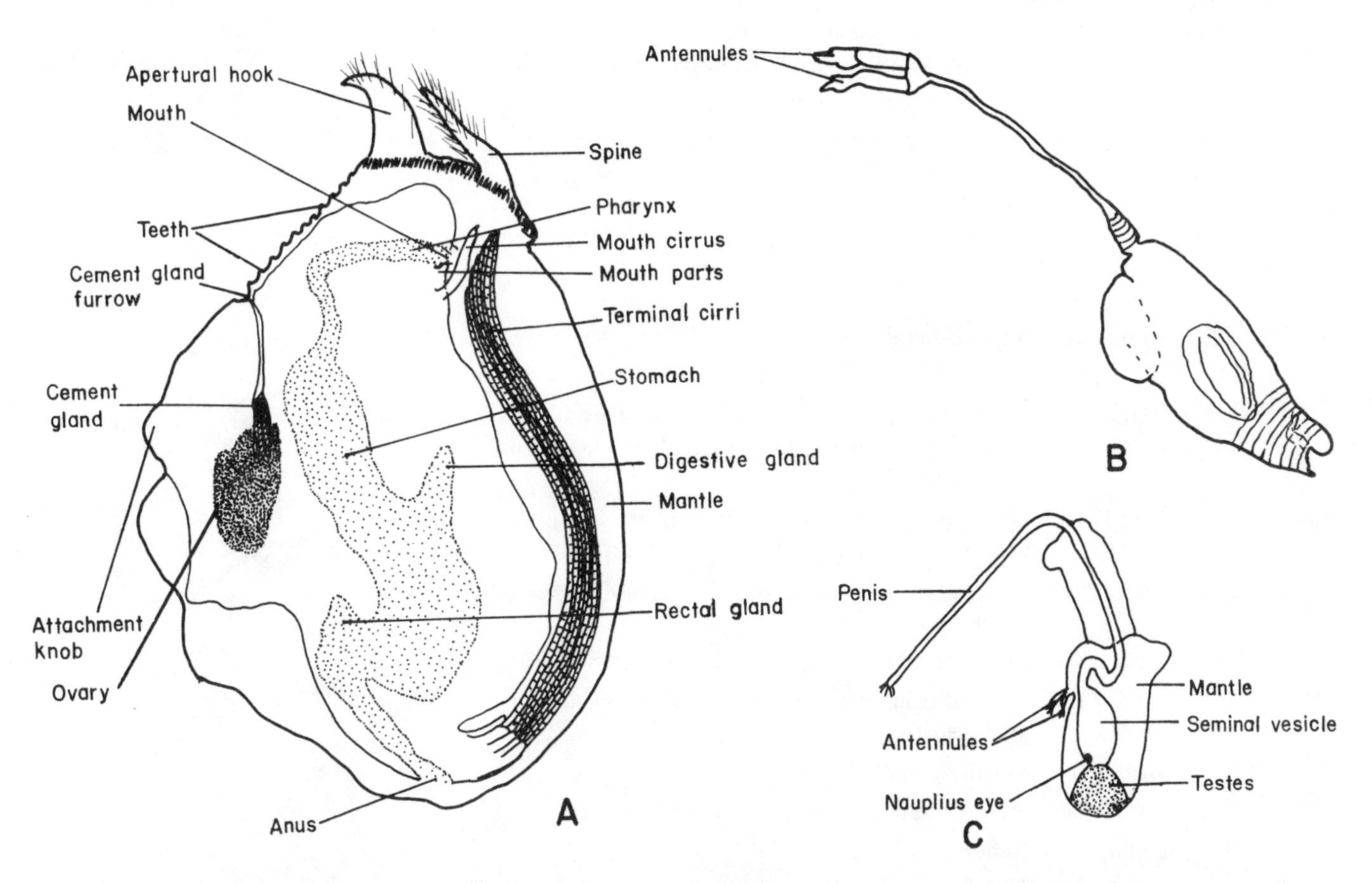

Figure 16 Cirripedia, Acrothoracica: A. Female, *Lithoglyptes* (lateral view); B. Male, *Lithoglyptes* [A, B after Tomlinson and Newman, 1960]; C. Male, *Trypetesa* [after Tomlinson, 1955].

sium hydroxide (10%) to remove the soft parts and then clearance in glycerin. If treated cleared specimens are available, you should examine the thoracic cirri, paying particular attention to the number of pairs and to the mouthparts and mouthcirri. The setation of the cirri is often of taxonomic significance, particularly the size and distribution of setae; however, it is not expected that you will be able to make specific identifications.

References

Batham, E. J., and J. T. Tomlinson, 1965. On *Cryptophialus melampygos* Berndt, a small boring barnacle of the order Acrothoracica abundant in some New Zealand molluscs. *Trans. Roy. Soc. N.Z.*, *7:* 141–154.

Berndt, W., 1906. Studien an bohrenden Cirripedien (Ordnung Acrothoracica Gruvel, Abdominalia Darwin). Pt. I: Die Cryptophialidae. *Arch. Biontol.*, *1:* 167–210.

———, 1907. Über das System der Acrothoracica. *Arch. Naturg.*, *73:* 287–289.

Krüger, P., 1940. Cirripedia. In H. G. Bronn, *Klassen und Ordnungen des Tierreichs*, 5(1) *3:* 1–560. Leipzig: Akad. Verl.

Newman, W. A., V. A. Zullo, and T. H. Withers, 1969. Cirripedia. In R. C. Moore (ed.), *Treatise on invertebrate paleontology*, Pt. R, Arthropoda, 4, *1:* R206–R295. Lawrence, Kansas: Geol. Soc. America and Univ. Kansas.

Schlaudt, C. M., and K. Young, 1960. Acrothoracic barnacles from the Texas Permian and Cretaceous. *J. Paleo.*, *34:* 903–907.

Tomlinson, J. T., 1955. The morphology of an acrothoracican barnacle, *Trypetesa lateralis*. *J. Morph.*, *96:* 97–122.

———, 1969a. The burrowing barnacles (Cirripedia: Order Acrothoracica). *Bull. U.S. Natl. Mus.*, 296: 1–162.

———, 1969b. Shell-burrowing barnacles. *Am. Zool.*, *9:* 837–840.

———, 1973. Distribution and structure of some burrowing barnacles, with four new species (Cirripedia: Acrothoracica). *Wasmann J. Biol.*, *31:* 263–288.

Tomlinson, J. T., and W. A. Newman, 1960. *Lithoglyptes spinatus*, a burrowing barnacle from Jamaica. *Proc. U.S. Natl. Mus.*, *112:* 517–526.

Turquier, Y., 1972. Contribution á la connaissance des Cirripèdes acrothoraciques. *Arch. Zool. Exp. Gèn.*, *113:* 499–551.

Utinomi, H., 1957. Studies on the Cirripedia Acrothoracica. 1. Biology and external morphology of the female of *Berndtia purpurea* Utinomi. *Publ. Seto Mar. Biol. Lab.*, *6:* 1–26.

———, 1961. Studies on the Cirripedia Acrothoracica. III. Development of the female and male of *Berndtia purpurea* Utinomi. *Publ. Seto Mar. Biol. Lab.*, *9:* 413–446.

———, 1963. Studies on the Cirripedia Acrothoracica. IV. Morphology of the female of *Balanodytes taiwanus* Utinomi. *Publ. Seto Mar. Biol. Lab.*, *11:* 57–73.

———, 1964. Studies on the Cirripedia Acrothoracica. V. Morphology of *Trypetesa habei* Utinomi. *Publ. Seto Mar. Biol. Lab.*, *12:* 117–132.

White, F., 1970. The chromosomes of *Trypetesa lampas* (Cirripedia, Acrothoracica) *Mar. Biol.*, *5:* 29–34.

ORDER THORACICA Darwin, 1854

Recent species	Approximately 800.
Size range	Up to 80 cm height; 20 cm diameter.
Carapace	Membranous "mantle" formed by bilateral folds of carapace; usually armed with calcareous plates; occasionally spinous or without plates or spines.
Eyes	Nauplius eye deeply embedded, rarely discernable in adult; nauplius and compound eyes present in cyprid.
Antennules	Vestigial in adult; uniramous in nauplius and cyprid; used for temporary attachment by cyprid.
Antennae	Absent in adult and cyprid; biramous in nauplius.
Mandibles	Without palp (mandibular palp shifted to labrum); with modified gnathobases.
Maxillulae	Modified gnathobase.
Maxillae	Modified gnathobase; pair sometimes fused.
Maxillipeds	None.
Thoracic appendages.	Usually 6 pairs of biramous cirri; in males, reduced, vestigial, or absent.
Abdominal appendages	None.

Telson	Absent.
Tagmata	Head and thorax.
Somites	Head with 5; thorax with 6.
Sexual characters	Openings of paired oviducts at bases of 1st cirri (1st thoracopods) in hermaphrodite and female; penis between bases of 6th cirri always present in hermaphrodite, sometimes in males.
Sexes	Rarely separate; usually functional hermaphrodites; occasionally with dwarf or complemental males attached to female or hermaphrodite respectively.
Larval development	Metamorphic; 6 naupliar substages → cyprid; occasionally naupliar substages passed through in egg.
Fossil record	U. Silurian to Recent.
Feeding types	Typically filter feeders; sometimes predators (free-living); occasionally absorption (parasites); males sometimes nonfeeding.
Habitat	Marine or brackish water, intertidal to abyssal, on rocks, pilings, shells, and other hard substrates, and on plants and other animals.
Distribution	Worldwide.

Thoracic cirripeds are attached permanently as adults. The mantle usually bears calcareous plates; the external surface, including the plates, is never molted. Six pairs of biramous appendages (cirri) are present and (except the 1st pair) are evenly distributed along the thorax. The 1st pair is closely associated with the mouthparts (trophi). Sexes usually are combined, rarely separate; males, when present, are reduced and associated with hermaphrodites as complemental males or with females as dwarf males. Thoracic cirripeds are primarily free-living, often commensal; occasionally they are symbiotic, rarely they are parasitic. Larval development is metamorphic and typically includes a nauplius stage with 6 substages and a cyprid stage. The nauplius is distinctive in that it can be distinguished from other crustacean nauplii by the presence of frontolateral horns. With metamorphosis, the broad dorsal shield of the nauplius becomes the bivalved carapace of the cyprid. In the cyprid the antennules are prehensile, the antennae absent, the mouthparts rudimentary, and the thoracic appendages natatory. Occasionally the nauplius stage is passed through in the egg and the young hatch as fully developed cyprids.

The Thoracica includes three suborders, each sufficiently distinct to be treated individually. A brief synopsis of major diagnostic characters precedes the discussion of each suborder.

SUBORDER LEPADOMORPHA Pilsbry, 1916

Calcareous plates	Usually present on capitulum, sometimes also on peduncle; capitulum with 5 or 6 principal plates; sometimes also with 1 or 2 paired laterals in upper whorl; sometimes also with secondary whorl of 3 pairs of laterals; with or without subrostrum and subcarina; less often also with basal whorls of smaller laterals; occasionally with fewer than 5 principal plates, with spines or without plates or spines.
Primordial plates	Chitinous principal plates except rostrum, occasionally 1 or 2 additional plates present in cyprid; retained at umbos of adult.
Basis	Absent.
Labrum	Bullate.
Caudal appendages	Usually present.

Basidorsal point (penis)	Absent.
Filamentary appendages	Usually present.
Sexes	Usually hermaphroditic, occasionally separate; sometimes with complemental or dwarf males.

The Lepadomorpha are distinguished from the sessile barnacles by being attached to the substrate by a stalk or peduncle rather than by a membranous or calcareous basis. The capitulum that typically is distinguished from the peduncle encloses the body and usually bears calcareous plates. The number and arrangement of the plates are diagnostic characters. There are 5 or 6 principal plates: paired scuta and terga, which border the occludent margin of the orifice, and an unpaired carina and sometimes a rostrum. In addition, there may be a second whorl of 9 pairs of lateral plates and sometimes several basal whorls of small plates.

The directions for the study of pedunculate morphology are based on representatives of the genera *Pollicipes* and *Lepas* (see Figure 17). *Pollicipes* is a member of the family Scalpellidae, which includes many diverse groups principally found in deep water, and is characterized by a capitulum with more than 5 principal plates and a peduncle with calcareous scales. In contrast, *Lepas* is a member of the family Lepadidae, which is characterized by a capitulum with 5 or fewer plates and a naked peduncle. If available, compare and contrast representatives of both genera. Note that the peduncle of *Pollicipes* is quite flexible although the integument is thick and covered with very small scales. The capitulum has several whorls of plates that decrease in size from above downward to the peduncle; the lowest whorl may have over 100 small plates. Identify the 6 principal plates. The rostrum is below the scutum in the midline. There is a secondary plate, a latus, between the scutum and tergum at the rostral level. In the next whorl identify the subrostrum and subcarina. Compare these capitular plates with those of *Lepas*, in which the plates of the capitulum include only the 5 primary plates, that is, paired scuta and terga and unpaired carina. The lepadid peduncle is equally as flexible, but the integument is thinner and unarmed.

By making two incisions on one side of the animal, the first to the side of the rostrum and the second to the side of the carina, the capitulum can be laid back to expose the body of the animal within (see Figure 18). The thoracic appendages (cirri) are the most apparent structures. They are orientated in such a manner that they can be extended through the orifice toward the rostral side of the animal. Directly in front of the cirri are the trophi. The adductor muscle appears as a dark, very prominent ovoid structure directly behind the rostrum and in front of the trophi. Below the pedicles of the cirri a series of *(Pollicipes)* or a few *(Lepas)* filamentary appendages can be observed. These appendages are not found in many other genera, nor even in all species of one genus. In *Lepas* they usually are limited to association with the 1st cirrus; their function is unknown. Remove the 6 cirri from one side in sequence while the animal is still within the capitulum, mounting them on a microscope slide. If a temporary mount is all that is desired, a medium such as glycerin is adequate; if you are making a permanent mount, a medium such as polyvinyl alcohol lactophenol is recommended. A method commonly used by cirripedologists is to mount the cirri from each side of the animal together under a single cover slip in the following manner:

Right side	I,	II,	III
	IV,	V,	VI
Left side	III,	II,	I
	VI,	V,	IV

By mounting the cirri in this way, you will be able to identify each cirrus. Observe that the anterior cirri (first 3 pairs) do not differ greatly from the posterior. Examine the pedicle (protopod) and rami (anterior and posterior, rather than endopod and exopod) of each cirrus. Are there differences in the setation of the cirri or of the two rami of one cirrus? The penis arises from between the bases of the 6th cirri. A pair of caudal appendages situated on each side of the anus are very small in *Pollicipes* and simple points continuous with the thorax in *Lepas*. Caudal appendages occur in all but a few genera and usually are more conspicuous, in some being long and multiarticulated. Free the specimen from the capitulum and remove the cirri from the other side before attempting to dissect out the mouthparts.

The trophi form a compact unit and can be removed, as such, from the rest of the body. First examine the trophi as a unit; the most apparent structure is the bullate labrum. Remove the paired mouthparts from the side opposite the labrum as follows: the maxillae (referred to by cirripedologists as the 2nd maxillae), the maxillulae (1st maxillae), the mandibles, and the labrum with palps. Mount the paired mouthparts with both members facing in the same direction so that both the outer and inner surfaces may be examined more easily. The maxillae are fused together basally and are best removed simultaneously, then separated when mounted.

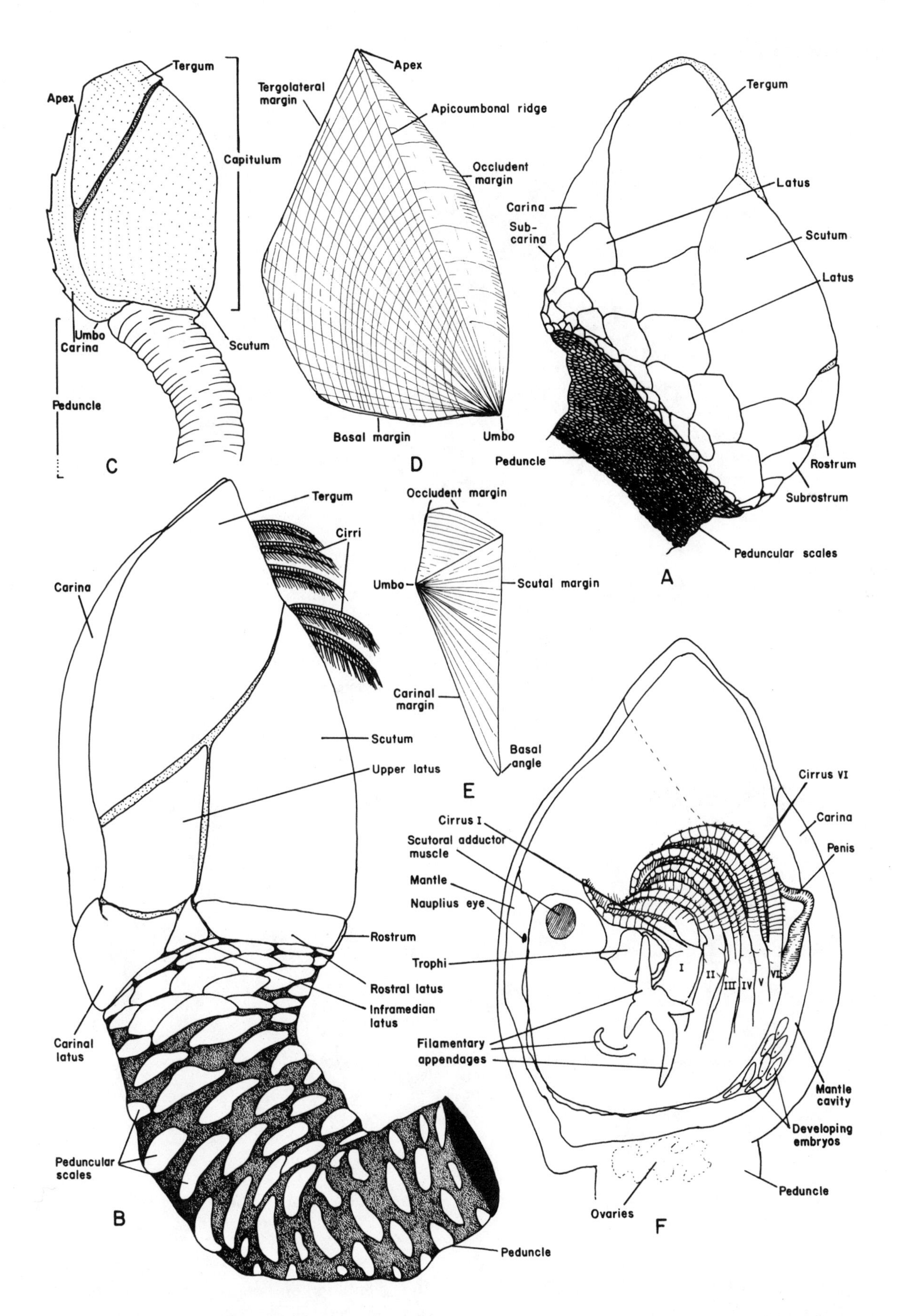

Figure 17 Cirripedia, Thoracica, Lepadomorpha: A. *Pollicipes* (lateral view); B. Scalpellid (lateral view); C. Lepadid (lateral view); D. Lepadid scutal valve; E. Lepadid tergal valve; F. Lateral view of animal within capitulum.

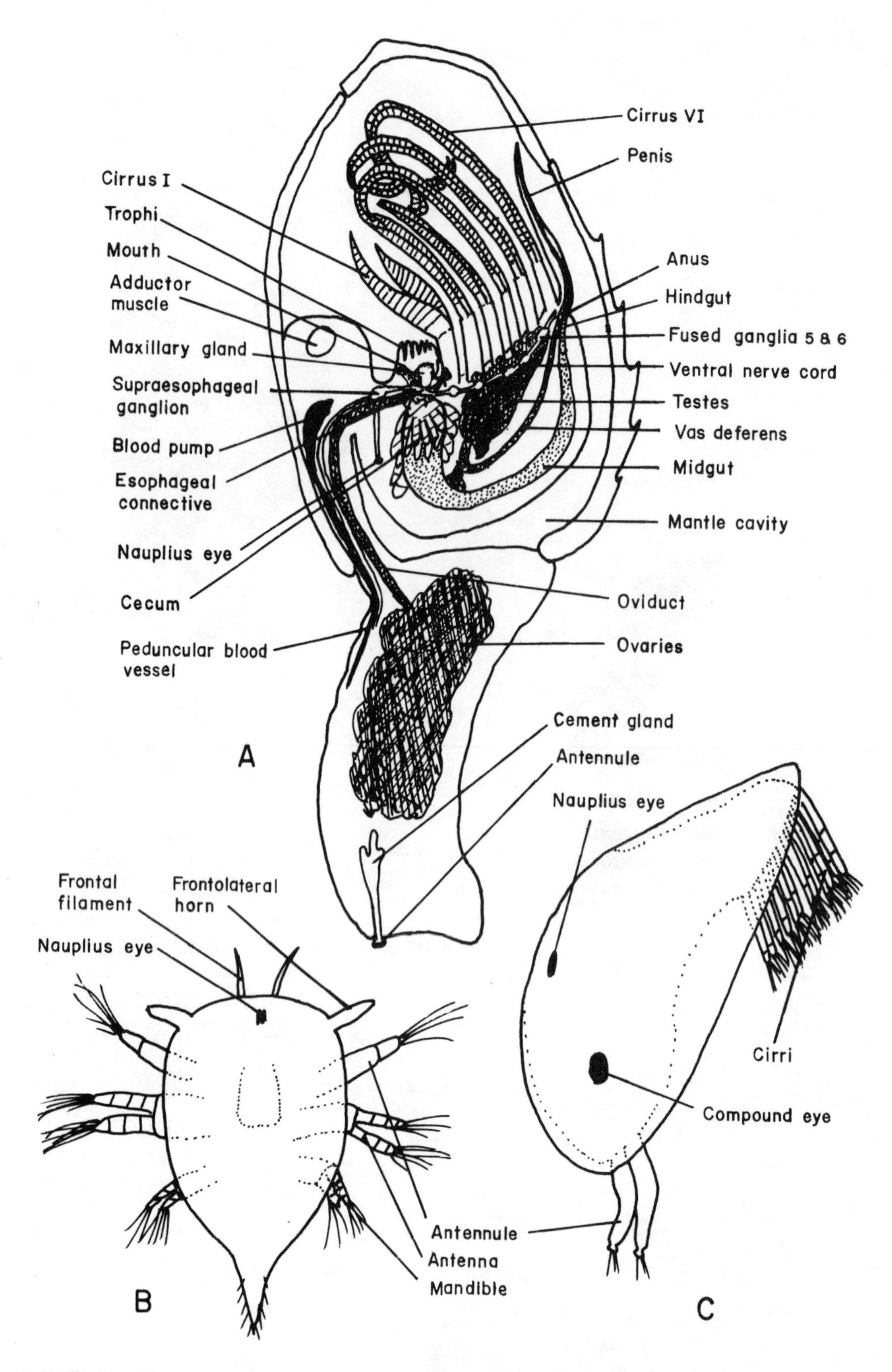

Figure 18 Cirripedia, Thoracica, Lepadomorpha: A. Diagrammatic lepadid with musculature removed to show major organ systems; B. Nauplius; C. Cyprid.

Remove the palps and much of the attached muscle and tissue from the labrum before mounting. The palps are attached to the lateral margins of the labrum rather than to the mandibles in all orders of the Cirripedia except the Acrothoracica.

The orientation of the cirriped body in the capitulum makes the study of the internal organ systems a little difficult. After removing the capitulum from one side, cut the peduncle on both the rostral and carinal sides and gently tease the integument away from the underlying muscles. Carefully remove the external membrane and superficial muscles that cover the body of the animal. In the peduncle remove the muscle fibers to expose the ovaries and cement glands, being careful not to destroy the ducts of the glands that follow the muscle wall to the antennules. The cement glands usually are quite small and may be difficult to distinguish in poorly preserved material. The peduncular space largely is filled by the tubules of the ovaries. Toward the midline of the peduncle rostrally locate the paired oviducts, and if possible trace one as it leads from the ovary, along the rostral side of the peduncle into the capitulum, and to the 1st cirrus. The oviduct terminates in glandular tissue within the pedicle of the cirrus that can be seen by removing the external membrane and the muscles of the pedicle.

In close proximity to the oviduct near the rostral margin of the capitulum identify the large membranous blood pump; it extends into the peduncle as the peduncular blood vessel. Cirripeds lack a true heart, and the blood pump, through alternate compression and release of compression, serves as the propulsory mechanism. Blood flows from the pump into the peduncle by way of the peduncular blood vessel and is carried back to the body through parenchymatous spaces into the blood sinuses in the mantle and body.

Often the most prominent structure in the body will be the greatly ramified testes. It may be necessary to remove the cirri from the exposed side of the body to be able to trace the vas deferens to the penis, as these tubes often are located near the midline of the body in close association with the midgut.

At the base of the labrum locate the mouth. Food is passed through the short esophagus or foregut into the U-shaped midgut, expanded anteriorly as a stomach, near which a pair of ceca are given off. Trace the midgut posteriorly; it will be difficult to distinguish the short hindgut from it as both are cuticularized. The anus opens near the base of the penis between the 6th pair of cirri. Adjacent to the esophagus is the end-sac of the maxillary gland with a duct opening on the maxilla.

Between the blood pump and the trophi locate the relatively small supraesophageal ganglion, which is connected with the ventral nerve cord and subesophageal ganglia by very long esophageal connectives. The ganglia of the 2nd to 4th cirri are distinct, whereas those of the 5th and 6th are fused in *Lepas*.

SUBORDER VERRUCOMORPHA Pilsbry, 1916

Calcareous plates	Asymmetrically arranged, forming wall and operculum; wall with 4 principal plates (rostrum, carina, fixed scutum, and fixed tergum); operculum with 2 principal plates (articulated scutum and tergum).
Primordial plates	Five chitinous plates (carina and paired scuta and terga).
Basis	Calcareous or membranous.
Labrum	Bullate.
Caudal appendages	Usually present.
Basidorsal point (penis)	Absent.
Filamentary appendages	Absent.
Sexes	Hermaphroditic.

These sessile barnacles exhibit pronounced asymmetry. In *Verruca*, the only Recent genus, the wall is composed of a rostrum and carina on one side and a fixed scutum and tergum on the other. The other scutum and tergum form the lidlike operculum. The basis is membranous or calcareous. In the characters derived from the body, *Verruca* resembles the lepadomorph cirripeds more than the balanomorphs, particularly in the mouthparts and the usual occurrence of caudal appendages. Verrucomorphs have a widespread distribution, primarily in deep water, usually on living organisms, less frequently on rocks.

Specimens of this suborder usually are not available for study; however, if some are, remove and clean the articulated scutum and tergum according to the instructions provided for the balanomorph opercular valves. Compare the verrucomorph shell with those of the balanomorphs, and, following the instructions given for the dissection and mounting of the lepadomorph animal, make a slide preparation of the appendages and mouthparts. What similarities and differences do you find between the verrucomorph and lepadomorph animals?

SUBORDER BALANOMORPHA Pilsbry, 1916

Calcareous plates	Symmetrically arranged, forming wall and usually operculum; wall with 2 principal plates (rostrum and carina) and 1–3 paired laterals; occasionally with plates wholly coalesced; rarely also with 1 or more basal whorls of small plates; operculum usually with 4 principal plates (i.e., articulated scutum and tergum on each side of aperture); occasionally opercular valves separate, fused, or absent.
Primordial plates	Absent.
Basis	Calcareous or membranous.
Labrum	Usually not bullate, notched; bullate in primitive members.
Caudal appendages	Present only occasionally.
Basidorsal point (penis)	Present only in Balanidae.
Filamentary appendages	Absent.
Sexes	Hermaphroditic; rarely with complemental males.

The Balanomorpha are differentiated from the other sessile suborder, the Verrucomorpha, by the symmetrical wall (see Figure 19). The 6 principal plates are homologous with those in the Lepadomorpha; that is, the rostrum and the carina in the shell wall and the paired scuta and terga that form the operculum. Basically the shell wall has, in addition, 3 pairs of lateral plates, but by loss or fusion of 1 or more plates, the shell typically has 6 or 4 plates; rarely the plates will be wholly coalesced. Each compartmental plate usually has 3 parts: a large central triangular area (paries, pl. parietes) and narrower segments on each side. If they are differentiated in structure from the paries, the one that overlaps the next compartment is called the radius, and the one that is overlapped, the ala. The carina and rostrum have alae on both sides; the rostrolaterals and compound rostrum have radii on both sides; and the carinolaterals and laterals have radii on one side and alae on the other. Shell walls with 8 plates are found only in the most primitive group; rarely there are also 1 or more basal whorls of small plates. In the two most primitive groups the parietes are solid and the mouthparts resemble those of the lepadomorphs. The parietes become progressively more complex with an inner lamina and tubes between it and the outer lamina in progressively more advanced taxa. In the most advanced, the radii also are tubiferous.

A large number of balanomorphs should be available for comparative study. The description and instructions are based on specimens of *Balanus*. Begin your examination with the shell. First determine which plates are overlapped on 1 or both sides. In *Balanus* the rostrum is compound (rostrolaterals fused, rostrum missing). If a specimen of *Chthamalus* (another balanomorph with 6 plates in the shell wall) is available, note that it has a true rostrum; it is the carinolateral plates that have been lost. Other diagnostic characters of the shell include the structure of the basis, the parietes, radii, and sheath. Many of the characters require removing the surface by filing or grinding, but you can determine whether the basis is calcareous or membranous, and whether the parietes are solid or tubiferous without special equipment. After you have removed the animal, examine the inner lamina and sheath.

Having completed an examination of the shell, remove the opercular valves by cutting the opercular membrane and underlying muscles with a scalpel. The terga are held in place partly by a strong muscle attaching to the shell near the base, in close proximity to the carina; the scuta similarly are attached by muscles inserting in the region of the rostrum. In removing the opercular valves be careful not to break off the tergal spur; it is an important character. Carefully detach the animal from the valves; before dissection of it, place the opercular valves in 5% sodium hypochlorite (bleach) to remove the attached tissues, as significant characters will be obscured otherwise.

The orientation of the animal of the balanomorph is very similar to that of a lepadomorph, and the instructions for dissection are approximately the same. Note that the anterior and posterior cirri are, for the most part, easily distinguishable. However, the cirri of the 4th, 5th, and 6th pairs resemble one another. The number of setae on the anterior margins of the 6th pair is important; therefore, take the time to mount the cirri as you did for the lepadomorphs. Care should be exercised in separating the 6th cirrus from the penis and base of the opposing member of the pair. It is advisable to remove the trophi as a unit before removing the cirri from the second side, as it is easy to inadvertently discard the trophi with the body remnants. After the remaining cirri have been removed, sever the penis at its base. Balanids have a penial structure not present in other cirripeds, a

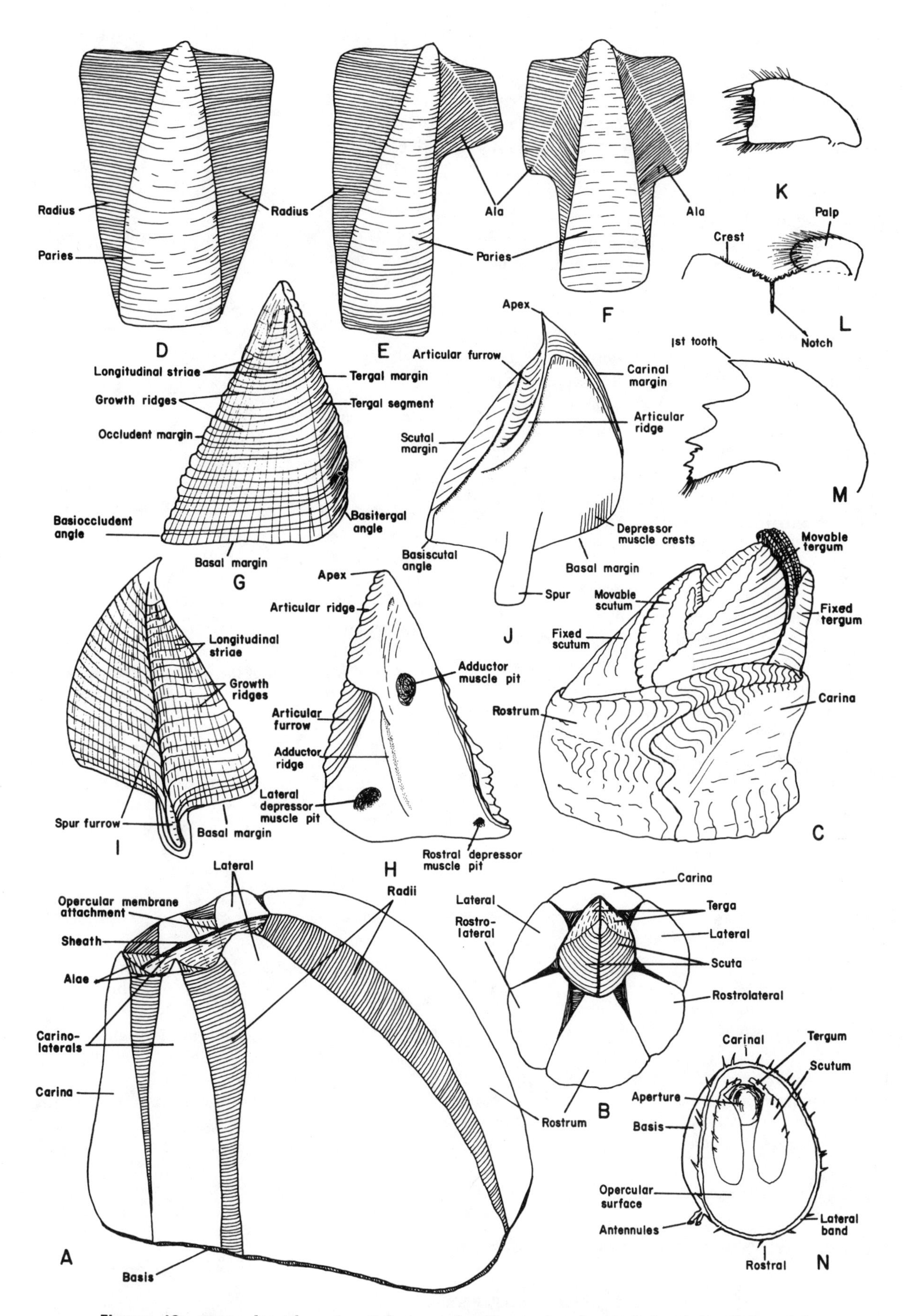

Figure 19 Cirripedia, Thoracica, Balanomorpha, Verrucomorpha: A. Balanid; B. Chthamalid; C. Verrucomorph; D—F, Parietes of shell. D. Paries with paired radii, e.g., fused rostrum (rostrum + rostrolaterals); E. Paries with radius and ala, e.g., lateral; F. Paries with paired alae, e.g., carina; G. Balanid scutum (external); H. Balanid scutum (internal); I. Balanid tergum (external); J. Balanid tergum (internal); K. Maxillule; L. Labrum; M. Mandible; N. Complemental male of balanid.

basidorsal point located just distal to the penis' attachment to the body. The basidorsal point often has taxonomic significance. Next mount the mouthparts following the instructions given for the lepadomorphs: maxillae, maxillulae, mandibles, palps, and labrum. Examine the mouthparts and cirri under moderately high magnification. How does the labrum differ from that of *Pollicipes* or *Lepas*? Are the other mouthparts similar to those in the lepadomorphs? How do they compare with the verrucomorphs, if specimens are available?

Returning to the opercular valves, rinse them in water after the tissue has been dissolved. Examine each tergum and scutum (the valves usually will become disarticulated in the bleach, but if they have not, gently separate them). Using the reference illustrations in Figure 19 as a guide, identify the various elements of the terga and scuta.

If slides of complemental males are available, examine them and compare their structures with those of the hermaphrodites. The characters most readily observable are those of the opercular valves, orifice, lateral band, and perhaps the antennules. If the penis is extended, observe the armature. The vestigial cirri or mouthparts can rarely be distinguished without careful dissection.

The major organ systems of the balanomorphs differ only slightly from those of the lepadomorphs. The most significant difference is in the location of the ovaries, which are still preoral but located in the basal and parietal portions of the mantle.

References

Barnard, K. H., 1924. Contributions to the crustacean fauna of South Africa. no. 7, Cirripedia. *Ann. S. Afr. Mus., 20:* 1–103.

Barnes, H., and W. Klepal, 1971. The structure of the pedicel of the penis in cirripedes and its relation to other taxonomic characters. *J. Exp. Mar. Biol. Ecol., 7:* 71–94.

Broch, H., 1922. Papers from Dr. Th. Mortensen's Pacific Expedition 1914–1916. 10. Studies on Pacific cirripeds. *Vidensk. Medd. Dansk Naturh. Foren. Kobenhavn, 73:* 215–358.

Dahl, E., 1963. Evolutionary lines among Recent Crustacea. In H. B. Whittington and W. D. I. Rolfe (eds.), *Phylogeny and evolution of Crustacea,* pp. 1–15. Spec. Publ. Mus. Comp. Zool. Harvard. Cambridge, Mass.: Harvard University Press.

Darwin, C., 1851. *A monograph on the sub-class Cirripedia, The Lepadidae; or pedunculated cirripedes.* 400 pp. London: Ray Soc.

———, 1854. *A monograph on the sub-class Cirripedia, The Balanidae (or sessile cirripedes); the Verrucidae, etc.* 684 pp. London: Ray Soc.

Henry, D. P., 1942. Studies on the sessile Cirripedia of the Pacific coast of North America. *Univ. Washington Publ. Oceanogr., 4:* 95–134.

———, 1958. Intertidal barnacles of Bermuda. *J. Mar. Res., 17:* 215–234.

Henry, D. P., and P. A. McLaughlin, 1967. A revision of the subgenus *Solidobalanus* Hoek (Cirripedia: Thoracica), with the description of a new species with complemental males. *Crustaceana, 12:* 43–58.

———, 1975. A revision of the *Balanus amphitrite* complex (Cirripedia: Thoracica). *Zool. Verhandl.*, no. 141, 254 pp.

Krüger, P., 1940. Cirripedia. In H. G. Bronn, *Klassen und Ordnungen des Tierreichs,* 5(1) 3(3): 1–560. Leipzig: Akad. Verl.

McLaughlin, P. A., and D. P. Henry, 1972. Comparative morphology of complemental males in four species of *Balanus* (Cirripedia: Thoracica). *Crustaceana, 22:* 13–30.

Newman, W. A., and A. Ross, 1971. Antarctic Cirripedia. *Am. Geophys. Un., Antarctic Res. Ser., 14:* 1–257.

———, 1976. Revision of the balanomorph barnacles; including a catalog of the species. *Mem. 9, San Diego Soc. Nat. Hist.*, 108 pp.

Newman, W. A., V. A. Zullo, and T. H. Withers, 1969. Cirripedia. In: R. C. Moore (ed.), *Treatise on invertebrate paleontology,* Pt. R, Arthropoda, 4, *1:* R206–R295. Lawrence, Kans.: Geol. Soc. America and Univ. Kansas.

Nilsson-Cantell, C. A., 1921. Cirripeden-Studien. Zur Kenntnis der Biologie, Anatomie und Systematik dieser Gruppe. *Zool. Bidr. Uppsala, 7:* 75–395.

Pilsbry, H. A., 1907. The barnacles (Cirripedia) contained in the collections of the U.S. National Museum. *U.S. Natl. Mus. Bull., 60:* 1–122.

———, 1916. The sessile barnacles (Cirripedia) contained in the collections of the U.S. National Museum. *U.S. Natl. Mus. Bull., 93:* 1–366.

Ross, A., 1969. Studies on the Tetraclitidae (Cirripedia: Thoracica): Revision of *Tetraclita. Trans. San Diego Soc. Nat. Hist., 15:* 237–251.

Tarasov, N. I., and G. B. Zevina, 1957. Usonogie raki (Cirripedia Thoracica) morei SSSR. *Fauna SSSR, Zool. Inst. Akad. Nauk, SSSR* n.s., 69, *6:* 1–268.

ORDER ASCOTHORACICA L.-Duthiers, 1880

Recent species	Approximately 30 in seven genera.
Size range	Up to 85 mm.
Carapace	Bivalved, open ventrally; occasionally absent.
Eyes	Nauplius eye in nauplius, with compound eyes in cyprid; both absent in adult.

Antennules	Uniramous; prehensile in cyprid and adult; often subchelate in adult.
Antennae	Biramous in nauplius; absent in cyprid and adult.
Mandibles	Without palp; part of protuberant oral cone modified for piercing and sucking.
Maxillulae	Part of protuberant oral cone.
Maxillae	Uniramous; part of protuberant oral cone; sometimes fused with paired stylet tips.
Maxillipeds	None.
Thoracic appendages	Three to 6 pairs of uni- or biramous thoracopods; sometimes reduced or absent.
Abdominal appendages	None.
Telson	Absent; caudal furca often present.
Tagmata	Head, thorax, and abdomen.
Somites	Head with 5 or 6 (1st thoracic somite sometimes fused to head); thorax usually with 5 or 6, occasionally obscure; abdomen with 4 or 5, or rarely coalesced with thorax.
Sexual characters	Female gonopore on conical papilla at base of 1st thoracopods; penis on 1st abdominal somite, at least in hermaphrodite.
Sexes	Typically separate; rarely hermaphroditic; males small.
Larval development	Nauplius → cyprid.
Fossil record	Cretaceous to Recent.
Feeding types	Parasitic; piercing and sucking.
Habitat	Marine; ecto- and endoparasites of coelenterates and echinoderms.
Distribution	Worldwide.

The Ascothoracica is the most primitive order of Cirripedia, as shown by the generalized form (see Figure 20). Prehensile antennules, a bivalved carapace, natatory thoracic appendages, and abdominal somites terminating in a furca are retained in the adult. [Bowman (1971) interprets the furca as uropods.] The sexes usually are separate, although hermaphroditism is known. Ascothoracicans are ecto- or endoparasites, and their mouthparts are adapted for piercing and sucking.

If infected coelenterates or echinoderms are available, a general study of ascothoracican morphology may be undertaken. Infection by *Ascothorax ophioctenis* produces a prominent swelling on the ophiuroid host; however, most species inhabit the gastric cavity of the host species and can be located only by the careful dissection of a large number of host specimens.

A thorough inspection of the female should be made for any attached males. If males are present, remove them to a separate container for subsequent examination. Carefully cut open the mantle of the female to reveal the body structure. In contrast to typical cirriped appendages, the thoracic appendages of the ascothoracicans are straight, natatory structures. Pay particular attention to the oral pyramid (mouth cone) and to the antennules, which usually are subchelate. In most species the abdominal segmentation can be distinguished and usually a pair of caudal rami is present. The external anatomy of the males of *Ascothorax ophioctenis* differs markedly from that of the female. The male has a bivalve carapace or mantle that scarcely covers the body and a pair of subchelate antennules extending anteriorly. Posteriorly, the abdomen and caudal furca are clearly distinguished. The degree of development of the appendages and the number present is correlated to the parasitic mode of the species.

References

Achituv, Y., 1971. *Dendrogaster asterinae* n. sp., an ascothoracid (Cirripedia) parasite of the starfish *Asterina burtoni* from the Gulf of Elat. *Crustaceana, 21:* 1–4.

Brattström, H., 1947. Ecology of the ascothoracid *Ulophysema oresundense*. *Lunds Univ. Arsskr. N.F.* (2) *43:* 4–75.

Caullery, M., 1952. *Parasitism and symbiosis*. Translated by A. M. Lysaght. xii + 321 pp. London: Sidgwick and Jackson.

Krüger, P., 1940. Ascothoracida. In H. G. Bronn, *Klassen und Ordnungen des Tierreichs*, 5(1) 3(4): 1–46. Leipzig: Akad. Verl.

Newman, W. A., V. A. Zullo, and T. H. Withers, 1969. Cirripedia. In R. C. Moore (ed.), *Treatise on invertebrate paleontology*, Pt. R, Arthropoda, 4, *1:* R206–R295. Lawrence, Kans.: Geol. Soc. America and Univ. Kansas.

Okada, Y. K., 1939. Les Cirripedes Ascothoraciques. *Trav. Sta. Zool. Wimereux, 13:* 489–514.

Pyefinch, A., 1939. Ascothoracica (Crustacea, Cirripedia). *John Murray Expedition, 1933–34. Scientific Reports, 5* (Zoology): 247–262.

Wagin, V. L., 1946. *Ascothorax ophioctenis* and the position of Ascothoracica in the system of the Entomostraca. *Acta Zool., 27:* 155–267.

———, 1964. On *Parascothorax synagogoides*, parasite on *Ophiura quadrispina* and on the geographical distribution of Ascothoracica. *Akad. Nauk Okeanol. Trudy Inst. Ocean. SSSR, 69:* 271–284 (in Russian).

ORDER RHIZOCEPHALA Müller, 1862

Recent species	Approximately 250 in ten genera.
Size range	As large as 10 × 5.5 × 3 cm.
Carapace	Absent.
Eyes	Nauplius eye in nauplius, with compound eye in cyprid; both absent in adult.
Antennules	Uniramous in cyprid; absent in adult.
Antennae	Biramous in nauplius; absent in cyprid and adult.
Mandibles	Absent.
Maxillulae	Absent.
Maxillae	Absent.
Maxillipeds	None.
Thoracic appendages	Absent.
Abdominal appendages	None.
Telson	Absent.
Tagmata	Not distinguishable.
Somites	Not distinguishable.
Sexual characters	Females with or without male-cell receptacle in mantle cavity.
Sexes	Usually hermaphroditic, occasionally separate, female and cyprid as functional male.
Larval development	Usually nauplius → cyprid → kentrogon; occasionally released as cyprid, with kentrogon.
Fossil record	Recent.
Feeding types	Parasitic; absorption through ramifying system.
Habitat	Marine; endo- or more rarely ectoparasites of Crustacea, particularly Decapoda.
Distribution	Worldwide.

The Rhizocephala are parasitic, greatly modified Cirripedia lacking appendages and a digestive tract in the adult. In part, the modification is so great that the affinity with the Cirripedia is determined only by the characteristic nauplius with frontal horns and/or the cyprid. Sexes until recently were thought to be combined; however, in some species the female invades the host and the spermatogenic material is provided by a male cyprid.

Very little, in terms of morphological studies, can be done with the rhizocephalans without histological sections. The various taxa, however, exhibit a variety of external reproductive capsules, and these should be examined. Rhizocephalans have been noted for the "parasitic castrating effect" they have on some hosts. Examine the host specimens and note any morphological changes that may have been induced by the rhizocephalan infection, such as the presence of female-type pleopods in males. Figure 20 illustrates the mode of reproduction of a species of *Peltogastrella*. The life cycle includes: (a) the female producing large eggs, which

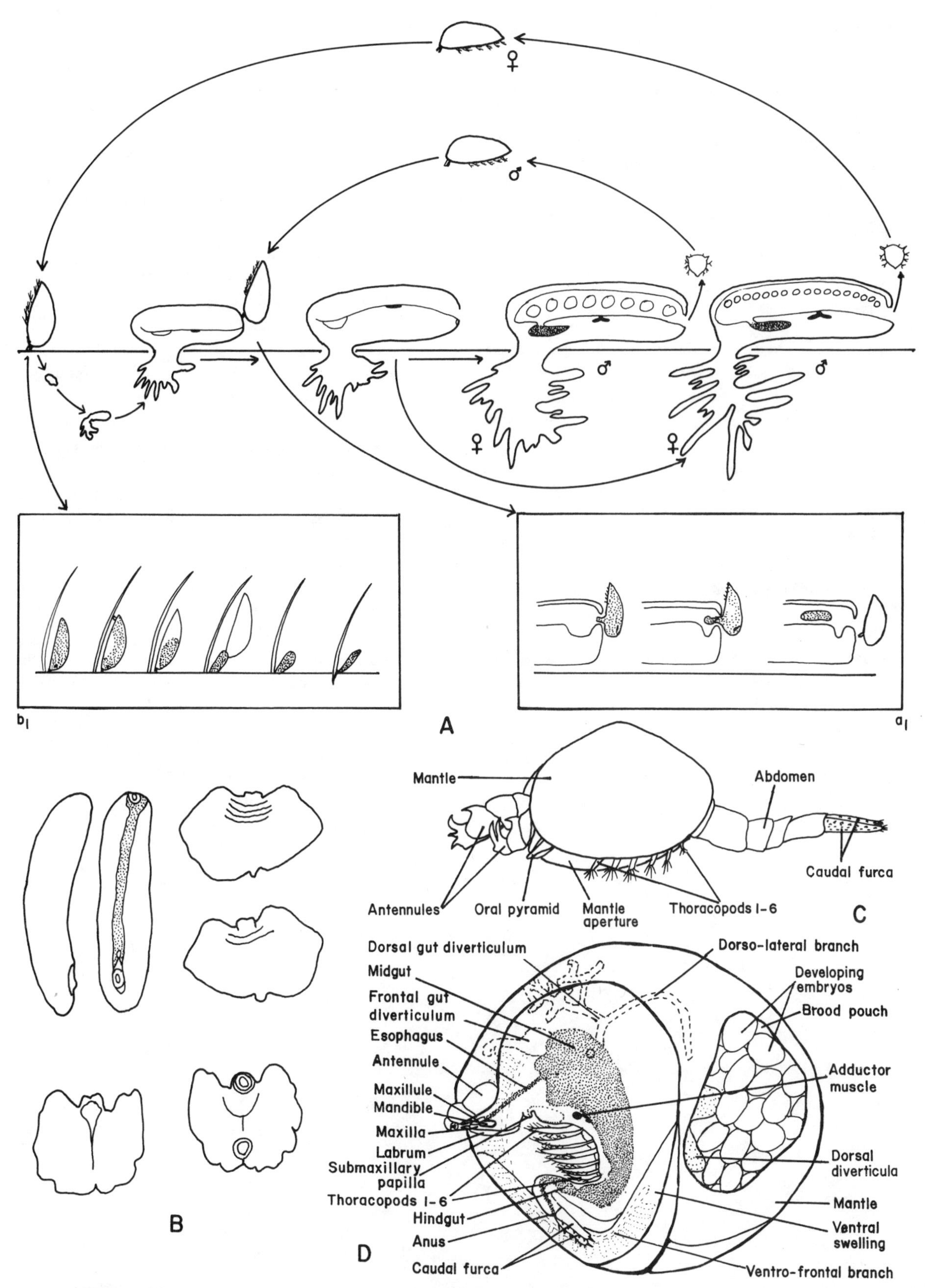

Figure 20 Cirripedia, Rhizocephala, Ascothoracica: A. Life cycle of a rhizocephalan parasite; a_1. Male spermatogenic cells injected into young female; b_1. Female cyprid attached at base of seta of host and casting cyprid shell prior to invasion of host [after Newman, Zullo, and Withers, 1969, from *Treatise on Invertebrate Paleontology*, courtesy of the Geological Society of America and University of Kansas]. B. Different forms of rhizocephalan reproductive sacs, *Peltogasterella, Sacculina, Lernaeodiscus* [after Boschma, 1950, 1952, 1959]; C. Ascothoracica, male; D. Ascothoracica, female [C, D after Wagin, 1946].

hatch as nauplii that become males; and (b) a female producing small eggs, which hatch as nauplii that become females. The male cyprid spermatogenic cells are injected into the young female and migrate to a receptacle (a_1); the female cyprid (kentrogon) attaches at the base of a seta of the host and casts the cyprid shell prior to invading the host's tissue (b_1).

References

Boschma, H., 1937. The species of the genus *Sacculina* (Crustacea Rhizocephala). *Zool. Meded., 19:* 187–328.

———, 1950. *Lernaeodiscus pusillus* nov. spec. A rhizocephalan parasite of a *Porcellana* from Egypt. *Bull. Brit. Mus. (Nat. Hist.), 1:* 61–65.

———, 1952. *Sacculina inconstans*, a new species of rhizocephalan parasite from the Gilbert Islands. *Koninkl. Nederl. Akad. Wetensch.*, Amsterdam (C) *55:* 1–6.

———, 1955. The described species of the family Sacculinidae. *Zool. Verhandl., 27:* 1–76.

———, 1959. The Crustacea Rhizocephala of Chile. Reports of the Lund University Chile Expedition, 1948–49. No. 37. *Lunds Univ. Arsskr. N.F.* (2) *56:* 3–20.

———, 1972. On the occurrence of *Carcinus maenas* (Linnaeus) and its parasite *Sacculina carcini* Thompson in Burma, with notes on the transport of crabs to new localities. *Zool. Meded., 47:* 145–155.

Caullery, M., 1952. *Parasitism and symbiosis.* Translated by A. M. Lysaght. xii + 321 pp. London: Sidgwick and Jackson.

Newman, W. A., V. A. Zullo, and T. H. Withers, 1969. Cirripedia. In R. C. Moore (ed.), *Treatise on invertebrate paleontology,* Pt. R, Arthropoda, 4, *1:* R206–R295. Lawrence, Kans.: Geol. Soc. America and Univ. Kansas.

Reinhard, E. G., 1942. Studies on the life history and host-parasite relationships of *Peltogaster paguri. Biol. Bull., 83:* 401–415.

———, 1944. Rhizocephalan parasites of hermit crabs from the northwest Pacific. *J. Wash. Acad. Sci., 34:* 49–58.

———, 1956. Parasitological reviews. Parasitic castration of Crustacea. *Parasitology, 5:* 79–107.

Reischman, P. G., 1959. Rhizocephala of the genus *Peltogasterella* from the coast of the state of Washington to the Bering Sea. *Koninkl. Neder. Akad. Wetensch.*, (C) *62:* 409–435.

Smith, G. W., 1906. Rhizocephala. *Fauna und Flora des Golfes von Neapel. Monogr. 29:* 1–123.

Veillet, A., 1945. Recherches sur le parasitisme des crabes et des Galathées par les Rhizocéphales et les Épicarides. *Ann. Inst. Océanogr. Paris, 22:* 193–341.

———, 1960. Observation de la fixation des larvaes mâles chez le Cirripède parasite *Septosaccus cuenoti* Duboseq. Dimorphisme des larves de Rhizocéphales. *Bull. Soc. Sci. Nancy, 19:* 109–112.

Yanagimachi, R., 1961. Studies on the sexual organization of the Rhizocephala. III. The mode of sex determination in *Peltogastrella. Biol. Bull., 120:* 272–283.

Yanagimachi, R., and N. Fujimake, 1967. Studies on the sexual organization of the Rhizocephala. IV. On the nature of "testis" of *Thompsonia. Annot. Zool. Jap., 18:* 130–134.

CLASS MALACOSTRACA Latreille, 1806

All remaining crustaceans are grouped in the class Malacostraca. Subclasses within this class include the Phyllocarida, Hoplocarida, and Eumalacostraca. The Phyllocarida and Hoplocarida are represented by the Recent orders Leptostraca and Stomatopoda respectively. The Eumalacostraca includes a number of superorders, orders, and suborders which, for discussion and study purposes, will be given equal attention.

General characters of the Malacostraca include paired compound eyes; usually biramous antennules and antennae, the latter with the exopod usually developed as a scaphocerite; and well-developed mandibles. Five to 8 pairs of uni- or biramous thoracic appendages and 5 pairs of biramous pleopods usually are present. The female gonopore(s) are invariable on the 6th thoracic somite or its appendages; the male gonopores are on the 8th or its appendages. The sexes usually are separate although protandry and protogyny are known.

SUBCLASS PHYLLOCARIDA Packard, 1879

ORDER LEPTOSTRACA Claus, 1880

Recent species	Eleven, in four genera: *Nebalia*, *Paranebalia*, *Nebaliopsis*, and *Nebaliella*.
Size range	0.8–4.0 cm.
Carapace	Bivalve, laterally compressed; lacking hinge or hinge line; enclosing thorax but not cephalon; rostrum articulated.
Eyes	Stalked, compound, except nonfunctional in *Nebaliella* and *Nebalia typhlops;* ocular scale sometimes present.
Antennules	Usually biramous; exopod scale or platelike, occasionally markedly reduced or absent.
Antennae	Uniramous, well developed; may be sexually dimorphic.
Mandibles	With palp; molar and incisor processes usually well developed, occasionally reduced.
Maxillulae	Paired endites and flagelliform palp; latter occasionally reduced.
Maxillae	Protopod usually with 4 endites; endopod and exopod usually well developed; occasionally endopod reduced, exopod vestigial.
Maxillipeds	None.
Thoracic appendages	Eight pairs of biramous foliaceous thoracopods, usually well developed, occasionally reduced; with or without epipods.
Abdominal appendages	Six pairs: 1st 4 pairs biramous, large, last 2 uniramous, reduced.
Telson	With caudal furca.
Tagmata	Head, thorax, and abdomen.
Somites	Head with 5, thorax with 8, abdomen with 7, excluding telson.
Sexual characters	Gonopores on coxae of 6th thoracopods of female, on 8th coxae of males. Antennules and antennae often sexually dimorphic.

Sexes	Separate.
Larval development	Epimorphic; usually hatch as postlarvae (manca stage); in *Nebaliopsis* as planktonic larvae.
Fossil record	Extensive, since Lower Cambrian. More than 50 fossil genera known; two orders known only from fossil record.
Feeding types	Generally filter feeder.
Habitat	Marine, usually between 10 and 400 m; most are benthic although one species is bathypelagic.
Distribution	Principally from Atlantic; one species from Puget Sound, Washington and one from Antarctic.

The Phyllocarida are represented by a single Recent order, the Leptostraca; it contains only the suborder Nebaliacea and family Nebaliidae with four currently recognized genera: *Nebalia*, *Paranebalia*, *Nebaliopsis*, and *Nebaliella*. Illustrations and the discussion of external morphology of this group are based primarily on specimens of *Paranebalia;* those of the internal anatomy are based on a composite. The few major morphological differences among the taxa will be pointed out. If an abundance of material is available, individual student dissections should be undertaken as these animals are easily manipulated, even when small in size. In cases where only very limited material is available, students should examine intact animals, then refer to slide preparations for the study of appendages, mouthparts, and internal organ systems.

The carapace is laterally compressed, forming a bivalvelike shell; however, no hinge or hinge line is present (see Figure 21). Anteriorly, the head is not covered by the carapace and the stalked compound eyes, as well as the articulated rostrum (also referred to as a rostral plate), are immediately apparent. The elongate antennules and antennae protrude from beneath the carapace. The number of articles in the antennular and antennal flagella are fewer in *Paranebalia* than in *Nebaliella* or *Nebalia* and generally also vary between the sexes. Although the virtually transparent carapace covers the entire thorax, it is attached only to the cephalon. The cephalic and thoracic appendages generally are visible beneath the carapace; however, for ease in distinguishing individual appendages, the carapace should be removed from one side of the body. After cutting the carapace along the dorsal midline, sever the prominent adductor muscle and remove one side of the carapace.

Before dissection, observe the thoracic and abdominal segmentation and the relationships of the appendages. Note that the abdomen bears pleopods on the anterior 6 somites but not on the 7th or the telson. [Bowman (1971) considers the abdomen 8-segmented; that is, the telson is lacking and the ultimate lobe is the anal somite. The caudal furca by his definition are uropods.]

Examine the stalked eyes; depending upon the genus, a scale or acicle may not be present at the base of each ocular peduncle. Remove the antennule and antenna from the exposed side of the body. The exopod of the antennule is extended as a denticulate plate or scalelike structure in *Paranebalia* and *Nebaliopsis,* reduced to a tuft of bristles in *Nebaliella,* and virtually absent in species of *Nebalia.* Sexual dimorphism is exhibited in the larger number of sensory bristles on the antennules and the considerably greater length and number of articles of the antennal flagella in males of some of the species. In a few species, the antennules and/or antennae of males are modified for grasping the female during copulation.

Remove one or more of the thoracopods. In *Paranebalia,* as figured, the thoracopods are unspecialized; the epipods are considerably reduced. In species of *Nebalia* both the exopods and epipods are very much flattened and inflated. In all species, except *Nebaliopsis typicus* G. O. Sars, the 8 thoracopods are very similar in structure. In *N. typicus* the 1st pair are slightly modified to act as interlimb space valves, and the subsequent 6 pairs are unsegmented and the epipods are represented by slight lobes on the outer margins; the last pair have very few setae. In most leptostracans, the tips of the thoracic endopods of the female carry long setae that curve inward toward the midline and form an effective brood pouch. The female gonopores open on the coxae of the 6th thoracopods but may be difficult to distinguish. The gonopores of the male open on low papillae on the coxae of the 8th pair of thoracopods.

Remove and examine one of the first 4 pairs and one of the last 2 pairs of pleopods. Observe that the endopods of the anterior pleopods are linked together by a small spinose structure, the appendix interna. The pleopods of the last 2 pairs are small, simple, and uniramous. The unsegmented caudal furca articulates freely with the telson.

Remove the mouthparts. You will notice that the mandible has a prominent molar process and palp but a weakly developed incisor process. The incisor process is

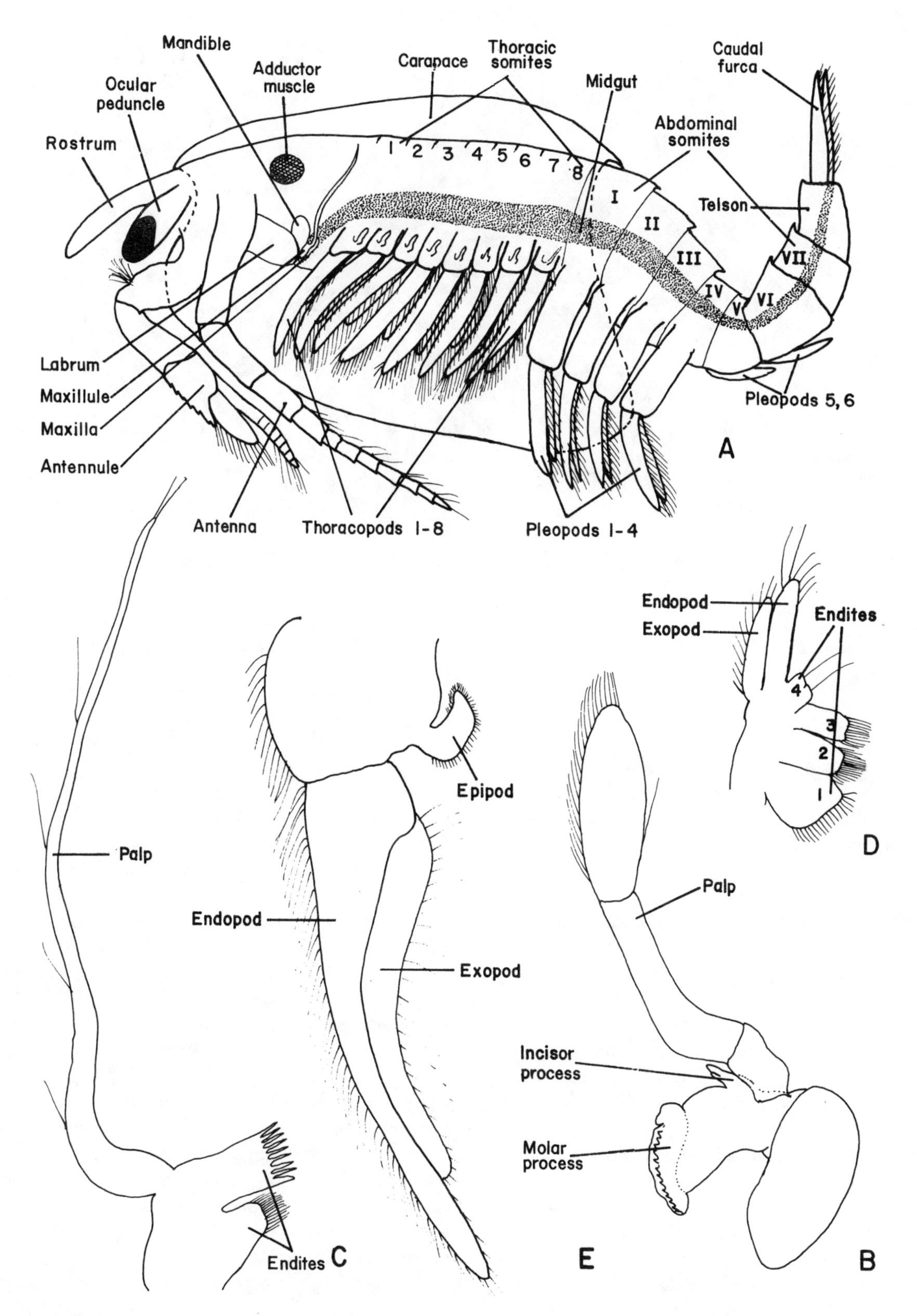

Figure 21 Phyllocarida: A. Lateral view of animal with left side of carapace removed; B. Mandible; C. Maxillule; D. Maxilla; E. Typical thoracopod.

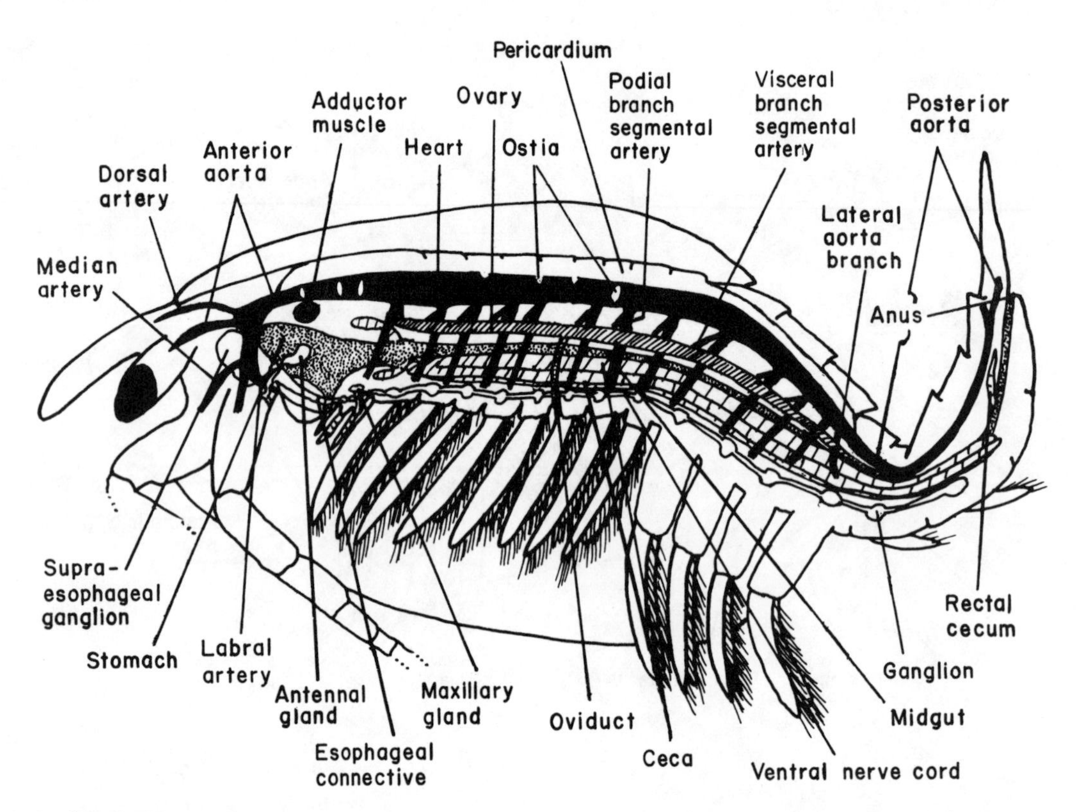

Figure 22 Phyllocarida: Diagrammatic leptostracan with musculature removed to show major organ systems.

also weakly developed in *Nebalia* and lacking in *Nebaliopsis;* however, it is well developed in *Nebaliella.* In all leptostracans, except *Nebaliopsis,* the maxillule has a well-developed, curved palp, which may represent a flagellate endopod. The maxilla usually has 4 endites and both an endopod and exopod as well; in *Nebaliopsis* the endopod is reduced and the exopod is vestigial.

The leptostracan body is quite muscular and these layers of muscle tissue must be removed before the major internal organ systems can be examined (see Figure 22). Begin dorsally in the midline and strip the muscles from each side of the body. Directly beneath the muscle layer is the tubular heart that extends from the cephalic region to the 4th abdominal somite surrounded by a pericaridial sinus. Identify 3 pairs of lateral ostia in the cephalic portion of the heart and the 3 dorsal and 1 large lateral pairs in the thoracic region. Follow the anterior aorta forward; it gives off branches ventrally to the antennules, antennae, and mouthparts and dorsally to the dorsal artery of the head and rostrum. In the thoracic region the heart gives off 8 pairs of segmental arteries; an additional 4 pair are given off in the abdomen. Each artery has a visceral branch that provides blood to the body organs and a podial branch that supplies the extrinsic muscles of the thoracopods and pleopods. Posteriorly identify the posterior aorta and its pair of lateral branches that supply blood to the posterior abdominal somites.

Laterally and slightly beneath the heart are the ovaries or testes. These are paired, long tubular organs that extend from the region of the stomach to the 3rd or 4th abdominal somite. In females the paired oviducts lead from the ovaries to the genital openings on the 6th thoracopods. In males the short vas deferens opens on small papillae on the coxae of the 8th thoracopods.

In most leptostracans food passes from the mouth through an esophagus or foregut into an enlarged cardiac stomach. The cardiac portion of the stomach is provided with masticatory ridges moved by muscles; the posterior or pyloric part of the stomach is provided with setose lobes and a dorsal groove that leads into the midgut as a chitinous funnel. Anteriorly the midgut gives off 1 or 2 pairs of dorsally and ventrally directed short ceca. Two or 3 elongate tubular ceca arise from a pair of ventral openings in the midgut and extend almost the entire length of the body. A short dorsal cecum arises from the midgut just anterior to the short cuticularized hindgut. The anus opens ventrally between the caudal rami. In *Nebaliopsis* there is no cardiac stomach, the midgut is very narrow, and only 1 large midgut cecum is present. In leptostracans excretion is accomplished by means of both antennal glands and maxillary glands.

Beneath the anterior aorta identify the large supraesophageal ganglion and the short connectives encircling the esophagus. Carefully remove the stomach, midgut, and ceca to expose the large ventral nerve cord and ganglia. Note that the nerve cord and ganglia terminate at the 6th abdominal somite. No ganglia are present in either the 7th somite or telson; however, 1 additional neuromere is present in the embryo and it has been assumed that the most posterior ganglion of the adult has resulted from the fusion of the two most posterior embryonic ganglia.

References

Brattegard, T., 1970. Leptostraca from shallow water in the Bahamas and southern Florida. *Sarsia, 44:* 1–8.

Cannon, H. G., 1927. On the feeding mechanism of *Nebalia bipes. Trans. Roy. Soc. Edinburgh, 55:* 355–369.

———, 1960. Leptostraca. In H. G. Bronn, *Klassen und Ordnungen des Tierreichs,* 2nd ed., 5(1) 4(1): 1–81. Leipzig: Akad. Verl.

Clark, A. E., 1932. *Nebaliella caboti* n. sp., with observations on other Nebaliacea. *Trans. Roy. Soc. Canad.,* (3) *26:* 217–235.

Hansen, H. J., 1920. Crustacea Malacostraca IV. *Danish Ingolf Exped., 3:* 1–86.

Hessler, R. R., and H. L. Sanders, 1965. Bathyal Leptostracea from the continental slope of the northeastern United States. *Crustaceana, 9:* 71–74.

Linder, F., 1943. Ueber *Nebaliopsis typica* G. O. Sars nebst einigen allgemeinen bermerkungen ueber die Leptostraken. *Dana Rept.* no. 25: 1–38.

Manton, S. M., 1934. On the embryology of the crustacean *Nebalia bipes. Phil. Trans. Roy. Soc. London,* (B) *223:* 163–238.

Packard, A. S., 1883. A monograph of the phyllopod crustaceans of North America, with remarks on the order Phyllocarida. *12th Ann. Rept. U.S. Geol. Geog. Surv. Terr. Wyom. Idaho,* pp. 295–457.

Sars, G. O., 1887. Report on the Phyllocarida collected by H.M.S. Challenger during the years 1873–76. *Challenger Sci. Rept. Zool., 14:* 1–38.

———, 1896. Phyllocarida and Phyllopoda. *Fauna Norveg., Vidensk. Selsk. Forhandl.* Kristiania. 140 pp.

Thiele, J., 1904. Die Leptostraken. *Wiss. Ergebn. Tiefsee-Exp. Valdivia, 8:* 1–26.

———, 1927. Leptostraca. In W. Kükenthal and Th. Krumbach (eds.), *Handbuch der Zoologie, 3:* 567–592. Berlin: Der Gruyter.

Verrill, A. E., 1923. Crustacea of Bermuda: Schizopoda, Cumacea, Stomatopoda, and Phyllocarida. *Trans. Conn. Acad. Arts Sci., 26:* 181–211.

Wakabara, Y., 1965. On *Nebalia* sp. from Brazil (Leptostraca). *Crustaceana, 9:* 245–248.

———, 1976. *Paranebalia fortunata* n. sp. from New Zealand (Crustacea, Leptostraca, Nebaliacea). *J. Roy. Soc. N. Z., 6:* 297–300.

Willemoës-Suhm, R., 1875. On some Atlantic Crustacea from the Challenger Expedition. *Trans. Linn. Soc., London* (2) *1:* 23–59.

SUBCLASS HOPLOCARIDA

ORDER STOMATOPODA Latreille, 1817

Recent species	Approximately 300 in four families.
Size range	5–55 cm.
Carapace	Well developed, laterally expanded; covering cephalon, except for acron and antennular somite, and anterior half of thorax.
Eyes	Stalked, compound; occasionally reduced.
Antennules	Peduncle with 3 segments; 3 flagella, dorsal and divided ventral.
Antennae	Peduncle with 2 segments; exopod with segment expanded as scaphocerite; endopod with flagellum.
Mandibles	With or without palp; well-developed toothed incisor and basal molar processes.
Maxillulae	Basal and coxal segments each with endite; endopod reduced; exopod absent.
Maxillae	With 4 segments, 1st and 2nd each with endite.
Maxillipeds	None.

Thoracic appendages	Eight pairs; first 5 subchelate, 2nd modified as powerful raptorial claw; all used in feeding; thoracopods 6–8 biramous, used in locomotion.
Abdominal appendages	Five pairs of biramous, laminar pleopods, often with gills; 6th somite with uropods.
Telson	Well developed, occasionally fused with 6th abdominal somite (pleotelson); usually well armed.
Tagmata	Head, thorax, and abdomen.
Somites	Head with 5, excluding acron; thorax with 8; abdomen with 6, excluding telson.
Sexual characters	Gonopores on 6th thoracic somite of female, 8th of male; male with well-developed penes. Armature of telson and size and armature of raptorial claw sexually dimorphic.
Sexes	Separate.
Larval development	Anamorphic; hatch as antizoea or pseudozoea; several pelagic substages before settlement.
Fossil record	Mississippian to Recent.
Feeding types	Predators.
Habitat	Usually marine, few reported from brackish water; usually inhabiting shallow water, but some deep-sea forms have been reported.
Distribution	Tropical and subtropical seas throughout world; sometimes temperate seas. In northern hemisphere as far north as Hokkaido, Japan; Massachusetts, U.S.A.; and southern North Sea. In southern hemisphere, South Australia and South Africa.

The figures and instructions for the study of the Stomatopoda are based on specimens of the family Squillidae; however, as the primary characters that distinguish the families are found in the armament of the telson, the use of species of other families will not invalidate the general discussion. You will notice immediately that stomatopods exhibit a number of characters that have not been present in the crustaceans that have been examined previously. Therefore, take time to examine the gross morphology of the animal before beginning your dissection (see Figure 23).

Examine the cephalon in dorsal view. Certain features of the ophthalmic area will be noticeably distinct. As in other stalk-eyed crustaceans, the eyes are situated on movable peduncles, the cornea broadly dilated. The ocular peduncles are attached to an articulated acron, also sometimes referred to as an ophthalmic somite, although it is doubtful that it is a true cephalic somite. It should be possible to distinguish 3 plates of the acron: the anteriormost plate, protruding in front of the basal margin of the ocular peduncles, is called the fastigial plate; the medial plate, at the bases of the peduncles, is the ocular plate; and the posterior plate, with dorsally directed small projections, the ocular scales, is the postocular plate.

Posterior to the acron is the antennular somite, also referred to as the posterior dorsal plate of the head; it is covered medially by the broad rostral plate or rostrum which articulates with the carapace. The antennular somite bears the 3-segmented antennules and, dorsolaterally, the antennular processes. The ventral flagellum of the antennule is subdivided into 2 flagella. The antennae arise lateral to the antennules and each consists of a 2-segmented peduncle, 2-segmented exopod, and 3-segmented endopod with flagellum. The terminal segment of the exopod is broadened and flattened, forming the antennal scale or scaphocerite.

The remaining cephalic somites and first 4 thoracic somites are covered dorsally by the carapace, which is broad and flattened. The surface of the carapace is variously sculptured in the different taxa but typically bears 3 grooves, that is, a transverse cervical groove on the posterior third and a pair of parallel, longitudinal gastric grooves. Usually there also is a median carina. A dorsal pit (median pore) between the cervical groove and the anterior bifurcation of the median carina may be easily distinguished in many species. The produced anterolateral angles may be armed or unarmed. In some species lateral and intermediate carinae between the gastric groove and the lateral margin on each side of the

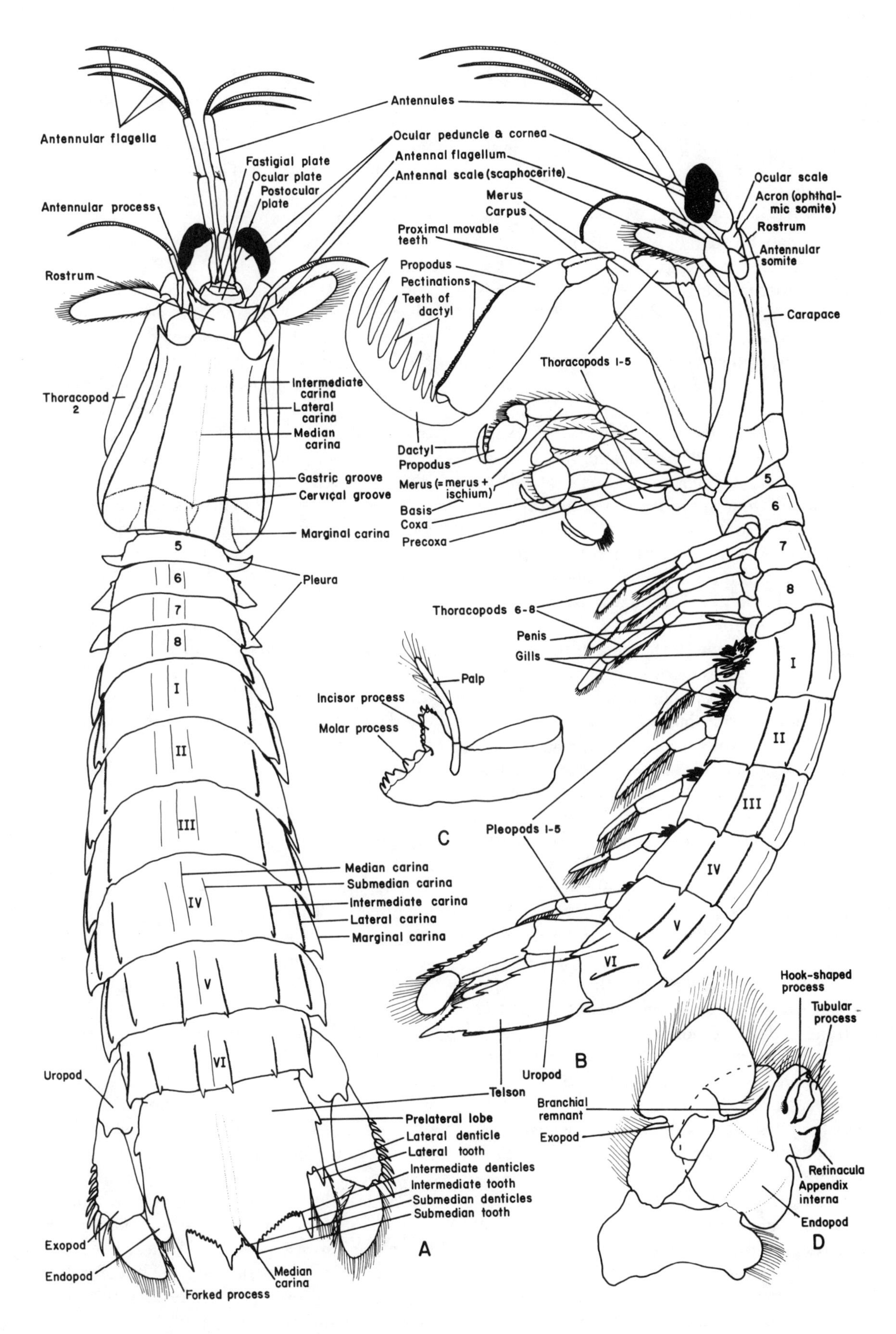

Figure 23 Stomatopoda: A. Whole animal (dorsal view); B. Male (lateral view); C. Mandible; D. 1st pleopod of male.

carapace, as well as a posterolateral marginal carina, also are present.

The last 4 thoracic somites are free (not covered by the carapace). They, as well as the first 5 abdominal somites, often have a number of longitudinal carinae, such as submedian, median, intermediate, lateral, or marginal. The 6th abdominal somite usually lacks marginal carinae. Identify the carinae present in your specimen(s). In some species secondary ridges may occur between the carinae. Not infrequently the carinae terminate in spines at the posterior margins of the somites. Denticles also may be present on the somites.

The telson usually is distinct from the 6th abdominal somite; however, fusion of the 6th somite and the telson occasionally does occur, resulting in a pleotelson. The telson usually is armed with teeth, denticles, or both. The teeth and denticles are identified according to their positions, such as lateral, intermediate, or submedian. Sketch the telson of your specimen(s) and identify the teeth and denticles present.

Place the specimen with its ventral surface up and examine the surfaces of the somites (the sternites), the appendages, and the telson. The first 5 pairs of thoracopods are modified for feeding, and the posterior pairs as walking legs. In older texts the anterior appendages often are referred to as maxillipeds; however, they are not equivalent to the maxillipeds of other crustaceans. Remove the anterior thoracopods from one side and observe their basic structure. All are subchelate and are composed of 7 segments. There is some disagreement about which are the fused segments; that is, whether the coxa and precoxa have fused or the ischium and merus have fused. Thus the thoracopods are thought to consist of either a coxa (fused), basis, ischium, merus, carpus, propodus, and dactyl, or a precoxa, coxa, basis, merus (fused), carpus, propodus, and dactyl. Notice how the dactyl is folded back on the propodus to form a subchela. Epipods may be present at the bases of the thoracopods. There is little difference in the 1st, 3rd, and 4th thoracopods except in size. The 2nd thoracopod is markedly stronger than the others, and the dactyl and propodus form a powerful and effective raptorial claw. The 5th thoracopod has a prominent tuft of stiff setae on the propodus that is used for cleaning the body. Remove one of the walking legs. You will observe that it is biramous. The protopod consists of 3 segments, the 2nd of which is elongate. The outer ramus consists of a single segment and, contrary to what should be expected, this ramus is the endopod. The inner ramus is the 2-segmented exopod; their relative positions have been reversed during development.

Five pairs of pleopods are present. They are biramous, broad, flattened, and provided with gills. In males the endopods of the 1st pair are modified markedly, and the second very slightly. Remove a 3rd pleopod, which may be considered typical, and examine the components. There is a basal protopod from which arise an endopod and exopod, each of which is lamellar and membraneous and obscurely divided into 2 segments. The basal segment of the exopod has 2 filamentous structures protruding from the internal margin. These are the gills. The distal segment of the endopod also has a structure protruding from the internal margin. This is an appendix interna that carries a group of coupling hooks or retinacula, which serve to attach the pair of pleopods together so that they may move in unison. Examine the 1st pleopod of a male specimen and note the modifications of the endopod to form the so-called petasma, particularly the hook-shaped and tubular processes. In addition to the 1st pleopods, males may be distinguished by a pair of tubular copulatory structures (penes) arising at the bases of the last pair of thoracic legs, and females by the presence of gonopores located submedially on the 6th thoracic sternite.

You should be able to distinguish the anal pore on the anteroventral surface of the telson and, directly posterior to it, a short post-anal carina. Strong ventral carinae are present medially on the abdominal sternal plates. The uropods (appendages of the 6th abdominal somite) typically consist of a 1-segmented protopod or basal segment, a 2-segmented exopod, and a 1-segmented endopod. The basal prolongation of the uropod is a flattened, usually bifurcate ventral projection, referred to as a forked process. This structure has considerable diagnostic significance at the generic and specific levels. The proximal segment of the exopod is provided with a series of movable spines.

Three pairs of cephalic appendages that constitute the mouthparts lie in close proximity to the broad epistome and above the bulbous labrum. Remove the mouthparts from one side. The maxillae are the most external of the mouthparts. The maxilla is somewhat different in appearance from those of most other crustaceans. It consists of 4 segments; the 1st and 2nd are provided with endites and that of the 2nd is bilobed. The orifice of the maxillary gland can be observed on the basal segment. The maxillule directly overlaps one lobe of the bulbous, bilobed labium. The maxillule consists of 2 segments, the coxa and basis, and each has a distinct endite. The exopod is absent and the endopod is represented by a very small segment arising near the base of the basal endite. It is sometimes referred to as the palp of the maxillule. The calcareous mandible is very large and has a prominently toothed incisor process and basal molar process. The incisor process lies over the labrum, but the molar process protrudes beneath the labrum. The mandibular palp may be 3-segmented or reduced to 1 or 2 segments. By probing the labrum ventrally in the midline, the mouth can be observed.

The position and the action of the mandibles is somewhat unusual, in that they have weak dorsal articulations and strong ventral articulations with small condyles on the posterior margins of the long lateral wings of the epistome that embrace the labrum. The fibers of the ventral muscles of the mandibles do not arise from a supporting ligament but are attached separately on large lateral expansions of an apodemal arch arising from premandibular invaginations between the epistome and the doublure of the carapace. A corresponding apodeme has not been observed in other crustaceans. Manipulation of the mandibles will show that the only movement of these jaws can be a partial rotation on their vertical epistomal hinges between the 2 points of articulation.

To examine the major aspects of the internal anatomy (see Figure 24), carefully free the carapace by cutting along the ventrolateral body wall where the laterally expanded carapace connects and posteriorly at its point of attachment. With a pair of sharp scissors, cut away the dorsal integument of the acron and the antennular somite, being careful not to damage the underlying structures. Then remove the remainder of the dorsal exoskeleton. A median and two lateral cuts with sharp pointed scissors will allow you to separate the exoskeleton from its attached muscles without major damage to the latter.

The posterior thorax and the abdomen are very well provided with thick layers of circular and longitudinal muscles. Examine these muscle layers and then carefully remove the dorsal portions to expose the underlying organs. Anteriorly the principal muscles are those of the antennae, antennules, and ocular peduncles. Carefully separate these muscles and trace each into its respective appendage. In the midline between the muscle pairs locate the anterior aorta. As the circulatory system in stomatopods is particularly significant from an evolutionary viewpoint, it should be studied in considerable detail. The anterior aorta is a large vessel that can be traced easily, even in noninjected specimens. At the level of the antennae it gives off a pair of large branches that are the common origins of the antennal and antennular arteries. In the acron the anterior aorta gives rise to the paired ophthalmic artery and smaller vessels that provide blood to the supraesophageal ganglion. After identifying the cephalic arteries, cut the overlying membrane and trace the anterior aorta posteriorly to its junction with the heart. At this level the paired lateral cephalic arteries arise. Trace the heart from its anterior end at the level of the 2nd thoracic somite to its termination in the 5th abdominal somite. The presence of paired segmental lateral arteries and ostia (the latter difficult to identify in noninjected specimens) are original in the Stomatopoda and considered to be a primitive condition. Trace several of the lateral arteries to their respective limbs; their small branches to the musculature and surrounding tissues will be difficult to distinguish. Exceptions to the segmental pattern are the 1st pair, which supplies blood to the mouthparts; the 2nd pair, which supplies both the 1st and 2nd thoracopods; the 9th pair, which supplies blood to the tissues and musculature of the 8th thoracic somite and the 1st abdominal somite; and the last pair, which serves both the last pair of pleopods and the uropods. Posterior to the heart the posterior aorta can be traced into the telson. Secondary branches of the lateral arteries connect with the large ventral subneural artery, which will not be observed until the digestive tract has been removed. Blood returning to the heart passes into the pleopodal gills through a large lacuna ventral to the gut. From the gills the blood flows into the pericardial sinus through segmentally arranged, dorsally directed sinuses. Details of the circulatory system will be difficult to observe in noninjected specimens.

Immediately beneath the heart are the reproductive organs. In females a pair of elongate ovaries are present with oviducts opening on the sternite of the 6th thoracic somite. In males a paired accessory gland is present anteriorly with ducts opening into the tubular copulatory appendages on the 8th thoracic somite. The testes extend from the 3rd abdominal somite to the telson, where the coiled pair fuse to form a single tube. Anteriorly from the testes trace the coiled paired vas deferens forward to their openings in the paired penes.

Return to the anterior part of the thorax and remove the muscles of the antennae and antennules. Observe the prominent anterior extension of the cardiac stomach. With careful teasing separate the stomach from the adjoining muscle bands. No clear distinction can be made between the esophagus and stomach in stomatopods. Near the anterior end of the heart observe the marked reduction in the size of the stomach. This is the pyloric part of the stomach that opens into the midgut and into a pair of digestive glands (ceca) through a structure called the *pars ampullaris*. Trace the midgut and one of the digestive glands posteriorly. In each segment the cecum is expanded into a large segmental pouch. In the telson the cecum branches into a series of tubules; the gut terminates at the anus, located anteroventrally on the telson. Cut the stomach free in the area of the mouth and at the beginning of the midgut, and remove and open it. Observe the structure of the filtering apparatus separating the cardiac and pyloric portions and the structure of the pyloric stomach itself. The former consists of a pair of sickle-shaped, finely toothed upper lateral plates, a pair of similarly shaped lower lateral plates, and a broad, ovate median plate. The pyloric stomach consists of a broad ventral filtering plate with anterior sclerite and right and left filtering plates with anterior sclerites. Each of the filtering plates is lined with rows of fine teeth that form a very efficient filtering apparatus.

With the cardiac stomach removed, the supraesophageal ganglion can readily be observed. Trace the nerves of the ocular peduncles, antennules, and antennae. The esophageal connectives in stomatopods are very elongate. Anteriorly the ventral nerve cord is enlarged by the fusion of the ganglia of the mouthparts and

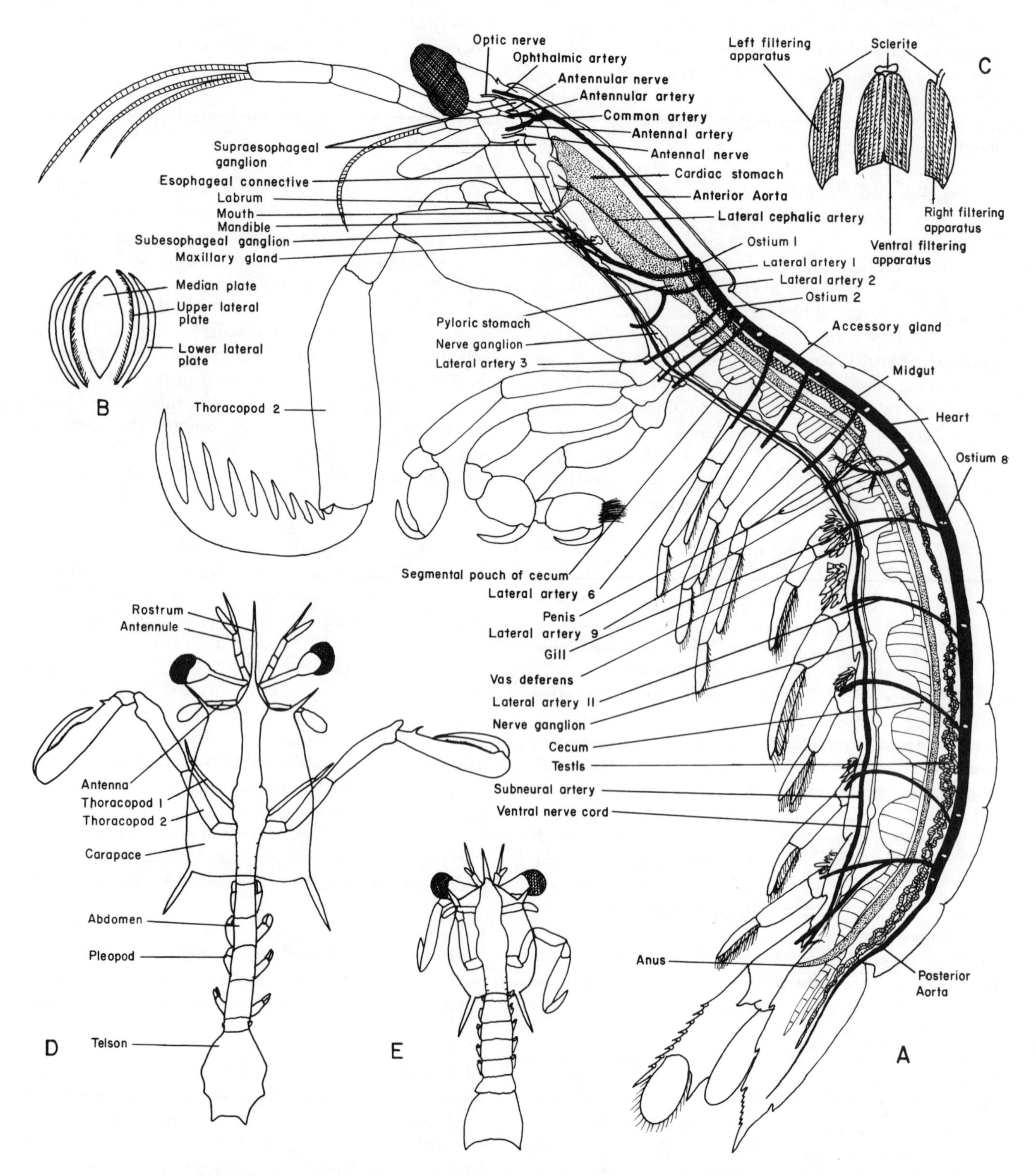

Figure 24 A. Diagrammatic stomatopod with musculature removed to show major organ systems; B. Filtering apparatus separating gastric from pyloric portions of stomach; C. Plates of pyloric stomach; D, E. Two forms of early stage pseudozoeae.

first 5 thoracomeres to form the very elongate subesophageal ganglion. Paired ganglia of the subsequent thoracic and abdominal somites appear fused as median ganglia separated by long connectives. Trace the nerve cord and identify the segmental ganglia. Beneath the nerve cord is the large subneural artery. Trace this blood vessel from its origin in the mandibular somite to its termination in the 6th abdominal somite.

Larval development in stomatopods is anamorphic and may be regular or irregular. Most stomatopods hatch as pseudozoeae, characterized by the presence of 4 or 5 functional pleopods, biramous antennules, and 2 uniramous thoracopods. In two genera, *Lysiosquilla* and *Coronida,* the larvae hatch as antizoea, characterized by the lack of pleopods (which develop progressively later), uniramous antennules, and 5 pairs of biramous thoracic appendages. The number of propelagic and pelagic larval substages appears to be variable among species and genera. Usually a single postlarval stage represents the transitional period between the free-swimming larvae and the benthic adults.

References

Balss, H., 1938. Stomatopoda. In H. G. Bronn, *Klassen und Ordnungen des Tierreichs,* 5(1)6(2): 1–173. Leipzig: Akad. Verl.

Barnard, K. H., 1950. Descriptive list of South African stomatopod Crustacea (mantis shrimp). *Ann. S. Afr. Mus., 38:* 838–864.

Blumstein, R., 1974. Stomatopod crustaceans from the Gulf of Tonkin with the description of new species. *Crustaceana, 26:* 113–126.

Chopra, B., 1939. Stomatopoda. *Sci. Rept. John Murray Expedition, 6:* 131–181.

Forest, J., 1973. Crustaceans. In W. Fisher (ed.), *FAO species identification sheets for fishery purposes.* Mediterranean and Black Sea (Fishing Area 37), 2, 45 pp.

Gurney, R., 1946. Notes on stomatopod larvae. *Proc. Zool. Soc. Lond., 116:* 133–175.

Holthuis, L. B., 1941. The Stomatopoda of the Snellius Expedition, 13. In: Biological results of the Snellius Expedition. *Temminckia, 6:* 241–294.

———, 1959. Stomatopod Crustacea of Suriname. *Stud. Fauna Suriname, 3:* 173–191.

———, 1967. Fam. Lysiosquillidae et Bathysquillidae. Stomatopoda I. In H. E. Gruner, and L. B. Holthuis (eds.), *Crustaceorum Catalogus,* ed. a, pars *1:* 1–28. Den Haag: W. Junk.

Holthius, L. B., and R. B. Manning, 1969. Stomatopoda. In R. C. Moore (ed.), *Treatise on invertebrate paleontology,* Pt. R, Arthropoda 4, *2:* R535–R552. Lawrence, Kans.: Geol. Soc. America and Univ. Kansas.

Kemp, S. W., 1913. An account of the Crustacea Stomatopoda of the Indo-Pacific region based on the collection in the Indian Museum. *Mem. Indian Mus., 4:* 1–217.

Makarov, R. R., 1977. Rotonogie raki v sborakh ekspyeditsii E/S "Akademik Knipovich". [Mantis shrimps (Crustacea: Hoplocarida: Stomatopoda) in collections of expeditions of the research vessel "Akademic Knipovich".] *Biologiya Morya,* no. 3: 14–23.

Manning, R. B., 1961. Sexual dimorphism in *Lysiosquilla scabricauda* (Lamarck) a stomatopod crustacean. *Quart. J. Fla. Acad. Sci., 24:* 101–107.

———, 1963. Preliminary revision of the genera *Pseudosquilla* and *Lysiosquilla* with descriptions of six new genera. *Bull. Mar. Sci. Gulf Carib., 13:* 308–328.

———, 1967. Review of the genus *Odontodactylus. Proc. U.S. Natl. Mus., 123:* 1–35.

———, 1968a. A revision of the family Squillidae, with the description of eight new genera. *Bull. Mar. Sci., 18:* 105–142.

———, 1968b. Stomatopod Crustacea from Madagascar. *Proc. U.S. Natl. Mus., 124:* 1–61.

———, 1969a. Notes on the *Gonodactylus* section of the family Gonodactylidae, with descriptions of four new genera and species. *Proc. Biol. Soc. Wash., 82:* 143–166.

———, 1969b. Stomatopod Crustacea from the western Atlantic. *Stud. Trop. Oceanogr.,* no. 8: 1–380. Miami, Florida: Univ. Miami Press.

———, 1970. The R/V "Pillsbury" deep-sea biological expedition to the Gulf of Guinea, 1964–65. 13. *Stud. Trop. Oceanogr.,* no. 4: 256–275.

———, 1972. Stomatopod Crustacea. Eastern Pacific expeditions of the New York Zoological Society. *Zoologica, N.Y., 56:* 95–113.

———, 1974. Stomatopods collected by Th. Mortensen in the eastern Pacific region (Crustacea, Stomatopoda). *Steenstrupia, 3:* 101–109.

———, 1977. A monograph of the West African stomatopod Crustacea. *Atlantide Rept.,* no. 12: 25–181.

———, 1978. Synopses of the Indo-West-Pacific species of *Lysiosquilla* Dana, 1852. *Smiths. Contr. Zool.,* no. 259: 1–16.

Schmitt, W. L., 1940. The stomatopods of the west coast of America based on collections made by the Allan Hancock Expeditions, 1933–1938. *Allan Hancock Found. Pacific Exped., 5:* 129–225.

Schram, F. R., 1973. On some phyllocarids and the origin of the Hoplocarida. *Fieldiana (Geologia), 26:* 77–94.

Serène, R., 1954. Observations biologiques sur les Stomatopodes. *Ann. Inst. Océanogr. Monaco, 29:* 1–93.

———, 1962. Révision du genre *Pseudosquilla* et définition de genres nouveaux. *Bull. Inst. Océanogr. Monaco,* no. 1241: 1–27.

Tirmizi, N. M. and R. B. Manning, 1968. Stomatopod Crustacea from West Pakistan. *Proc. U.S. Natl. Mus., 125:* 1–48.

Williamson, D. I., 1967. On a collection of planktonic Decapoda and Stomatopoda (Crustacea) from the Mediterranean coast of Israel. *Bull. Sea Fish. Res. Stn. Israel,* no. 45: 32–64.

SUBCLASS EUMALACOSTRACA Grobben, 1892

The Eumalacostraca includes the three superorders Syncarida, Peracarida, and Eucarida. Collectively eumalacostracans can be distinguished from phyllocarids by the absence of a bivalve carapace and the lack of a 7th abdominal somite. Characters that distinguish the Eumalacostraca from the Hoplocarida are the absence of an acron and the lack of 5 pairs of subchelate, feeding thoracopods, the 2nd of which is extremely powerful and raptorial.

Superorder SYNCARIDA Packard, 1885

The Syncarida is a relatively small taxon possessing a unique combination of characters, the most outstanding of which is the complete absence of any vestige of a carapace or cephalic shield. The body plan is simple and quite fundamental; the biramous thoracopods are considered to represent the original primitive condition. The three orders, Anaspidacea, Stygocaridacea, and Bathynellacea, are discussed individually.

ORDER ANASPIDACEA Calman, 1904

Recent species	Seven or eight in three families.
Size range	2–50 mm.
Carapace	Absent.
Eyes	Stalked or sessile, rarely absent.
Antennules	Biramous; with statocyst.
Antennae	Uni- or biramous; latter with scaphocerite.
Mandibles	With palp.
Maxillulae	With 2 endites and vestige of (?) endopod.
Maxillae	With 2 endites and bilobed terminal segment.
Maxillipeds	One pair.
Thoracic appendages	Seven pairs of thoracopods; 1st through 5th biramous, with epipods; 6th and 7th with or without exopods; 6th sometimes without epipods; 7th without epipods.
Abdominal appendages	Five pairs of pleopods usually with endopods reduced or absent; endopods of 1st and 2nd pleopods modified as copulatory structures (gonopophyses) in males. Uropods well developed, biramous, spatulate, or slender.
Telson	Sometimes spatulate.
Tagmata	Head, thorax, and abdomen.
Somites	Head with 5 + 1 thoracic (maxilliped); thorax with 7; abdomen with 6, excluding telson.
Sexual characters	Gonopores on 6th thoracic somite of female, 8th of male. Endopods of 1st and 2nd pleopods modified as copulatory structures in males; females with spermatheca.
Sexes	Separate.
Larval development	Epimorphic.
Fossil record	Triassic to Recent.
Feeding types	Primarily filter feeders on detritus and algae; sometimes predators.
Habitat	Freshwater mountain lakes and streams.
Distribution	Tasmania and southern Australia.

The restricted distribution (Tasmania and Australia) of Recent species makes it probable that no more than demonstration materials of anaspidaceans will be available for study. However, this is a generally basic group of higher crustaceans and its morphological patterns will provide a basis for subsequent modifications. The discussion and instructions for study are based on specimens of *Anaspides tasmaniae* Thomson; but major differences between this species and others in the order will be pointed out.

Observe the whole animal from both dorsal and ventral aspects (see Figure 25). Dorsally, the cephalon is produced into a broad, ventrally deflected rostrum; it may be considerably reduced in other taxa. A pair of prominent stalked eyes project from beneath the rostrum; in *A. tasmaniae* they are directed almost laterally, but in other taxa more anteriorly. Anaspidids all have stalked eyes, but the eyes in the koonugids are sessile; *Micraspides* is blind. A small tubercle, commonly found in mysids, has been reported on the eyestalk of *Allanaspides*. The antennular peduncles are very broad and the flagella biramous. A slitlike opening of the statocyst may be observed on the 1st peduncular segment. The antennal flagellum is moderately long; the exopod is developed as a scaphocerite in the Anaspididae but is absent in the Koonugidae. The cephalon and 1st thoracic somite are fused, but a mandibular or cervical groove or sulcus is very prominent in *Anaspides* and *Allanaspides*. It is less pronounced in *Paranaspides* and virtually obscured in *Koonuga*. In the midline, anterior to this groove, a small pigmented area can be observed. It is frequently referred to as a "4-celled sense organ"; its function is unknown. Some carcinologists have suggested that it might be similar to or homologous with the dorsal organ of some nonmalacostracans. The thorax consists of 7 somites (2–8), each with a pair of thoracopods. The first 5 pairs (thoracopods 2–6) are biramous and usually have a pair of epipods developed. The exopod of thoracopod 7 is reduced or absent in the Anaspididae and absent in the Koonugidae; 1 or 2 epipods are present. The last pair of thoracopods are uniramous and without epipods in all taxa.

The 6 somites of the abdomen are very well developed; the telson usually is spatulate. Five pairs of biramous pleopods are present in the Anaspididae, but endopods are lacking in the Koonugidae, except for the first 2 pairs in males. The first 2 pairs of pleopods are modified for reproductive purposes in all males. The biramous uropods are spatulate in *Anaspides*, and together with the telson form a tailfan. In other taxa the uropods may be styliform or reduced but are always biramous.

Examine the ventral aspects of the whole animal before making a detailed study of the appendages and mouthparts. In males the vas deferens open through a pair of oblique slits on the sternite of the 8th thoracomere in *Anaspides* or through a single opening in *Allanaspides*. In addition to paired gonopores that open on the medial faces of the coxae of the 6th thoracopods, females have the anterior margin of the 8th thoracic sternite produced to form a sperm receptacle or spermatheca. In females particularly, the coxal segments of the thoracopods have setose medial lobes projecting in toward the midline.

Examine a thoracopod from one of the anterior pairs and compare its structure with that of the 7th and 8th. Typically, each thoracopod consists of a 2-segmented protopod, an exopod and endopod. A pair of epipods arise laterally from the coxa. The endopod, in contrast to most malacostracans, is 6-segmented; that is, preischium, ischium, merus, carpus, propodus, and dactyl. Basis-preischium fusion is progressive in the thoracopods of *Allanaspides*. The exopod consists of a basal stipe and an annulated terminal segment. What differences from this typical pattern do you find in the last 2 pairs of thoracopods? Examine the first 2 pleopodal endopods of the male. In *A. tasmaniae* the remaining 3 pairs in males and all 5 pairs in females have reduced, 1-segmented endopods. Endopods are lacking, except for the modified first 2 pairs in males, in *Allanaspides*.

Examine slide preparations of the mouthparts. The mandible is large, with well-developed incisor and molar processes; the mandibular palp has 3 segments. The maxillule has 2 endites and a small lobe that may be a vestige of either the endopod or exopod. The maxilla has 2 small endites and 1 that is large and bilobed. The maxilliped is not appreciably modified from a typical thoracopod, except for 2 articulated endites (gnathobases) on the coxa. These structures are absent in *Koonuga*, but the maxilliped is distinguishable by its much stouter appearance.

From a carefully dissected specimen, in sagittal view it will be possible to examine the major organ systems (see Figure 26). Dorsally the heart, which lies in close proximity to the very thin exoskeleton, extends from the 1st thoracic to the 4th abdominal somite. A single pair of ostia is present at the level of the 3rd thoracomere. Anterior from the heart, the prominent anterior aorta extends into the cephalon. Also arising anteriorly from the heart is the 1st of 7 pairs of lateral arteries. Almost immediately each lateral artery gives off a branch to the viscera, the remaining branch continuing to supply blood to the appendages. The 2nd pair of lateral arteries arises at the level of the 8th thoracomere. The lateral branch of only one of the pair is well developed, the descending aorta, which supplies blood to the thoracopods via the supraneural artery. The 3rd pair of lateral arteries provides blood to the body tissues, while the remaining 4 pairs provide blood to the pleopods and uropods.

The digestive system consists of an esophagus, stomach, midgut, hindgut, and ceca. The stomach is divided into cardiac and pyloric parts; a pars ampullaris at

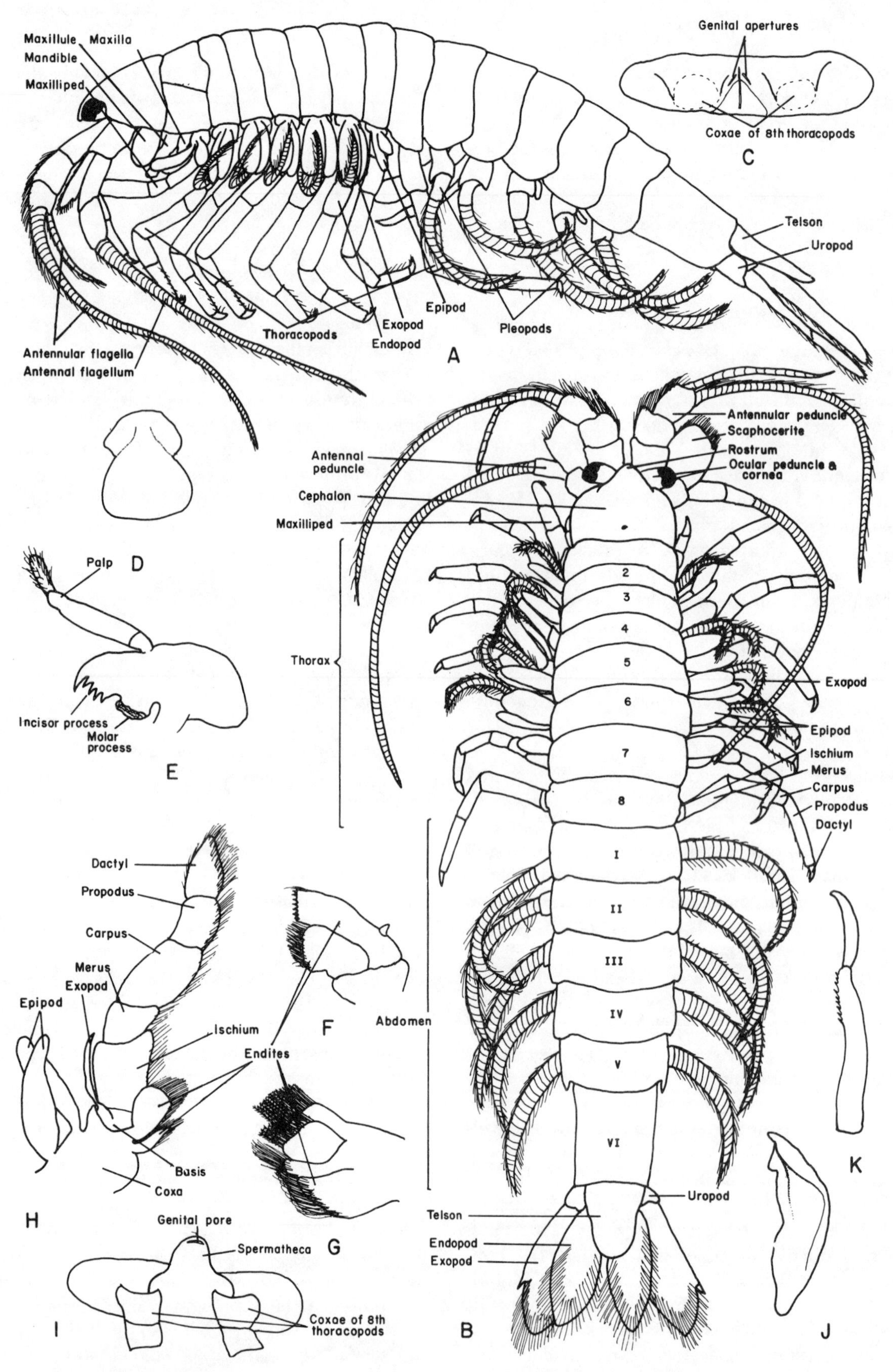

Figure 25 Anaspidacea: A. Whole animal (lateral view); B. Whole animal (dorsal view); C. 8th sternite of male; D. Labrum; E. Mandible; F. Maxillule; G. Maxilla; H. Maxilliped; I. 8th sternite of female; J. Endopod of 1st male pleopod; K. Endopod of 2nd male pleopod.

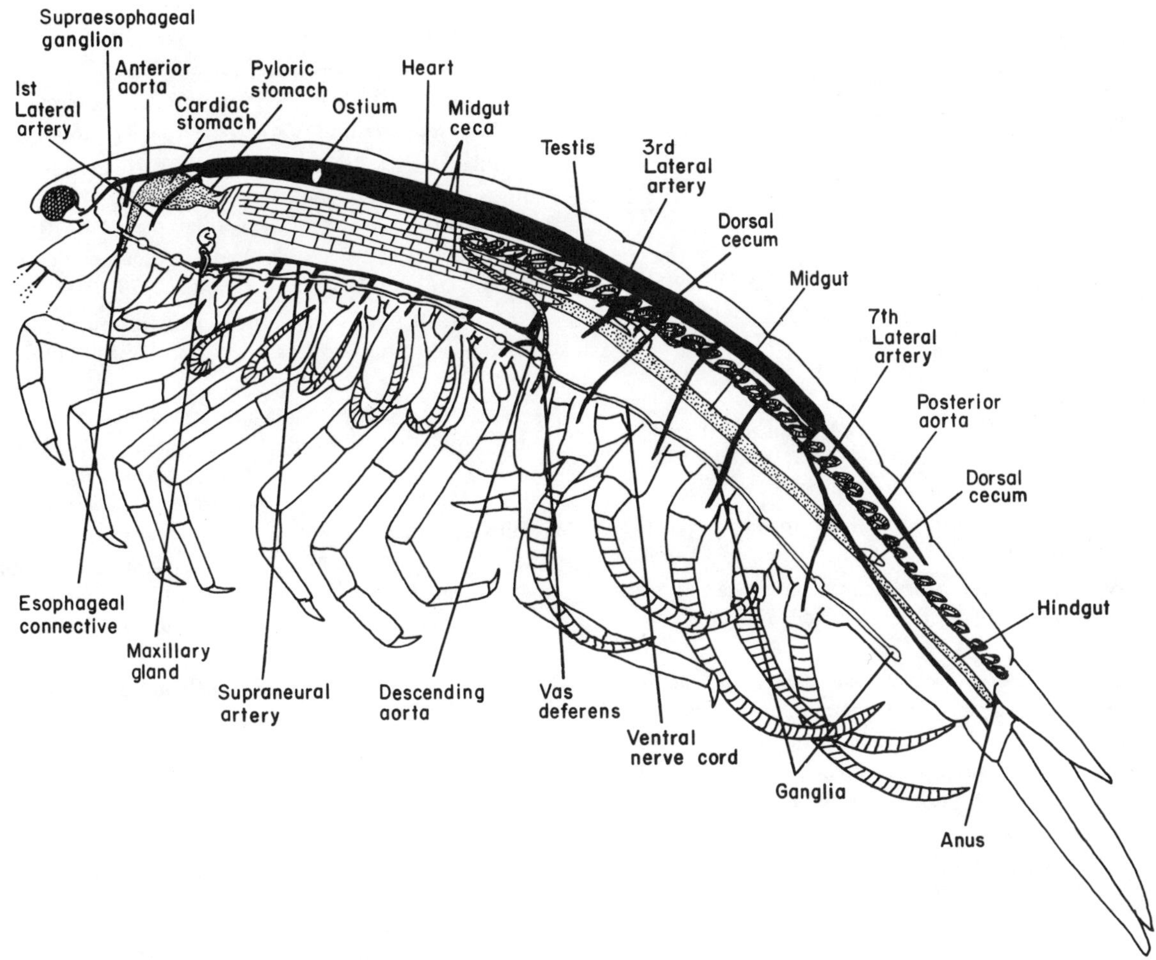

Figure 26 Anaspidacea: Diagrammatic *Anaspides* with musculature removed to show major organ systems.

the entrance of the ceca is homologous with that of stomatopods. In *Anaspides* the numerous ceca are thin tubules extending the length of the thorax. At the level of the 1st abdominal somite a short, anteriorly directed dorsal cecum arises from the midgut, and a second dorsal cecum often is present at the level of the 5th somite. A change in the muscular wall of the gut at the level of the 6th somite suggests the beginning of the hindgut. The anus opens at the base of the telson. Excretion is by means of a maxillary gland, the tubules of which are easily recognized.

The large supraesophageal ganglion is located anterodorsally in the cephalon and the prominent optic, antennular, and antennal nerves and the esophageal connectives leading from it can be distinguished without difficulty. In contrast to other malacostracans, no subesophageal ganglion is present; the ganglia of the mouthparts all are distinct. The ventral nerve cord, lying directly beneath the supraneural artery in the thorax, consists, in *A. tasmaniea,* of a broad band of parallel fibers with ganglia in each somite.

The reproductive system in the male consists of paired, thin coiled testes that extend from the 5th thoracic somite to the telson. Anteriorly from the testes paired vas deferens lead to the gonopores on the 8th thoracic sternite. In *Allanaspides* the vas deferens unite before opening into the median gonopore. In the female the ovaries are a pair of broad tubular structures that lie on either side of the midgut. Although typically extending from the 6th thoracic to the 6th abdominal somite, they often extend anteriorly to the region of the 2nd thoracomere. A pair of oviducts lead from the ovaries to the gonopores on the coxae of the 6th thoracopods. The sternite of the 8th thoracomere is produced anteriorly to form a spermatheca.

Development in anaspidaceans is epimorphic, with the young hatching with the full complement of body somites and appearing quite similar to the adult. At the time of hatching, young *Anaspides* have sessile eyes, a vestigial nauplius eye, and a deeply cleft telson; the endopods of the pleopods have not developed. The complete adult form is attained in successive molts.

References

Brooks, H. K., 1962. On the fossil Anaspidacea, with a revision of the classification of the Syncarida. *Crustaceana 4:* 229–242.

Calman, W. T., 1896. On the genus *Anaspides* and its affinities with certain fossil Crustacea. *Trans. Roy. Soc. Edinburgh, 38:* 787–802.

———, 1915. Opossum shrimps. *Nature*, London, *168:* 924–925.

Cannon, H. G., and S. M. Manton, 1929. On the feeding mechanism of the syncarid Crustacea. *Trans. Roy. Soc. Edinburgh, 56:* 175–189.

Chappuis, P. A., 1927. Anaspidacea. In W. Kükenthal and Th. Krumbach, *Handbuch der Zoologie, 3:* 593–606. Berlin: Der Gruyter.

Hansen, H. J., 1925. *Studies on Arthropoda, 2:* 1–176. Copenhagen: Gyldendalsk.

Hickman, V. V., 1937. The embryology of the syncarid crustacean *Anaspides tasmaniae. Pap. Roy. Soc. Tasmania, 1936:* 1–36.

Manton, S. M., 1930. Notes on the habits and feeding mechanisms of *Anaspides* and *Paranaspides* (Crustacea, Syncarida). *Proc. Zool. Soc. London, 1930:* 791–800, 1079.

Nicholls, G. E., 1921. *Micraspides calmani* a new syncaridan from the west coast of Tasmania. *J. Linn. Soc. Zool., 37:* 473–488.

———, 1947. On the Tasmanian Syncarida. *Rec. Victoria Mus., 2:* 9–16.

Noodt, W., 1963. Anaspidacea in der südlichen Neotropis. *Verhandl. Deutschen Zool. Ges., 26:* 568–578.

———, 1964. Natürliches System und Biogeographie der Syncarida. *Gewässer und Abwässer, 37/38:* 77–186.

Schminke, H. K., 1974. *Psammaspides williamsi* gen. n., sp. n., ein Vertreter einer neuen Familie mesopsammaler Anaspidacea (Crustacea, Syncarida). *Zool. Scr., 3:* 177–183.

Siewing, R., 1954. Verwandtschaftsbeziehungen der Anaspidaceen. *Verhandl. Deutschen Zool. Ges., 18:* 240–252.

———, 1959. Syncarida. In H. G. Bronn, *Klassen und Ordnungen des Tierreichs*, 5 (1) 4(2): 1–121. Leipzig: Akad. Verl.

———, 1963. Studies in malacostracan morphology: Results and problems. In H. B. Whittington and W. D. I. Rolfe (eds.), *Phylogeny and evolution of Crustacea*, pp. 85–110. Spec. Publ. Mus. Comp. Zool., Harvard. Cambridge, Mass.: Harvard University Press.

Smith, G. W., 1908. Preliminary account of the habits and structure of *Anaspides* with remarks on some freshwater Crustacea from Tasmania. *Proc. Roy. Soc. London*, (B) *80:* 465–473.

———, 1909. On the Anaspidacea, living and fossil. *Quart. J. Micro. Sci.*, n.s. *53:* 489–578.

Swain, R., I. S. Wilson, J. L. Hickman, and J. E. Ong, 1970. *Allanaspides helonomus* gen. et sp. nov. (Crustacea: Syncarida) from Tasmania. *Rec. Victoria Mus.*, no. 35: 1–7.

Williams, W. D., 1965. Zoological notes on Tasmanian Syncarida. *Int. Rev. Ges. Hydrobiol., 50:* 95–126.

ORDER STYGOCARIDACEA Noodt, 1964

Recent species	Five in two genera, *Stygocaris* and *Parastygocaris.*
Size range	2.0–4.2 mm.
Carapace	Absent.
Eyes	Absent.
Antennules	Biramous; with statocyst.
Antennae	Uniramous; nephropore absent.
Mandibles	Without palp; dentate setae (penicillae) present; incisor and molar processes well developed.
Maxillulae	With 2 endites.
Maxillae	With 4 enditic lobes.
Maxillipeds	One pair.
Thoracic appendages	Seven thoracopods, 2–8; 2–6 with or without exopods; 7 and 8 uniramous; 2–7 with 2 epipods.
Abdominal appendages	Pleopods usually absent; 1st and 2nd pairs present in males, forming petasma. Uropods biramous.
Telson	With rudimentary caudal furca.
Tagmata	Head, thorax, and abdomen.
Somites	Head with 5 + 1 thoracic (maxilliped); thorax with 7; abdomen with 6, excluding telson.

Sexual characters	Gonopores on 6th thoracic somite of female, on 8th of male; female also with spermatheca between bases of 8th thoracopods; male with petasma; head, antennules, and antennae sometimes exhibiting sexual dimorphism.
Sexes	Separate.
Larval development	(?) Anamorphic.
Fossil record	Permian to Recent.
Feeding types	Probably grazers on sand grains.
Habitat	Interstitial in fresh water.
Distribution	South America and New Zealand.

The Stygocaridacea is a very small order, with only four described species in a single family, all from interstitial habitats. It is probable that material will not be available for detailed study of this group; thus description and illustrations will be used to explain the general morphology of the order (see Figure 27).

The body is elongate and vermiform, with little distinction between thoracic and abdominal somites, except in the appendages. A carapace and eyes are lacking. The rostrum is a thin, platelike structure either quadrate or strap-shaped, occasionally deeply cleft. The cephalon is fused with the 1st thoracic somite. The cephalic appendages include the biramous antennules, each with a well-developed statocyst on the 1st segment, the uniramous antennae, mouthparts, and 1 pair of maxillipeds. There are 7 pairs of thoracopods. Exopods are present on thoracopods 2 to 6 in species of *Parastygocaris* but are absent in *Stygocaris*. A pair of epipods is present on each thoracopod except the last pair in all species.

Abdominal appendages are absent except for the modified first 2 pairs in males. These appendages form a petasma, which is distinctly structured in each species. The uropods have a basal protopod, a 1-segmented endopod, and usually a 2-segmented exopod. Remnants of a caudal furca are present as small setose protuberances from the terminal margin of the telson. [Bowman (1971) interprets these protuberances as lobes of a slightly incised telson.]

The mandible has a well-developed incisor process with a toothlike structure directly beneath it. One or 2 penicillae usually are present between these structures and the prominent molar process. The maxillule consists of a pair of endites; the maxilla has 4 enditic lobes. The maxilliped is provided with a setose endite and a 7-segmented endopod.

The larval development of stygocaridaceans is not known, but is assumed to be similar to that of the bathynellaceans.

References

Brooks, H. K., 1969. *Syncarida*. In R. C. Moore (ed.), *Treatise on invertebrate paleontology,* Part R Arthropoda, 4, *1:* R345–R359. Lawrence, Kans.: Geol. Soc. America and Univ. Kansas.

Gordon, I., 1964. On the mandible of the Stygocaridae and some other Eumalacostraca, with special reference to the lacinia mobilis. *Crustaceana, 7:* 150–157.

Morimoto, Y., 1977. A new *Stygocaris* (Syncarida Stygocarididae) from New Zealand. *Bull. Nat. Sci. Mus., A3:* 19–24.

Noodt, W., 1963. *Anaspidacea* (Crustacea, Syncarida) in der südlichen Neotropis. *Verhandl. Deutsche Zool. Ges. Wien 1962:* 568–578.

———, 1965. Natürliches System und Biogeographie der Syncarida. *Gewässer und Abwässer,* 37/38 (1964): 77–186.

———, 1970. Eidonomie der Stygocaridacea, einer Gruppe interstitieller Syncarida (Malacostraca). *Crustaceana, 19:* 227–244.

Schiminke, H. K., and W. Noodt, 1968. Discovery of Bathynellacea, Stygocaridacea and other interstitial Crustacea in New Zealand. *Naturwissenschaften, 54:* 184–185.

ORDER BATHYNELLACEA Chappuis, 1915

Recent species	Approximately 69 in three families: Bathynellidae, Parabathynellidae, and Leptobathynellidae.
Size range	0.5–5.4 mm.
Carapace	Absent.

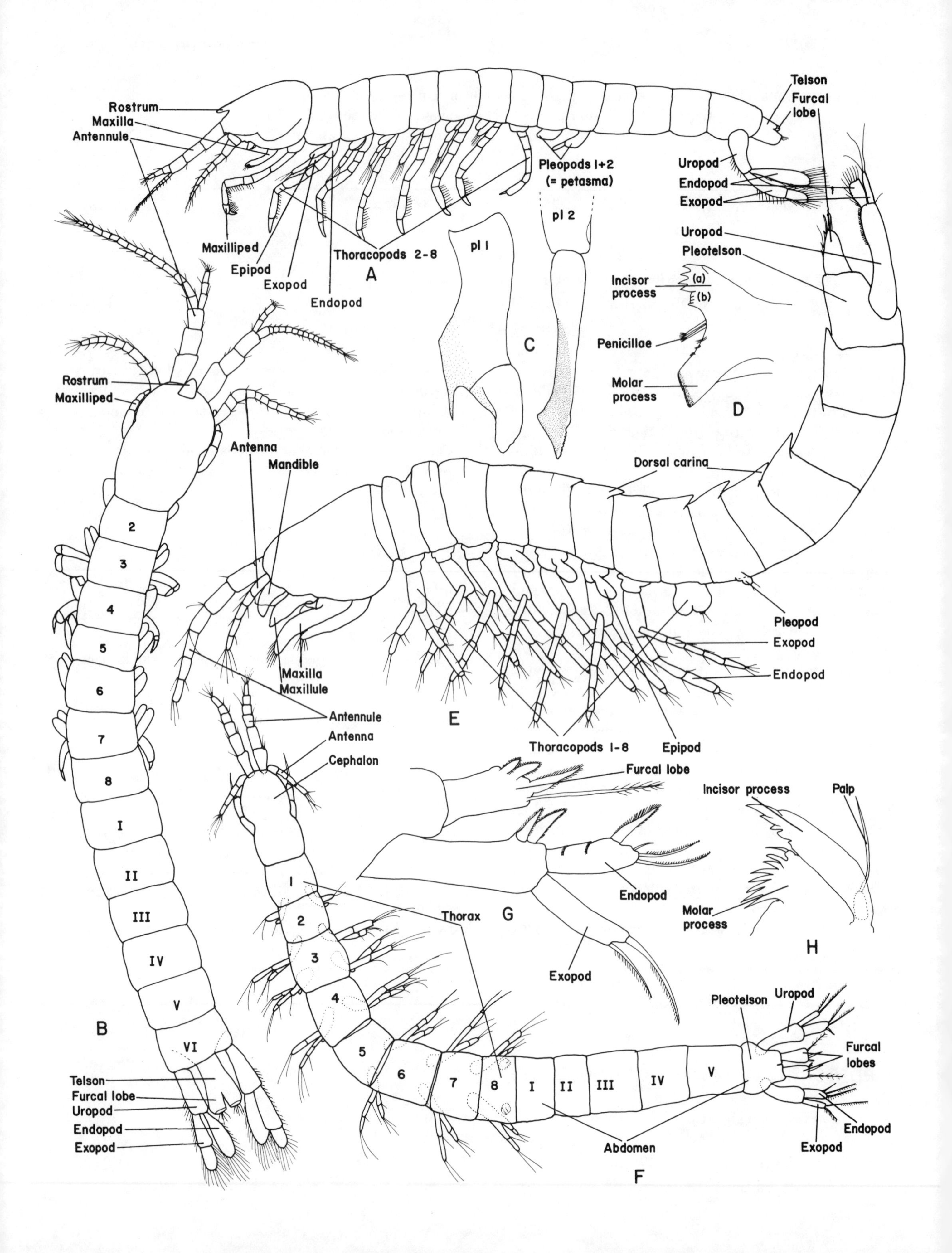
Rostrum
Maxilla
Antennule
Maxilliped
Epipod
Exopod
Endopod
Thoracopods 2-8
A
Pleopods 1+2
(= petasma)
Telson
Furcal
lobe
Uropod
Endopod
Exopod
pl 1
pl 2
C
Uropod
Pleotelson
Incisor
process
(a)
(b)
Penicillae
Molar
process
D
Rostrum
Maxilliped
Antenna
Mandible
2
3
4
5
6
7
8
I
II
III
IV
V
VI
B
Telson
Furcal lobe
Uropod
Endopod
Exopod
Dorsal carina
Pleopod
Exopod
Endopod
Maxilla
Maxillule
Antennule
E
Thoracopods 1-8
Epipod
Antenna
Cephalon
Furcal lobe
Endopod
G
Exopod
Incisor process
Palp
Molar
process
H
Thorax
1
Pleotelson
Uropod
Furcal
lobes
Endopod
Exopod
Abdomen
F

Figure 27 Stygocaridacea, Bathynellacea: A—D, Stygocaridacea; E—H, Bathynellacea. A. *Stygocaris* (lateral view) [after Kaestner, 1970]; B. *Parastygocaris* (dorsal view) [after Brooks, 1969, from *Treatise on Invertebrate Paleontology,* courtesy of the Geological Society of America and University of Kansas]; C. Petasma of *Stygocaris gomezmillasi* Noodt [after Noodt, 1970]; D. Mandible of *Parastygocaris andina* Noodt [after Gordon, 1964]; E. *Allobathynella* (lateral view) [after Ueno, 1961]; F. *Leptobathynella* (dorsal view) [after Noodt, 1972]; G. Left uropod and furcal lobe of telson [after Noodt, 1972]; H. Mandible of *Parabathynella fagei* Delamare Deboutteville and Chappuis [after Gordon, 1964].

Eyes	Absent.
Antennules	Uniramous; without statocyst.
Antennae	Uni- or biramous; nephropore absent.
Mandibles	With palp; incisor and molar processes well developed.
Maxillulae	With 2 basal endites.
Maxillae	With 2 or 3 endites.
Maxillipeds	None.
Thoracic appendages	Usually 7 biramous thoracopods, 1–7; 8th reduced or vestigial, occasionally absent; 1 or 2 epipods frequently present on thoracopods 1–7.
Abdominal appendages	Usually 1st or 1st and 2nd pairs of pleopods; occasionally absent. Uropods biramous.
Telson	Fused with 6th abdominal somite (pleotelson); with furcal lobes.
Tagmata	Head, thorax, and abdomen.
Somites	Head with 5; thorax with 8; abdomen with 5, excluding pleotelson.
Sexual characters	Gonopores on 6th thoracic somite of female, on 8th thoracopod of male; 8th thoracopods sexually dimorphic.
Sexes	Separate.
Larval development	Anamorphic.
Fossil record	Recent.
Feeding types	Detritus feeders.
Habitat	Interstitial in fresh water.
Distribution	Africa, Madagascar, southwestern Asia, Japan, New Zealand, South America, United States, and Europe.

Like the rather closely related stygocaridaceans, species of Bathynellacea are very small interstitial crustaceans. They are, however, somewhat better known than the former order, with some 69 species within the three families. The typically small size and the relative paucity of specimens in teaching collections make it necessary to study this group from demonstration materials. If these are not available, at least familiarize yourself with the general morphology from Figure 27 and the discussion.

As in other syncarids, the carapace is lacking. The eyes also are absent, as is a rostral plate. The cephalon is not fused with the first thoracic somite; however, the number of body somites is the same as in the stygocaridaceans as a result of the fusion of the 6th abdominal somite with the telson to form a pleotelson. [Bowman (1971) interprets the abdominal segmentation as 6 unfused somites, with the terminal lobes representing a divided telson rather than caudal rami.]

The cephalon generally is more elongate than any of the individual body somites. The antennules are uniramous; the antennae have a small exopod in the Bathynellidae, a rudimentary exopod in the Leptobathynellidae, and lack an exopod in the Parabathynellidae. The 1st thoracopod is not modified as a maxilliped.

Eight pairs of thoracopods often are present; however, this number sometimes is reduced to 7 and occasionally to 6. The 8th pair, when present, is markedly smaller and usually modified as gonopods in males. Typically both endopods and exopods are present and most thoracopods also carry epipods, which may be reduced in size. Pleopods may be absent, or present only on the 1st or the 1st and 2nd abdominal somites. The pleotelson has a pair of biramous uropods and a pair of setose or spinose caudal lobes (caudal rami).

The mandible is well developed, with an incisor process, molar process, and 1- to 3-segmented mandibular palp that frequently is prehensile. The maxillule typically has a pair of endites, and the maxilla 2 or 3 endites. Paragnaths often also are present.

A newly hatched bathynellid has antennules, antennae, mouthparts, and 1st thoracopods, but only 10 body somites. Following the first molt 11 somites are present in addition to the 2nd thoracopods and limb buds of the 3rd and 4th thoracic and 1st abdominal appendages. Usually 6 juvenile substages occur, and one or two additional molts are required for the animal to reach sexual maturity.

References

Brooks, H. K., 1969. Syncarida. In R. C. Moore (ed.), *Treatise on invertebrate paleontology,* Part R Arthropoda, 4, *1:* R345–R359. Lawrence, Kans.: Geol. Soc. America and Univ. Kansas.

Calman, W. T., 1917. Notes on the morphology of *Bathynella* and some allied Crustacea. *Quart. J. Micro. Sci., 62:* 489–514.

Chappuis, P. A., 1915. *Bathynella natans* und ihre Stellung im System. *Zool. Jahrb. (Syst.), 40:* 147–176.

———, 1948. Le développement larvaire de *Bathynella. Bull. Soc. Sci. Cluj, 10:* 305–309.

Dancau, D., and E. Serban, 1963. Sur une nouvelle *Parabathynella* de Roumaine, *Parabathynella motasi* nov. sp. *Crustaceana, 5:* 241–250.

Delamare Deboutteville, C., 1961a. Nouvelles récoltes de Syncarides et compléments systématiques. *Ann. Spéol., 16:* 217–222.

———, 1961b. Présence d'un Syncaride d'un genre nouveau dans les eaux interstitielles des lacs de la Patagonie andine, et remarques biogéographiques. *Compt. Rend. Séanc. Acad. Sci., 251:* 1038–1039.

Delamare Deboutteville, C., and P. A. Chappuis, 1953. Les Bathynelles de France et d'Espagne avec diagnoses d'especes et de formes nouvelles. *Vie et Milieu, 4:* 114–115.

Delamare Deboutteville, C., N. Coineau, and E. Serban, 1975. Découverte de la famille des Parabathynellidae (Bathynellacea) en Amerique du Nord: *Texanobathynella bowmani* n. g. n. sp. *C. R. Acad. Sci., Paris D280:* 2223–2226.

Delamare Deboutteville, C., and Ch. Roland, 1963. Syncarides. *Austrobathynella patagonica* Delamare. In *Biologie de l'Amerique Australe, 2:* 55–62. Paris.

Gordon, I., 1964. On the mandible of the Stygocaridae and some other Eumalacostraca with special reference to the lacinia mobilis. *Crustaceana, 7:* 150–157.

Husmann, S., 1965. Morphologische, ökologische und verbreitungsgeschichtliche Studien über die Bathynellen (Crustacea, Syncarida) des Niederrhein-Grundwasserstromes bei Krefeld. *Gewässer und Abwässer,* 37/38 (1964): 46–76.

———, 1973. *Bathynella stammeri* Jakobi, 1954 (Syncarida) aus den stygohithral des Alpen; Studien zu Morphologies, Ökologie und Verbreitungsgeschichte. *Crustaceana, 25:* 21–34.

Jakobi, H., 1954. Biologie, Entwicklungsgeschichte und Systematik von *Bathynella natans. Zool. Jahrb. Abt. Syst., 83:* 1–184.

———, 1958. Ein neues Genus der Grundwasserfamilie Bathynellidae (Grobben) aus der Dünenzone der Insel Santa Catarina, Südbrasilien. *Dusenia, 8:* 25–36.

Noodt, W., 1965. Natürliches System und Biogeographie der Syncarida. *Gewässer und Abwässer, 37/38* (1964): 77–186.

———, 1967. Biogeographie der Bathynellacea. *Symp. Crustacea, Ernakulum, Mar. Biol. Assoc. India, 1:* 411–417.

———, 1972. Brasilianische Grundwasser—Crustacea, 2. *Nannobathynella, Leptobathynella* und *Parabathynella* aus der Serra do Mar von São Paulo (Malacostraca, Syncarida). *Crustaceana, 23:* 152–164.

Noodt, W., and M. H. Galhano, 1969. Studien an Crustacea subterranea (Isopoda, Syncarida, Copepoda) aus dem Norden Portugals. *Publ. Inst. Zool. "Dr. A. Nobre," 107:* 1–75.

Por, F. D., 1968. *Parabathynella calmani* n. sp. (Syncarida, Bathynellacea) from Israel. *Crustaceana, 14:* 151–154.

Schminke, H. K., 1976. Systematische Untersuchungen an Grundwasserkrebsen—eine Bestandsaufnahme (mit der Beschreibungen zweier neuer Gattungen der Familie Parabathynellidae (Bathynellacea). *Int. J. Speleol., 8:* 195–216.

Schminke, H. K., and W. Noodt, 1968. Discovery of Bathynellacea, Stygocaridacea and other interstitial Crustacea in New Zealand. *Naturwissenschaften, 54:* 184–185.

Serban, E., 1975. Sur les *Bathynella* de Roumanie: *B. (B.) boyrsi* Serban, *B. (B.) vaducrisensis* n. sp., *B. (B.) plesai* Serban et *B. (B.) motrensis* Serban (Bathynellacea, Bathynellidae I partie). *Int. J. Speleol., 7:* 357–398.

———, 1977. Sur péréiopodes de *Iberobathynella* cf. *fagei* de Majorque (Bathynellacea, Parabathynellidae). *Crustaceana, 33:* 1–16.

Serban, E., and N. Coineau, 1975. Sur les Bathynellidae (Podophallocarida, Bathynellacea) d'Afrique du Sud. Les genres *Transvaalthynella* nov. et *Transkeithynella* nov. *Ann. Spéléol., 30:* 137–165.

Siewing, R., 1956a. *Thermobathynella amyxi* nov. spec. aus dem Brackwasser der Amazonasmüdung. *Kieler Meeresforsch., 12:* 114–119.

———, 1956b. Morphologie der Malacostraca. *Zool. Jahrb. Abt. Anat., 75:* 39–176.

———, 1963. Studies in malacostracan morphology: Results and problems. In H. B. Whittington and W. D. I. Rolfe (eds.), *Phylogeny and evolution of Crustacea,* pp. 85–110. Spec. Publ. Mus. Comp. Zool., Harvard. Cambridge, Mass.: Harvard University Press.

Ueno, M., 1961. A new Japanese bathynellid. *Crustaceana, 2:* 85–88.

Superorder PERACARIDA Calman, 1904

General characters of the Peracarida include a carapace that is never fused to all the thoracomeres (the carapace in isopods and amphipods secondarily has been lost); an antenna typically with a 3-segmented protopod; and a mandible usually with a lacinia mobilis. At least the 1st pair of thoracopods is modified as a pair of maxillipeds; the pleopods lack an appendix interna; and females have a brood pouch usually formed by medial lamellar outgrowths of the coxae of some of the thoracopods.

Seven orders are included in the Peracarida: Mysidacea, Thermosbaenacea, Spelaeogriphacea, Cumacea, Tanaidacea, Isopoda, and Amphipoda. Of these, the suborders of the Isopoda and Amphipoda are sufficiently distinctive and important that they each warrant specific consideration. Therefore, diagnoses of these taxa are presented as diagnoses of the individual suborders.

ORDER MYSIDACEA Boas, 1883

Recent species	Approximately 450.
Size range	10–350 mm.
Carapace	Well developed; covering head and most of thorax; fused with first 3, rarely also 4th, thoracic somites.
Eyes	Stalked, compound.
Antennules	Biramous.
Antennae	Biramous; exopod scalelike.
Mandibles	With well-developed palp.
Maxillulae	Usually with pair of endites and 1 exite; occasionally also with endopod.
Maxillae	With endopod, endites; usually also with exopod.
Maxillipeds	One or 2 pairs.
Thoracic appendages	Usually 6 or 7 pairs; exopods usually present.
Abdominal appendages	Biramous when present, often reduced or absent in females; uropods frequently with statocysts.
Telson	With uropods forming well-developed tailfan.
Tagmata	Cephalothorax and abdomen.
Somites	Head with 5 + 1 or 2 thoracic (maxillipeds); thorax with 6 or 7; abdomen with 6, excluding telson.
Sexual characters	Gonopores on coxae of 6th thoracopods of female, on genital papillae or penes of 8th thoracomere of male; pleopods often absent in females, sometimes modified as gonopods in male; female with oostegites.
Sexes	Separate.
Larval development	Epimorphic; released as miniature adults.
Fossil record	Mississippian to Recent.
Feeding types	Filter feeders or predators.
Habitat	Primarily marine, some brackish and freshwater species; pelagic, epibenthic, or sometimes benthic.
Distribution	Worldwide.

The Mysidacea include two suborders, the Lophogastrida and the Mysida. The Lophogastrida is considered the more primitive suborder; its representatives generally inhabit deep water and may attain a large size. The description of lophogastrid mysidaceans has been based primarily on characters exhibited by the genera

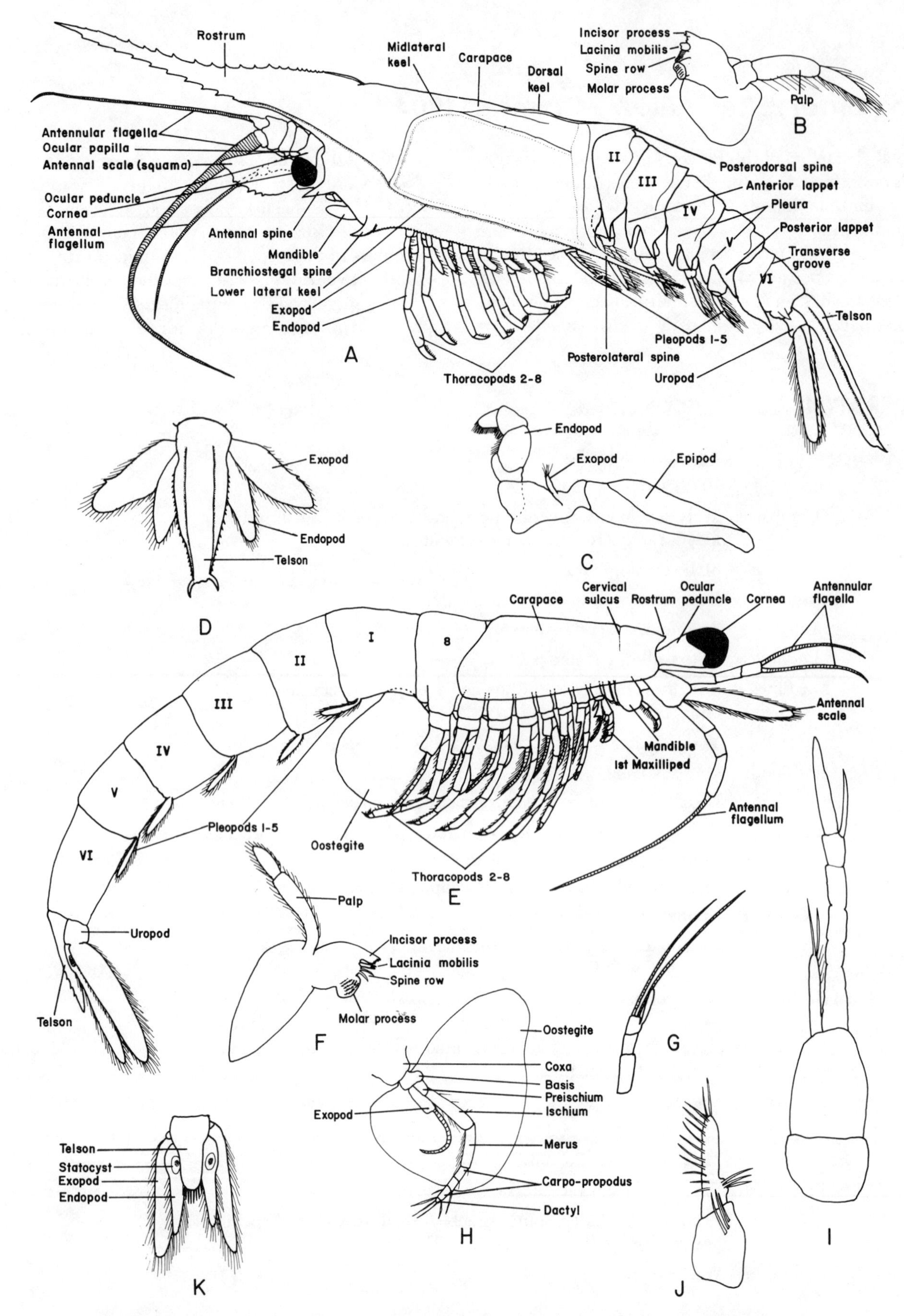

Figure 28 Mysidacea: A—D, Lophogastrida; E—K, Mysida. A. Whole animal (lateral view); B. Mandible; C. Maxilliped; D. Telson and uropods; E. Whole animal (lateral view); F. Mandible; G. Antennule of male; H. Thoracopod with oostegite; I. 1st pleopod of male; J. 2nd pleopod of male; K. Uropods and telson.

Lophogaster and *Gnathophausia*, as these taxa are more frequently available for study (see Figure 28). The carapace is very well developed, usually ornamented with dorsal and lateral keels and marginal and posteromarginal spines, and is produced anteriorly into a very prominent, elongate rostrum. At least in some species of *Lophogaster*, the rostrum is often more elongate in adult females than in males and immature specimens. The eyes are usually well developed, even if occasionally of small size. The ocular peduncle, at least in species of *Gnathophausia*, carries a moderately prominent papilla. The antennular peduncle is 3-segmented; in *Lophogaster* the 3rd segment has a scalelike process that often is referred to as an antennular lamella. The antennular flagella are paired; 1 flagellum often is much stronger than the other. The exopod of the antenna is developed into a prominent scale.

In the lophogastrids the 1st thoracopod is modified as a maxilliped, and the 2nd through 8th are well developed. The form of these latter thoracopods is one of the characters used to separate the two families of Lophogastrida. In the Lophogastridae all 7 thoracopods are similar in structure; in the Eucopiidae the 2nd to 4th are short and subchelate, the 5th to 7th are long and subchelate, and the 8th is typical in structure. Gills of 3 or 4 branches are present on the 2nd to 7th thoracopods; when present on the 8th, gills are reduced or rudimentary. Females have oostegites on all 7 pairs of thoracopods.

The abdomen of lophogastrids is well developed and the tergites are laterally produced to form prominent pleura, which, in *Gnathophausia* at least, are subdivided into anterior and posterior lappets. The 6th abdominal somite is transversely grooved, which has led to speculation that this somite primitively may have represented a fusion of the 6th with a 7th somite. Five pairs of biramous pleopods are present in both sexes and usually are not modified in males, although some modifications have been reported in species of *Eucopia*. The uropods and telson are well developed and form a tailfan. The telson frequently carries a pair of terminal spines. No statocysts are present on the endopods of the uropods in lophogastrids.

The mouthparts consist of paired mandibles, maxillulae, and maxillae in addition to the pair of maxillipeds. The left mandible bears a lacinia mobilis between the incisor and molar processes; the right has a fixed cusp in this position. The maxillule has a large, swollen proximal endite with few setae; the distal endite has a row of blunt spines. No palp is present. The maxilla is provided with 2 endites (not subdivided in *Lophogaster*), an exite, and a palp.

If specimens are available for dissection, remove, examine, and illustrate the mouthparts, maxilliped, and 1 thoracopod. Pay particular attention to the structure of the right and left mandibles. Compare the mouthparts and appendages of the lophogastrids with those of the mysids and the euphausiids, a superficially similar group of planktonic crustaceans at one time classified with the Mysidacea. Differences in internal organ systems will be discussed with the Mysida.

The description and instructions for the study of the suborder Mysida have been based on a composite of species. If you have been able to examine specimens of lophogastrids, the differences in carapace shape and development will be apparent immediately. In typical mysids the carapace is much more weakly developed and does not cover the bases of the thoracopods laterally. A rostrum may be present but is never appreciably elongate. Begin your examination of the external morphology by observing the animal from a dorsal view. The stalked eyes are prominent but there is no movable acron or ophthalmic somite such as was observed in the stomatopods. In the anterior portion of the carapace locate the transverse cervical sulcus. It is possible that this groove may be homologous with the cervical groove of anaspidaceans. Frequently not all of the thoracic somites are covered by the carapace; lift the posterior margin of the carapace and see if you can determine the number of thoracic somites to which it is fused. Even in preserved specimens numerous chromatophores usually can be observed, and these serve as an aid in the recognition of many mysids. In lateral view, observe the antennules, antennae, and appendages. The flagella of the antennule frequently are relatively short and of equal strength; the 3rd peduncular segment may have an accessory lobe in males. The exopod of the antenna is produced into a prominent scale, which may be diagnostic both in its shape and in its segmentation. The lateral part of the mandible frequently can be recognized without dissection. The 1st or in some cases the 1st and 2nd thoracopods are modified as maxillipeds. Thoracopods 3 to 8 may vary among themselves depending upon the taxa under study. After you have completed your general examination of the specimen, remove several thoracopods and identify the components. As in anaspidaceans, the endopod will consist of 6 segments, a preischium being present in addition to the typical components. In some taxa the carpus and propodus may be fused and may be multiarticulate in addition, the number of carpopropodal articles varying from 1 to 8. Exopods may be present or absent; epipods are present on the 1st thoracopod only. In females, 2 or 3 posterior thoracopods have oostegites developed.

The abdominal tergites of typical mysids are not laterally produced to form pleura. Pleopods generally are reduced or absent in females; in males they sometimes are normally developed, biramous, highly modified, or by a combination of conditions. Depending upon the taxon, the 1st, 2nd, 3rd, and 5th may be well developed or reduced; usually the 4th is modified, presumably for

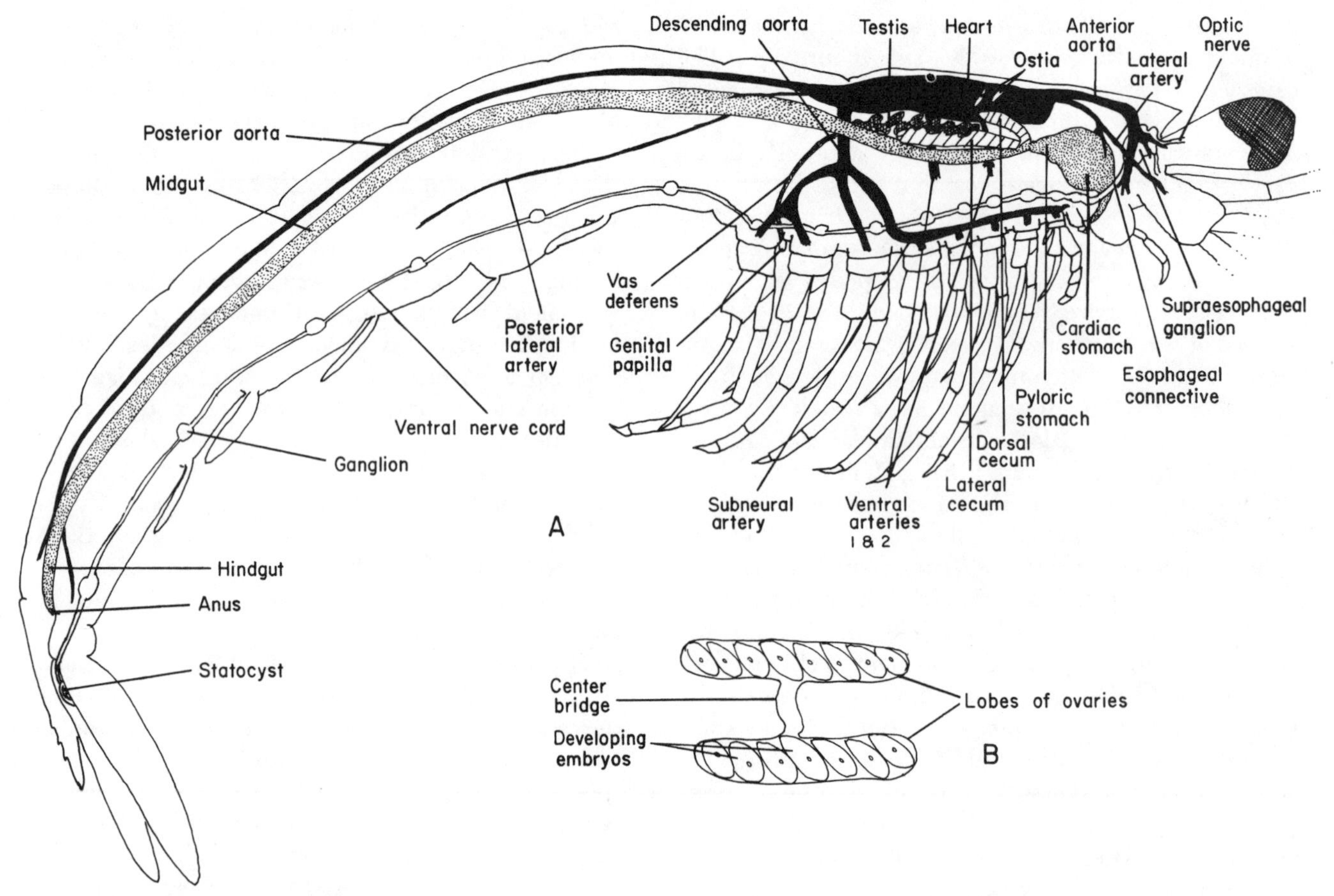

Figure 29 Mysidacea: A. Diagrammatic mysid with musculature removed to show major organ systems. B. Diagrammatic mysid ovaries.

copulation. The uropods have a prominent statocyst on each endopod. The telson is well developed, and with the uropods forms a tailfan.

Remove and examine the mouthparts. A mandibular palp is well developed. How do the mandibles compare with those of the Lophogastrida? The maxillule consists usually of 2 endites. The maxilla has an exopod, 2-segmented endopod, and paired endites, the distal one bilobed. Identify the epipod of the maxilliped. Compare the maxilliped with one of the anterior thoracopods and determine the feeding adaptations of this appendage.

The major internal organ systems of mysidaceans exhibit marked differences between the two suborders, particularly the circulatory system. The description is based on the condition found in mysids, but differences in lophogastrids will be pointed out (see Figure 29). Begin your study by teasing the carapace free of the anterior thoracic somites. With fine scissors or a microscalpel cut the integument of the abdomen and carefully remove the exoskeleton from one side. The mysid body is quite muscular, particularly in the abdominal region; this musculature must be removed from one side to expose the organs.

Dorsally in the cephalothorax locate the heart. It is moderately elongate, but rarely extending beyond the cephalothorax. In mysids 2 pair of ostia are present at the level of the 4th and 3rd thoracomeres; in lophogastrids a 3rd pair is present at the level of the 2nd thoracomere. Anteriorly a pair of lateral cephalic arteries and the unpaired anterior aorta lead from the heart and provide blood to the cephalic appendages, nervous system, and stomach. In mysids 3 unpaired ventral arteries arise from the heart medially. The first 2 provide blood to the gut sinus; the 3rd, the large descending aorta, divides into 3 branches. The 1st branch continues anteriorly as the subneural artery with branches to the first 5 pairs of thoracopods. The 2nd branch provides blood to the 6th thoracopods, and the 3rd branch supplies the 7th and 8th thoracopods. In lophogastrids 9 pairs of lateral arteries are present. The 1st pair are the lateral cephalic arteries also present in mysids; the 8th pair constitute the descending aorta (right branch) that contributes to the gut sinus and the subneural artery (left branch) that supplies blood to all the thoracopods. The 9th pair are the posterior lateral arteries leading posteriorly with the posterior aorta into the abdomen. In mysids these are the only vessels providing blood to the tissues and appendages; in lophogastrids a pair of lateral branches from the posterior aorta lead to each of the paired pleopods. Blood returns to the heart via sinuses.

The digestive system consists of an esophagus, stomach, midgut, hindgut, and a variable number of ceca. The stomach is divided into an enlarged cardiac and smaller pyloric part. At the junction of the pyloric stomach and midgut frequently an unpaired dorsal cecum is present; 1 to 5 pairs of short to moderately long ceca opening through a pair of common ducts also may be present. The midgut is elongate; the cuticularized hindgut quite short, terminating in an anus on the ventral surface of the telson. The excretory organ in mysids is the antennal gland; in lophogastrids both antennal and maxillary glands are present.

The nervous system exhibits some variation among different taxa. The supraesophageal ganglion always is well developed; a subesophageal ganglion is present in lophogastrids, but apparently not in mysids. The ventral nerve cord in *Mysis* consists of a coalesced ribbon of ganglia and connectives extending the length of the cephalothorax; in *Boreomysis* individual pairs of cephalothoracic ganglia can be distinguished. Six pairs of abdominal ganglia always are present. In mysids the statocysts on the endopods of the uropods are innervated from the ventral nerve cord.

The female reproductive system in mysids differs from that of lophogastrids in that the ovaries in the former are connected by a narrow to moderately well-developed bridge. Each ovary is a moderately short tubule; posteriorly an oviduct leads to a gonopore on the coxa of the 6th thoracopod. The testes are paired coiled structures located in the cephalothorax that are easily identified by their pyriform follicles. A pair of vas deferens leads to genital papillae on the sternite of the 8th thoracomere. Oostegites form a brood pouch in females.

Larval development in the Mysidacea is epimorphic. Embryos hatch in the brood pouch of the female and remain there until all appendages have developed. The young, when released, resemble miniature adults.

References

Almeida Prado, M. S., de, 1974. Sistemática dos Mysidacea (Crustacea) na região de Cananéia. *Bol. Inst. Oceanogr., 23:* 47–87.

Bacescu, M., 1955. Mysidacea. *Faune Rep. Popil. Române, 4:* 3–122.

———, 1968a. Études de quelques Leptomysini (Crustacea Mysidacea) des eaux du Brézil et Cuba; description d'un genre et de cinq autres taxons nouveau. *Ann. Mus. Stor. Nat. Genova, 77:* 232–249.

———, 1968b. Contributions to the knowledge of the Gastrosaccinae psammobionte of the tropical America, with the description of a new genus (*Bowmaniella* n. g.) and three new species of its frame. *Trav. Mus. Hist. Nat. Gr. Antipa, 8:* 355–373.

Banner, A. H., 1948. A taxonomic study of the Mysidacea and Euphausiacea (Crustacea) of the northeastern Pacific, Pts. 1 and 2. *Trans. Roy. Canad. Inst., 26:* 345–399; *27:* 65–125.

———, 1953. On a new genus and species of mysid from southern Louisiana. *Tulane Stud. Zool., 1:* 3–8.

Brattegard, T., 1969. Marine biological investigations in the Bahamas. 10. Mysidacea from shallow water in the Bahamas and southern Florida. Pt. 1. *Sarsia, 39:* 17–106.

———, 1970. Marine biological investigations in the Bahamas. 10. Mysidacea from shallow waters in the Bahamas and southern Florida. Pt. 2. *Sarsia, 41:* 1–35.

———, 1973. Mysidacea from shallow water on the Caribbean coast of Colombia. *Sarsia, 54:* 1–65.

———, 1974a. Additional Mysidacea from shallow water on the Caribbean coast of Colombia. *Sarsia, 57:* 47–85.

———, 1974b. Mysidacea from shallow water on the Caribbean coast of Panama. *Sarsia, 57:* 87–107.

———, 1975. Shallow-water Mysidacea from the Lesser Antilles and other Caribbean regions. *Stud. Fauna Curaçao Carib. Isl., 47:* 102–115.

Fage, L., 1940. Sur le déterminisme des caractéres sexuels secondaires des Lophogastrides. *C. R. Acad. Sci., 211:* 335–337.

———, 1941. Mysidacea: Lophogastrida I. *Dana Rept.*, no. 19, 52 pp.

———, 1942. Mysidacea: Lophogastrida II. *Dana Rept.*, no. 23, 67 pp.

Gordan, J., 1957. A bibliography of the order Mysidacea. *Bull. Amer. Mus. Nat. Hist., 112:* 279–394.

Hansen, H. J., 1910. The Schizopoda of the "Siboga" Expedition. *Siboga-Exped., 37:* 1–123.

Ii, N., 1964. Mysidae (Crustacea). *Fauna Japonica.* Tokyo: Biogeographical Society of Japan. 610 pp.

Nouvel, H., 1940. Observations su la sexualité d'un Mysiacé, *Heteromysis armoricana* n. sp. *Bull. Inst. Oceanogr. Monaco*, no. 789, 11 pp.

———, 1942a. Sur la systématique des espèces du genre *Eucopia* Dana 1852 (Crustacea Mysidacea). *Bull. Inst. Oceanogr. Monaco*, no. 818, 10 pp.

———, 1942b. Sur la sexualité des Mysidacés du genre *Eucopia* (caractères sexuels secondaires, taille et maturité sexuelle, anomalies et action possible d'un Epicaride). *Bull. Inst. Oceanogr. Monaco*, no. 820, 11 pp.

Ortmann, A. E., 1905. Schizopods of the Hawaiian Islands collected by the steamer "Albatross" in 1902. *Bull. U.S. Fish. Comm., 23:* 961–973.

———, 1907. Schizopod crustaceans in the U.S. National Museum. The families Lophogastridae and Eucopiidae. *Proc. U.S. Natl. Mus., 31:* 23–54.

Tattersall, O. S., 1960. Notes on mysidacean crustaceans of the genus *Lophogaster* in the U.S. National Museum. *Proc. U.S. Natl. Mus., 112:* 527–547.

Tattersall, O. S., 1967. A survey of the genus *Heteromysis* (Crustacea: Mysidacea) with descriptions of five new species from tropical coastal waters of the Pacific and Indian Oceans, with a key for the identification of the known species of the genus. *Trans. Zool. Soc. London, 31:* 157–193.

Tattersall, W. M., 1922. Indian Mysidacea. *Rec. Indian Mus., 24:* 445–504.

———, 1926. Crustaceans of the orders Euphausiacea and Mysidacea from the western Atlantic. *Proc. U.S. Natl. Mus., 69:* 1–28.

———, 1933. Euphausiacea and Mysidacea from western Canada. *Contr. Can. Biol. Fish., 8:* 1–25.

———, 1939. The Mysidacea of eastern Canadian waters. *J. Fish. Res. Bd. Canad., 4:* 281–286.

———, 1951. A review of the Mysidacea of the United States National Museum. *Bull. U.S. Natl. Mus.,* 201, 292 pp.

Tattersall, W. M., and O. S. Tattersall, 1951. *British Mysidacea.* London: Ray Soc., *135:* 1–360.

Wittmann, K. J., 1977. Modification of association and swarming in North Adriatic Mysidacea in relation to habitat and interacting species. In B. F. Keegan, P. O. Ceidigh, and P. J. S. Boaden, *Biology of benthic organisms,* pp. 613–620. Oxford, New York: Pergamon Press.

Zimmer, C., 1927. Mysidacea. In W. Kükenthal and Th. Krumbach, *Handbuch der Zoologie, 3:* 607–650. Berlin: Der Gruyter.

———, 1933. Mysidacea. In G. Grimpe and E. Wagner, *Die Tierwelt der Nord-und Ostee,* 10(23) *4:* 70–120. Leipzig: Akad. Verl.

ORDER THERMOSBAENACEA Monod, 1927

Recent species	Eight in four genera: *Monodella, Thermosbaena, Halosbaena,* and *Limnosbaena.*
Size range	1.5–4.0 mm.
Carapace	Present; short, fused to 1st thoracic somite and covering adjacent 2 or occasionally 3 somites; dorsally enlarged to form brood pouch in females.
Eyes	Greatly reduced or absent; sometimes discernible only in sections.
Antennules	Biramous; sometimes with aesthetascs.
Antennae	Uniramous; flagella with few articles.
Mandibles	With palp; molar and incisor processes; lacinia mobilis well developed.
Maxillulae	With 2 endites and palp.
Maxillae	With 3 endites, palp, and rudimentary lobe possibly representing exopod.
Maxillipeds	One pair; uni- or biramous; sometimes sexually dimorphic.
Thoracic appendages	Biramous; 5–7 pairs; epipods absent.
Abdominal appendages	Two anterior pairs of uniramous pleopods; uropods biramous.
Telson	Well developed; freely articulated or fused with ultimate abdominal somite (pleotelson).
Tagmata	Head, thorax, and abdomen.
Somites	Head with 5 + 1 thoracic (maxilliped); thorax with 7; abdomen with 6, excluding telson, or 5, excluding pleotelson.
Sexual characters	Gonopores on 6th thoracic somite of female, on 8th of male; female with dorsal brood pouch, without oostegites; male with paired penes. Maxilliped sometimes sexually dimorphic.
Sexes	Separate.
Larval development	Epimorphic; manca stage; when released sometimes without 6th and 7th thoracopods.
Fossil record	Recent.

Feeding types	Plant detritus feeders (although mandibles have biting capabilities).
Habitat	Thermal springs, fresh and brackish subterranean lakes, and brackish coastal interstitial areas.
Distribution	Southern Italy, Dead Sea, Tunisia, West Indies, and Hays County, Texas.

The Thermosbaenacea is a small order from a generally restricted habitat. Representatives of this order were known only from brackish waters with temperatures in excess of 30°C until a species from fresh cool water recently was described from Texas.

The small size, as well as the lack of reference specimens, restricts the classroom study of this very interesting group. Examine demonstration slides if available or, if not, study the general morphological characters of the order from the illustration in Figure 30. The two families have some major differences that will be pointed out.

The carapace, which is fused with the 1st thoracic somite, covers the cephalon and in some taxa several thoracic somites; in females it is expanded to provide a dorsal brood pouch. Eyes usually are absent or can be distinguished only in serial sections. The antennules are well developed, biramous appendages; the antennae are uniramous and considerably smaller. The 1st pair of thoracopods are modified as maxillipeds.

In the Monodellidae thoracopods 2–8 are well developed biramous appendages; the first 4 pairs are much shorter than the last 3. In the Thermosbaenidae only thoracopods 2–6 are present; they also are biramous but are relatively small and generally equal in length. The endopods have been interpreted as frequently having a fused or partially fused basis-ischium in the maxilliped and a fused ischium-merus in some thoracopods. Often the dactyl terminates in an elongate claw referred to as an ungulus.

Six abdominal somites and a telson are present in the Monodellidae; 5 abdominal somites and a pleotelson occur in the Thermosbaenidae. In both families the uropods each consist of a protopod, a 1-segmented endopod, and a 2-segmented exopod; the uropods of the monodellids usually are much larger.

Of the mouthparts the mandible perhaps is the most structurally complex. It consists of a well-developed molar process bearing a series of fine guide setules at the anterior angle. The incisor process and lacinia mobilis are borne on a keellike projection. The relatively broad space separating these structures from the molar process is occupied by a row of "lifting spines," also referred to in certain other taxa as a setal row. A very well-developed mandibular palp also is present. The maxillule consists of 2 endites and a palp; the maxilla has 3 endites, a palp, and a rudimentary structure considered by some investigators to represent the exopod.

The pair of 1st maxillipeds are sexually dimorphic in some species. In these the maxillipeds each consist of 2 endites, an exopodal palp, and an epipod in the female; the male endopod is a 5-segmented prehensile structure presumably functioning as a clasping structure.

Larval development has been observed in *Thermosbaena* and *Monodella*. In the former, the young hatch as miniature adults; in the latter, hatching occurs at the manca stage and the young pass through a postmarsupial stage during which the 6th and 7th thoracopods and the pleopods develop.

The thermosbaenaceans have been related, by various carcinologists, to the Hoplocarida, Syncarida, and Peracarida. What similarities and differences do you find among these orders and to which, if any, would you relate the Thermosbaenacea?

References

Barker, D., 1956. The morphology, reproduction and behaviour of *Thermosbaena mirabilis* Monod. *Intern. Congr. Zool.*, *14*, 1953, pp. 503–504, Copenhagen.

———, 1958. On *Thermosbaena mirabilis* Monod and the distribution and systematics of the Thermosbaenacea. *Proc. Centen. Bicenten. Congr. Biol.*, pp. 253–258, Singapore.

———, 1959. The distribution and systematic position of the Thermosbaenacea. *Hydrobiologia*, *13:* 209–235.

———, 1962. A study of *Thermosbaena mirabilis* (Malacostraca, Peracarida) and its reproduction. *Quart. J. Micro. Sci.*, *103:* 261–286.

Bruun, A. F., 1939. Observations on *Thermosbaena mirabilis* Monod from the hot springs of El Hamma, Tunisia. *Vidensk. Medd. Naturh. Foren. Kjøb.*, *103:* 493–501.

Delamare Deboutteville, C., 1960. *Biologie des eaux souterraines littorales et continentales.* 740 pp. Paris: Hermann.

Fryer, G., 1964. Studies on the functional morphology and feeding mechanism of *Monodella argentarii* Stella (Crustacea: Thermosbaenacea). *Trans. Roy. Soc. Edinburgh*, *64*(4): 49–90.

Karaman, S., 1953. Ueber einen Vertreter des Ordnung Thermosbaenacea (Crustacea Percarida) aus Jugoslavien, *Monodella halophila* n. sp., *Acta Adriat.*, *5:* 55–67.

Maguire, B., Jr., 1964. Crustacea: A primitive Mediterranean group also occurs in North America. *Science*, *146:* 931–932.

———, 1965. *Monodella texana* n. sp., an extension of the range of the crustacean order Thermosbaenacea to the western hemisphere. *Crustaceana*, *9*(2): 149–154.

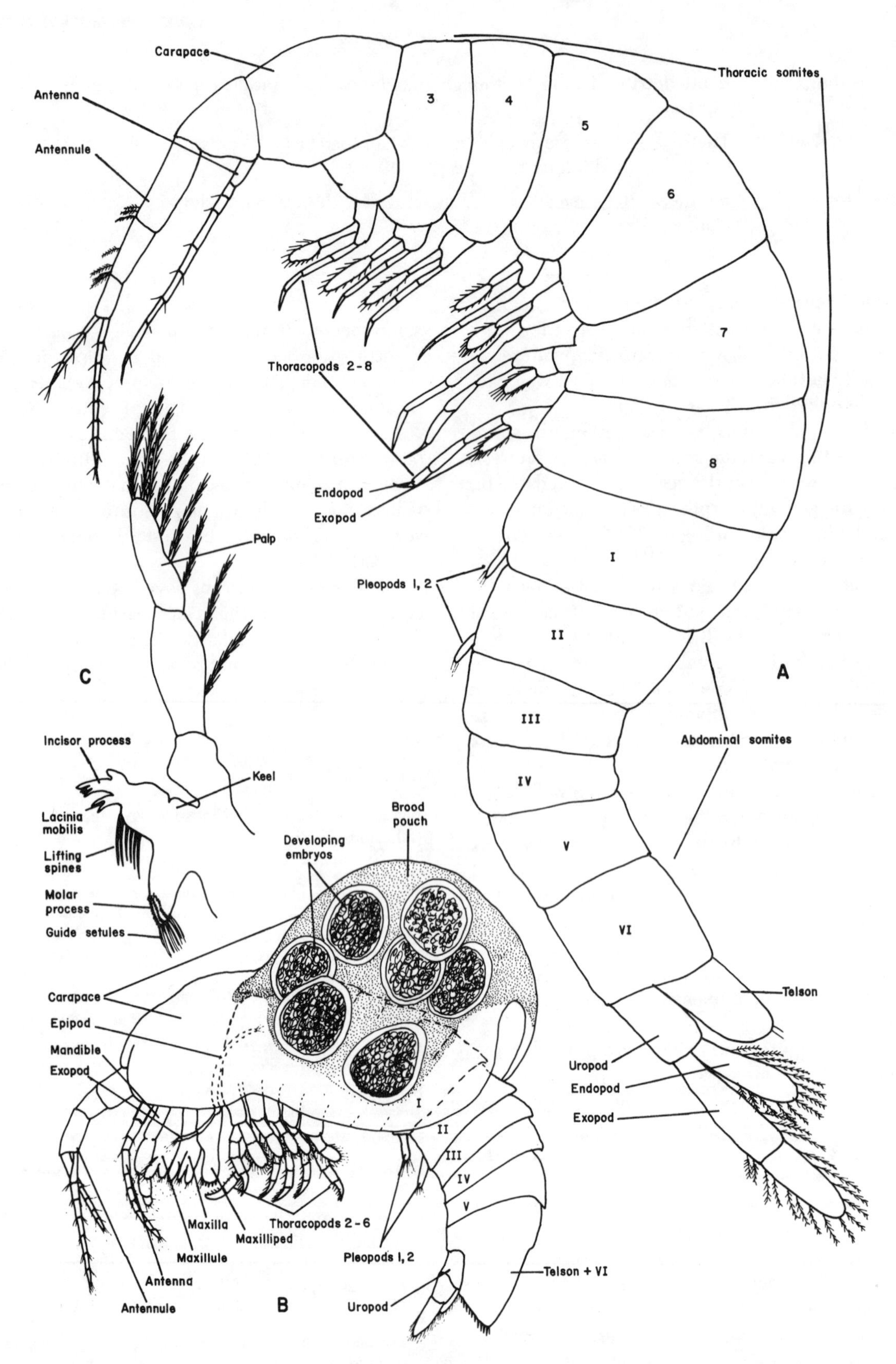

Figure 30 Thermosbaenacea: A. *Monodella* (lateral view) [after C. Delamare Deboutteville: *Biologie des eaux souterraines littorales et continentales* (Hermann, Paris, 1960)]; B. *Thermosbaena*, ovigerous female (lateral view) [after Barker, 1962]; C. Mandible of *Monodella relicta* Por [after Por, 1962].

Monod, T., 1924. Sur un type nouveau de Malacostrace: *Thermosbaena mirabilis* nov. gen., nov. sp. *Bull. Soc. Zool. Fr.*, *49:* 58–68.

———, 1927a. *Thermosbaena mirabilis* Monod. Remarques sur la morphologie et sa position systématique. *Faune Colon. Franc.*, *1:* 29–51.

———, 1927b. Nouvelles observations sur la morphologie de *Thermosbaena mirabilis*. *Bull. Soc. Zool. Fr.*, *52:* 196–200.

———, 1940. Thermosbaenacea. In H. G. Bronn, *Klassen und Ordnungen des Tierreichs*, 5,(1), 4, 24 pp. Leipzig: Akad. Verl.

Por, F. D., 1962. Un nouveau Thermosbaenace, *Monodella relicta* n. sp. dans la depression de la Mer Morte. *Crustaceana*, *3*(4): 304–310.

Ruffo, S., 1949a. *Monodella stygicola* n. gen., n. sp. nuovo crostaceao Thermosbaenaceo delle acque sotteranee della Penisola Salentina. *Arch Zool. Ital.*, *34:* 31–48.

———, 1949b. Sur *Monodella stygicola* Ruffo des eaux souterraines de l'Italie meridionale, deuxiéme espèce connue de l'ordre des Thermosbaenacés (Malacostraca, Peracarida). *Hydrobiologica*, *2*(1): 56–63.

Siewing, R., 1958. Anatomie und Histologie von *Thermosbaena mirabilis*. Ein Beitrag zur Phylogene der Reihe Pancarida (Thermosbaenacea). *Abh. Math.-Nat. Kl. Akad. Wiss. Mainz*, 7, 197–270.

———, 1963. Studies in malacostracan morphology: Results and problems. In H. B. Whittington and W. D. I. Rolfe (eds.), *Phylogeny and evolution of Crustacea*, pp. 85–103. Spec. Publ. Mus. Comp. Zool., Harvard. Cambridge, Mass.: Harvard University Press.

Stella, E., 1951a. *Monodella argentarii* n. sp. di Thermosbaenacaeo (Crustacea, Peracarida) limnotroglobio di Monte Argentario. *Arch. Zool., Torino*, *36:* 1–15.

———, 1951b. Notizie biologische su *Monodella argentarii* Stella, Thermosbenaceo della acque di una grotta di Monte Argentario. *Boll. Zool.*, *18* (4–6): 227–233.

———, 1953. Sur *Monodella argentarii* Stella, espèce de Crustacé Thermosbaenacé des eaux d'une grotte de l'Italie Centrale. *Hydrobiologia*, *5*(1–2): 1–11.

———, 1959. Ulteriori osservazioni sulla riproduzione e lo sviluppo di *Monodella argentarii* (Pancarida, Thermosbaenacea). *Riv. Biol., Perugia*, *51* (1): 121–144.

Stella, E., and F. Baschieri Salvadori, 1953. La fauna acquatica della grotta "di punta degli Stretti" (Monte Argentario). *Arch. Zool. Ital.*, *38:* 441–483.

Stock, J. H., 1976. A new genus and two new species of the crustacean order Thermosbaenacea from the West Indies. *Bijd. Dierk.*, *46:* 47–70.

Taramelli, E., 1954. La posizione sistematica die Thermosbenacei quale results dall studio anatomico di *Monodella argentarii* Stella. *Monit. Zool. Ital.*, *62:* 9–27.

Zimmer, C., 1927. *Thermosbaena mirabilis* Monod. In W. Kükenthal and Th. Krumbach, *Handbuch der Zoologie*, *3:* 809–811, 1078. Berlin: Der Gruyter.

ORDER SPELAEOGRIPHACEA Gordon, 1957

Recent species	One.
Size range	7.5–8.6 mm.
Carapace	Present; short; fused dorsally to 1st and covering most of 2nd thoracic somite.
Eyes	Stalked, nonfunctional.
Antennules	Biramous.
Antennae	Biramous; exopod scalelike.
Mandibles	With palp; lacinia mobilis well developed.
Maxillulae	With 2 endites, innermost slightly bilobed.
Maxillae	With 3 endites, 3rd strongly bilobed.
Maxillipeds	One pair.
Thoracic appendages	Seven pairs; anterior 3 pairs with ventilatory exopods; thoracopods 5–7 with gills.
Abdominal appendages	Pleopods 1–4 biramous, well developed; 5th vestigial; uropods biramous, broad.
Telson	Present.
Tagmata	Head, thorax, and abdomen.
Somites	Head with 5 + 1 thoracic (maxilliped); thorax with 7; abdomen with 6, excluding telson.

Sexual characters	Gonopores at bases of 6th thoracopods of female, 8th of male; female with 5 pairs of oostegites on thoracopods 2–6; male with pair of penes on coxae of 8th thoracopods, antennules sexually dimorphic.
Sexes	Separate.
Larval development	(?) Epimorphic.
Fossil record	Mississippian, Recent.
Feeding types	Apparently detritus feeders.
Habitat	Freshwater stream, bottom dwelling, and pelagic.
Distribution	Bats Cave, Table Mountain, South Africa.

As the Spelaeogriphacea is a monotypic order represented by only a single species, known only from the type locality, study material of this group will usually not be available. However, from the description and illustrations provided you should be able to draw some general comparisons between this order and others within the Peracarida (see Figure 31).

The carapace, which is fused with the 1st thoracic somite, is short and laterally expanded to enclose the branchial cavity and effectively conceal the mouthparts. Anteriorly, the carapace is produced into a broadly triangular rostrum. A cervical groove is present, but frequently is indistinct. A pair of ocular lobes is present, although spelaeogriphaceans lack functional eyes. The cephalon bears a pair of rather closely set, short antennules, a pair of more robust antennae with scalelike exopods, and mouthparts, including a pair of maxillipeds.

The 2nd through 8th thoracic somites are free, although the 2nd is partially covered by the carapace. The 7 pairs of thoracic appendages are simple and ambulatory, and are divisible into 2 types. The anterior 3 pairs each bear ventilatory exopods; the exopods of thoracopods 5–7 are reduced and modified as gills, and those of the last pair usually are absent. The 3 anterior pairs usually are more robust in males. Males have a simple penial process on the coxa of each 8th thoracopod.

The abdomen is elongate, usually exceeding half the total body length; the somites bear pleura that are quite delicate and not easily distinguished, except on the 5th somite. The first 4 pairs of pleopods are well developed and natatory; the 5th pair is reduced and concealed by the epimere of the somite. The uropods are elongate. The telson is short and terminally rounded.

Females have oostegites on the first 5 pairs of thoracopods, and it is assumed that development is epimorphic; however, juvenile stages have not been observed.

References

Gordon, I., 1957. On *Spelaeogriphus*, a new cavernicolous crustacean from South Africa. *Bull. Brit. Mus. (Nat. Hist.)*, *5:* 31–47.

———, 1960. On a *Stygiomysis* from the West Indies, with a note on *Spelaeogriphus* (Crustacea, Peracarida). *Bull. Brit. Mus. (Nat. Hist.)*, *6:* 285–323.

Grindley, J. R., and R. R. Hessler, 1971. The respiratory mechanism of *Spelaeogriphus* and its phylogenetic significance (Spelaeogriphacea). *Crustaceana*, *20:* 141–144.

ORDER CUMACEA Kröyer, 1846

Recent species	Approximately 800.
Size range	0.5–4.0 mm.
Carapace	Well developed; covering anterior part of thorax.
Eyes	Sessile, compound, fused, usually dorsomedial; sometimes absent.
Antennules	Uni- or biramous.
Antennae	Uniramous; short in female and juvenile, well developed in mature male.
Mandibles	Without palp; lacinia mobilis well developed.
Maxillulae	With 2 endites and palp.

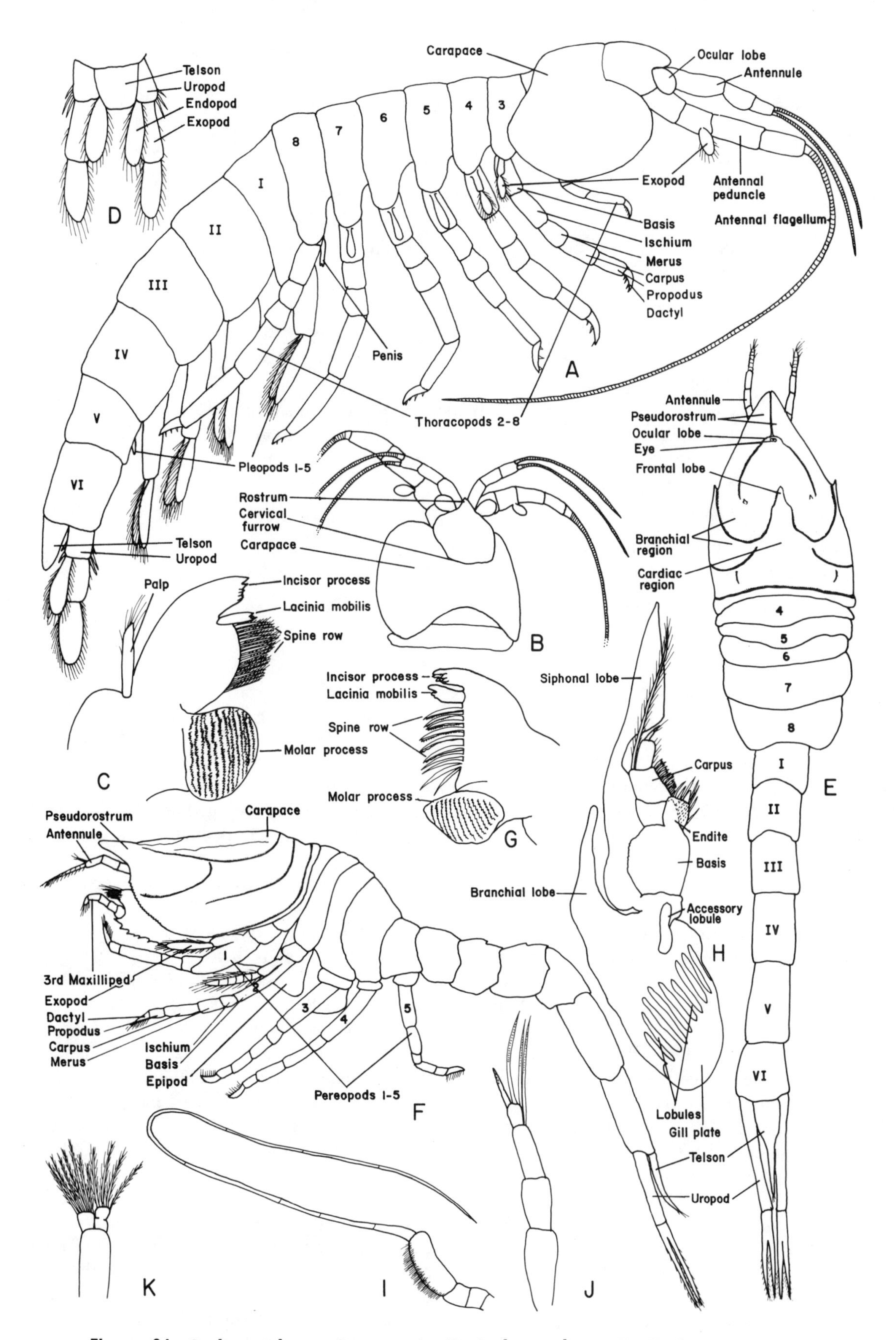

Figure 31 Spelaeogriphacea, Cumacea: A—D, Spelaeogriphacea; E—K, Cumacea. A. *Spelaeogriphus lepidops* Gordon (lateral view) [after Gordon, 1957]; B. Dorsal view of carapace and cephalic appendages [after Kaestner, 1970]; C. Mandible [after Gordon, 1957]; D. Telson and uropods [after Kaestner, 1970]; E. Whole animal (dorsal view); F. Female (lateral view); G. Left mandible; H. 1st maxilliped; I. Antenna of male; J. 1st pleopod of male; K. 2nd pleopod of male.

Maxillae	With pair of endites and lamellar exopod.
Maxillipeds	Three pairs.
Thoracic appendages	Five pairs of pereopods, with or without exopods.
Abdominal appendages	Absent in female; present in varying numbers in male; uropods elongate, biramous, styliform.
Telson	Present; occasionally fused with 6th abdominal somite (pleotelson).
Tagmata	Cephalothorax and abdomen.
Somites	Head with 5 + 3 thoracic (maxillipeds); thorax with 5 [= cephalothorax]; abdomen with 6, excluding telson, or 5, excluding pleotelson.
Sexual characters	Gonopores on coxae of 3rd pereopods of female, on sternite of last thoracic somite (8th) of male; female with oostegites forming marsupium or brood pouch; male with numerous secondary sexual characters.
Sexes	Separate.
Larval development	Epimorphic; manca stage → postlarva.
Fossil record	Permian to Recent.
Feeding types	Grazers on sand grains or filter feeders.
Habitat	Primarily marine, few brackish and freshwater species; burrowers.
Distribution	Worldwide.

The general body shape of cumaceans is characteristic of the order; all members have an inflated cephalothorax and a very slender abdomen with styliform uropods (see Figure 31). The telson frequently is reduced in size or fused with the 6th abdominal somite to form a pleotelson. Pleopods are absent in females but may be present in variable numbers in males. A very specialized character is the development of the anterior carapace. It is produced in 2 prominent plates to form a structure called a pseudorostrum that covers the anterior extension of the branchial cavity. Within the cavity, or channel, the greatly developed exopods of the 1st maxillipeds unite to form a siphon. The antennules usually each have a single flagellum, or in some taxa an inner flagellum may be reduced or vestigial. The antennae exhibit sexual dimorphism; in the female they are very small or vestigial and in the male they are well developed and the flagella may exceed the length of the body.

The last 3 to 5 cephalothoracic somites are not covered by the carapace; however, in the female, the oostegites often appear to be continuations of the carapace. The pereopods usually are similar in structure and generally are considered to be fossorial in function. The 1st pereopods usually are strongly developed; the 2nd pereopods sometimes have a reduced number of segments. Remove the appendages from one side of the body so that the distinctive structure of the carapace and branchial chamber can be observed. The 3 pairs of maxillipeds lie in the midline. Laterally the carapace is fused to the body wall to form a relatively closed branchial chamber. The epipod or branchial lobe of the 1st maxilliped is considerably enlarged and lies in the branchial chamber. Usually it is furnished with respiratory processes or it may only be vascularized. Notice the narrow channel anteriorly that is covered by the pseudorostrum. Observe the development of the exopod of the maxilliped that forms a siphonal structure capable of being extended in front of the head.

The abdomen is elongate and considerably narrower than the cephalothorax. The telson may be free or fused with the 6th abdominal somite to form a pleotelson. [Bowman (1971) considers that certain families (e.g., Nannastacidae, Leuconidae, Bodotriidae) lack telsons rather than having pleotelsons.] Pleopods are absent in all females and occasionally also in males. When present, they vary in number according to taxa. Uropods are present in both sexes and usually are elongate and styliform.

The mouthparts of most cumaceans are small and rather difficult to remove. If sufficient magnification is available, remove and examine not only the maxillipeds but the other mouthparts as well. The 2nd maxilliped lacks an exopod, although the female usually has a small plate bearing a fringe of setae that projects backward into the marsupial chamber; it is thought to aid in keeping the eggs and embryos in motion. The maxilla has a pair of endites. The endopod of the maxillule is usually retroverted. The mandible lacks a palp, but has a lacinia mobilis and spine or setal row between the incisor and molar processes.

Larval development in cumaceans is epimorphic. After hatching, the young remain in the brood pouch of the female until the manca or postlarval stage is reached. At this stage they resemble the adult cumacean but lack the last pair of pereopods. Secondary sexual characters (e.g., the male antennae, exopods of the pereopods and pleopods, and the female oostegites) develop gradually with successive molts.

References

Bacescu, M., 1951. Cumacea. *Fauna republicii populare Romane, Bucharest, 4:* 1–91.

Calman, W. T., 1905. Cumacea. *Fish. Ireland, Sci. Invest.*, 1904. *1:* 1–52.

———, 1912. The Crustacea of the order Cumacea in the collection of the United States National Museum. *Proc. U.S. Natl. Mus., 41:* 603–676.

Fage, L., 1940. Les Cumacés de la Méditerranée. Remarques systematiques et biologiques. *Bull. Inst. Oceanogr.*, 783: 1–14.

———, 1951. Cumacés. *Faune de France, 54:* 1–136. Paris: Paul Lechevalier.

Gamô, S., 1967. Studies on the Cumacea (Crustacea, Malacostraca) of Japan. *Publ. Seto Mar. Biol. Lab., 15:* 245–274.

Hale, H. M., 1937. Cumacea and Nebaliacea. *Rep. B.A.N.Z. Antarct. Res. Exped.* 1929–31, *4b:* 37–56.

———, 1941. Australian Cumacea, 8. Codotriidae. *Trans. Roy. Soc. S. Australia, 68:* 225–285.

———, 1945. Australia Cumacea, 9. Nannastacidae. *Rec. S. Australian Mus., 8:* 145–218.

Hart, J. F. L., 1930. Some Cumacea of the Vancouver Island region. *Contr. Canad. Biol. Fish.*, n.s., *6:* 1–18.

———, 1939. Cumacea and Decapoda of the western Canadian Arctic region, 1936–37. *Canad. J. Res., 17:* 62–67.

Jones, N. S., 1969. The systematics and distribution of Cumacea from depths exceeding 200 meters. *Galathea Rep., 10:* 99–180.

———, 1973. Some new Cumacea from deep water in the Atlantic. *Crustaceana, 25:* 297–319.

———, 1974. *Campylaspis* species (Crustacea: Cumacea) from the deep Atlantic. *Bull. Brit. Mus. (Nat. Hist.) Zool., 27:* 249–300.

Jones, N. S., and H. L. Sanders, 1972. Distribution of Cumacea in the deep Atlantic. *Deep-Sea Res., 19:* 737–745.

Le Loeuff, P., and A. Intes, 1972. Les Cumacés du plateau continental de Côte d'Ivoire. *Cah. O.R.S.T.O.M., Ser. Oceanogr., 10:* 19–46.

Lie, U., 1969. Cumacea from Puget Sound and off the northwestern coast of Washington, with descriptions of two new species. *Crustaceana, 17:* 19–30.

Lomakina, N. B., 1955. Cumacea of Far East Seas. *Trudy Zool. Inst. Leningr., 18:* 112–165 (in Russian).

———, 1958. Kumovye raki (Cumacea) morei SSSR. *Akad. Nauk SSSR, Izdavaemye Zool. Inst. Akad. Nauk SSSR,* 66: 1–301.

———, 1968. Cumacea of the Antarctic region. *Issled. Faun. Morei, 6:* 97–140 (in Russian).

Oelze, A., 1931. Beiträge zur Anatomie von *Diastylis rathkei* Kr. *Zool. Jahrb. Anat., 54:* 235–294.

Reyss, D., 1972. Résultats scientifiques de la campagne du N.O. 'Jean Charcot' en Méditerranée Occidentale, Mai-Juin-Juillet 1970. *Crustaceana,* suppl. *3:* 362–377.

———, 1973. Distribution of Cumacea in the deep Mediterranean. *Deep-Sea Res., 20:* 1119–1123.

———, 1974. Contribution à l'étude des Cumacés de profondeur de l'Atlantique Nord: le genre *Makrokylindrus* Stebbing. *Crustaceana, 26:* 5–28.

Stebbing, Th. R. R., 1912. The Sympoda. *Ann. S. Afr. Mus., 10:* 129–176.

———, 1913. Cumacea. *Das Tierreich, 39:* 1–210.

Watling, L., 1977. Two new genera and a new subfamily of Bodotriidae (Crustacea: Cumacea) from eastern North America. *Proc. Biol. Soc. Wash., 89:* 593–598.

Zimmer, C., 1941. Cumacea. In H. G. Bronn, *Klassen und Ordnungen des Tierreichs,* 5(1) *4:* 1–222. Leipzig: Akad. Verl.

———, 1943. Über neue und weniger bekannte Cumaceen. *Zool. Anz., 141:* 148–167.

———, 1944. Cumaceen des tropischen Westatlantika. *Zool. Anz., 144:* 121–137.

ORDER TANAIDACEA Dana, 1853

Recent species	Approximately 800.
Size range	5–20 mm.
Carapace	Short; fused dorsally with first 2 thoracic somites.
Eyes	Compound, usually on produced lobe; frequently absent.
Antennules	Uni- or biramous.

Antennae	Uni- or biramous.
Mandibles	With or without palp; lacinia mobilis sometimes only on left mandible.
Maxillulae	With endopod and 1 or 2 endites; occasionally absent.
Maxillae	Developed or vestigial.
Maxillipeds	One pair.
Thoracic appendages	Usually 7 pairs, uni- or biramous; 1st pair often chelate.
Abdominal appendages	Often 5 pairs, biramous; sometimes reduced, both in size and number; occasionally absent; uropods small, slender.
Telson	Present; fused with 6th or occasionally 5th and 6th abdominal somites (pleotelson).
Tagmata	Head, thorax (pereon), and abdomen (pleon).
Somites	Head with 5 + 1 thoracic (maxilliped); thorax with 7; abdomen with 4 or 5, excluding pleotelson.
Sexual characters	Gonopores on coxae of 6th thoracopods of female, on 8th thoracic somite of male; female with oostegites forming marsupium; male with single or pair of genital cones. Chelipeds sexually dimorphic (greatly enlarged in mature males).
Sexes	Separate or hermaphroditic; gonochoristic and protogynous females.
Larval development	Epimorphic; manca stage → postlarva.
Fossil record	(?) Mississippian, Pennsylvanian to Recent.
Feeding types	Usually filter feeders on detritus and plankton; sometimes predators.
Habitat	Primarily marine, occasionally brackish and freshwater; burrowers or tube dwellers, occasionally inhabiting gastropod shells.
Distribution	Worldwide.

The Tanaidacea is an order of small peracaridan crustaceans that until recently was classified with the Isopoda. The extreme degree of sexual dimorphism exhibited among taxa of the Tanaidacea has resulted in considerable taxonomic chaos. The situation is further complicated by the probability that both protandry and protogyny exist in this group. Another source of difficulty in interpreting morphological characters comes from the fact that a series of structurally different stages in both juveniles and adults occur in most species.

The description of the morphology and instructions for the study of tanaidaceans have been based on species of *Pagurapseudes* and *Leptochelia;* however, major differences between these and other representatives of the two suborders Monokonophora and Dikonophora will be pointed out (see Figure 32).

Eyes frequently are lacking; when present they usually are set on small ocular lobes. The carapace covers the cephalon, which is fused with the 1st thoracic somite. The carapace is also fused with the 2nd thoracic somite and is laterally expanded to enclose the branchial chambers. Both the antennules and antennae may be uni- or biramous; the number of peduncular segments varies among taxa. In the Monokonophora the antennal scale (exopod), when present, is small. The thorax, also referred to as the pereon, contains 7 somites, of which the posterior 6 are free. Disagreement among carcinologists exists as to the numbering of the somites and their accompanying appendages. As previously indicated, in this text the numbering of somites refers to the actual morphological somites; thus in tanaidaceans, the first free somite is the 3rd thoracic somite (the 1st is fused to the cephalon and the 2nd is covered by and fused to the carapace). Traditionally, when a maxilliped is present on the 1st true thoracic somite and the number 1 is assigned to a somite other than the morphological 1st, the pair of appendages associated with the former somite is also assigned the number 1. Some recent carcinologists have changed the designation of thoracic appendages in the tanaidaceans. The chelipeds (appendages of the 2nd thoracic somite) have been referred to simply as the chelipeds and the appendages of the following somite

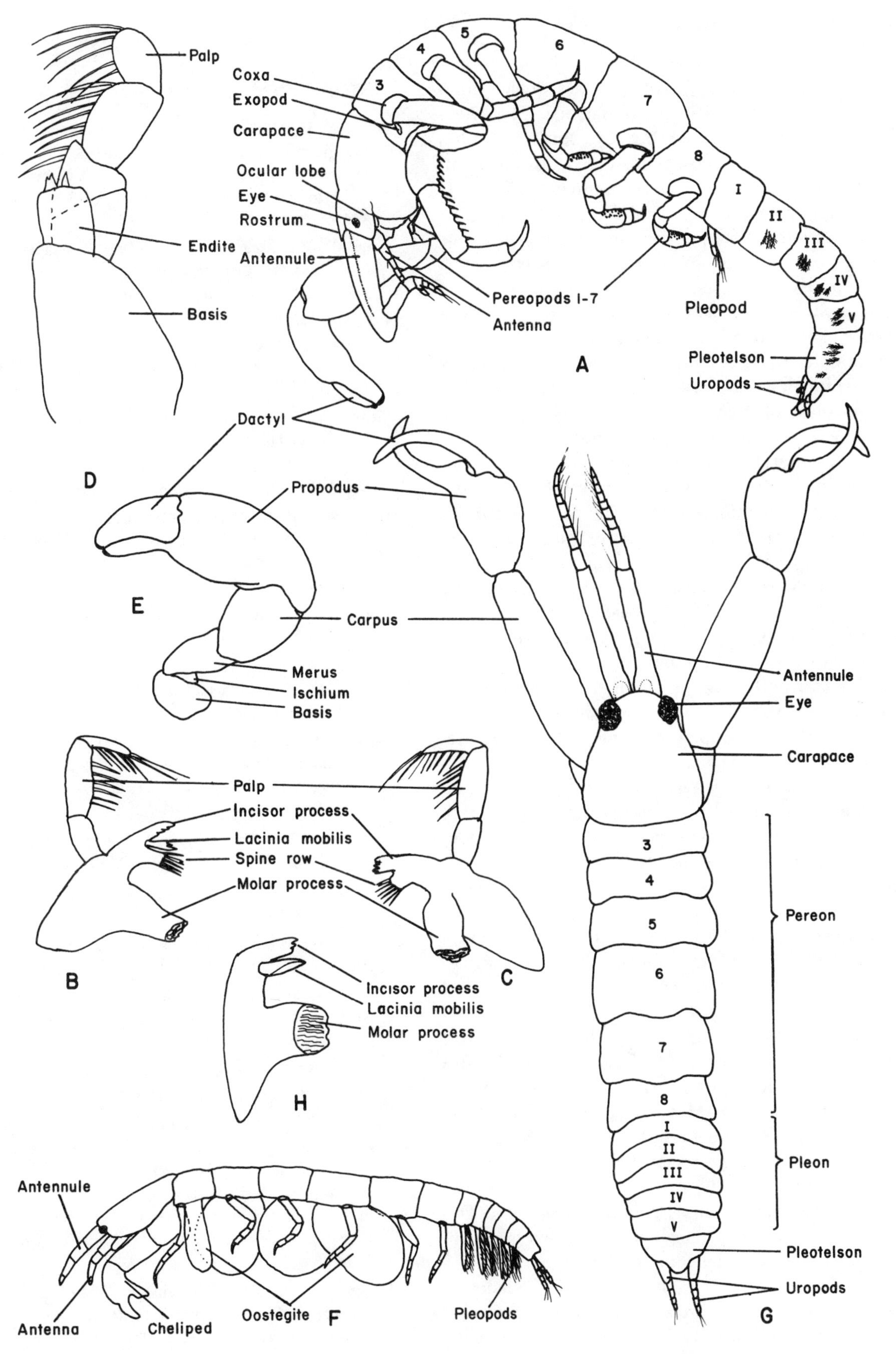

Figure 32 Tanaidacea: A—E, Monokonophora *(Pagurapseudes);* F—H, Dikonophora *(Leptochelia).* A. Whole animal (lateral view); B. Left mandible; C. Right mandible; D. Maxilliped; E. Right cheliped of male; F. Female (lateral view); G. Mature male (dorsal view); H. Left mandible.

(3rd thoracic somite) as the 1st pereopods. Such a numbering system is not used in this manual. The chelipeds are considered to be the 1st pereopods (2nd thoracopods); and the subsequent pereopods are numbered 2 through 7. The chelipeds are usually moderately well developed in juveniles and females but become massive, and often asymmetrical, in mature males. The 2nd pereopods often are stronger than the succeeding 5 pairs. Exopods are present on the 1st and 2nd pairs of pereopods in the Monokonophora but absent from the posterior pairs; exopods are lacking on all pereopods in the Dikonophora. The 2nd pereopods in the Kalliapseudidae often have sensory structures either on the dactyls or in place of them; the remaining pereopods also often carry sensory structures. Examine the chelipeds, 1 or 2 of the anterior pereopods, and 1 of the posterior pairs of each sex. In species of *Pagurapseudes,* the last 3 pairs are distinctly different from the preceding pairs; each carries a rasp of chitinous scales or spines, similar to the propodal rasps of hermit crabs.

The abdomen usually consists of 4 or 5 slender, usually compressed, somites and a pleotelson (telson + 1 or 2 pleomeres). Biramous pleopods usually are well developed on the 5 pleomeres, but in some taxa, such as *Pagurapseudes* or *Tanais,* the number of pleopods may be reduced to 3 or fewer; occasionally they are completely lacking. The uropods are uni- or biramous, depending upon the taxon.

The mouthparts are variable among the tanaidacean taxa. The labrum usually is a large bulbous structure lying in close proximity to the epistome, which occasionally carries a prominent spine. The mandible is well developed, often with a palp, incisor process, lacinia mobilis, spine row, and molar process. However, the palp is absent in the Dikonophora and not infrequently the lacinia mobilis is present only on the left mandible. Be sure to examine the appendages from both sides of the animal.

Mature females have genital apertures (gonopores) opening on the coxae of the 6th thoracopods (5th pereopods). On the ventral surface of the 8th thoracic somite, a genital cone (Monokonophora) or a pair of cones (Dikonophora) are present in males. Successive stages in the morphological series have been reported for both sexes. In females these include 1 or 2 preparatory stages prior to stages of brood pouch development. Depending upon the taxon, females may develop from 1 to 4 pairs of oostegites, which form the brood pouch.

Larval development is epimorphic. Newly hatched embryos are held in the brood pouch for 1 or 2 weeks, during which time limb development takes place. They are released at the 1st submanca stage lacking only the last thoracic appendages and pleopods. With the molt to the 2nd manca substage these remaining appendages appear and are completely developed with the next molt to the 1st juvenile or neuter stage.

References

Bacescu, M., 1960. Cîteva animale necunoscute înca în marea Neagra si descrierea unor Malacostracei noi *(Elaphognathia monodi* n. sp. si *Pontotanais boracei* n. gen. n. sp.) provenind din apele pontice prebosforice. *Rev. Biol.* (Roumaine), *5:* 107–124.

Dennell, R., 1939. On the feeding mechanism of *Apseudes talpa* and the evolution of the precaridan feeding mechanisms. *Trans. Roy. Soc. Edinburgh, 59:* 57–78.

Gardiner, L. F., 1976. The systematics, postmarsupial development and ecology of the deep-sea family Neotanaidae (Crustacean: Tanaidacea). *Smiths. Contr. Zool.,* no. 170: 1–265.

Greve, L., 1964. The records of *Leptognathia dentifera* G. O. Sars (Tanaidacea). *Sarsia,* no. 15, p. 71.

———, 1965. Tanaidacea from Trondheimsfjorden. *Kgl. Norsk. Vid. Selskabs Forhandl.* (B) *38:* 140–143.

Gutu, M., 1972. Phylogenetic and systematic considerations upon the Monokonophora (Crustacea, Tanaidacea) with the suggestions of a new family and several new subfamilies. *Rev. Roum. Biol. Zool., 17:* 297–305.

Kudinova-Pasternak, R. K., 1965. Deep-sea Tanaidacea from the Bougainville trench of the Pacific. *Crustaceana, 8:* 75–91.

———, 1970. Tanaidacea kurilo-Kamchatskogo zheloba. *Akad. Nauk SSSR Trudy Inst. Oceanol., 86:* 341 – 381.

———, 1973. Tanaidacea (Crustacea, Malacostraca), sobrannye na nis "Vityaz" v raionax aleutskogo zheloba zaliba alyaska. *Akad. Nauk SSSR, Trudy Inst. Oceanol., 91:* 141–168.

Lang, K., 1949. Contribution to the systematics and synonomies of Tanaidacea. *Ark. Zool. 42A:* 1–14.

———, 1953. The postmarsupial development of the Tanaidacea. *Ark. Zool.,* (2) *4:* 409–422.

———, 1955. Tanaidacea from tropical West Africa. *Atlantide Rept., 3:* 57–81.

———, 1956. Kalliapseudidae, a new family of Tanaidacea. *Bertil Hanstrom, Zoological papers in honor of his 65th birthday,* Nov. 20, 1965: 205–225.

———, 1957. Tanaidacea from Canada and Alaska. *Contr. Dép. Pech.,* Quebec, *52:* 1–54.

———, 1967. Taxonomische und phylogenetische Untersuchungen über die Tanaidaceen. 3. Der Umfang der Familien Tanaidae Sars, Lang und Paratanaidae Lang nebst Bemerkungen über den taxonomischen Wert der Mandibeln und Maxillulae. Dazu eine taxonomische-monographische Darstellung der Gattung *Tanaopsis* Sars. *Ark. Zool.,* (2) *19:* 343–368.

Menzies, R. J., 1953. The apseudid Chelifera of the eastern tropical and north temperate Pacific Ocean. *Bull. Mus. Comp. Zool. Harvard, 107:* 443–496.

Miller, M. A., 1968. Isopoda and Tanaidacea from buoys in coastal waters of the continental United States, Hawaii, and the Bahamas (Crustacea). *Proc. U.S. Natl. Mus., 125:* 1–53.

Richardson, H., 1905. Descriptions of a new genus of Isopoda belonging to the family Tanaidae and of a new species of

Tanais, both from Monterey Bay, California. *Proc. U.S. Natl. Mus.*, *28:* 367–370.

Shiino, S. M., 1965. Tanaidacea from the Bismarck Archipelago. *Vidensk. Medd. Dansk Naturh. Foren.*, *128:* 177–203.

Sieg, J., 1976. Zum natürlichen System der Dikonophora Lang (Crustacea, Tanaidacea). *Mitt. Zool. Syst. Evolutionsforsch*, *14:* 177–198.

———, 1977. Taxonomische Monografie der Familie Pseudotanaidae (Crustacea, Tanaidacea). *Mitt. Zool. Mus. Berlin*, *53:* 1–109.

Siewing, R., 1953. Morphologische Untersuchungen an Tanaidaceen und Lophogastriden. *S. Wiss. Zool.*, *157:* 333–426.

Wolff, T., 1956a. Crustacea Tanaidacea from depths exceeding 6000 meters. *Galathea Rept.*, *2:* 187–241.

———, 1956b. Six new abyssal species of *Neotanais* (Crustacea Tanaidacea). *Vidensk. Medd. Dansk Naturh. Foren.*, *118:* 41–52.

Zimmer, C., 1926. Tanaidacea. In W. Kükenthal and Th. Krumbach, *Handbuch der Zoologie*, 3(1) *10:* 683–696. Berlin: Der Gruyter.

ORDER ISOPODA Latreille, 1817

The Isopoda includes the suborders Valvifera, Anthuridea, Flabellifera, Microcerberidea, Asellota, Phreatoicidea, Gnathiidea, Oniscoidea, and Epicaridea; the latter is exclusively parasitic. Although there are major differences among the other eight suborders, their overall general morphology is sufficiently similar so that they do not need to be discussed separately. The major points of difference are indicated in the tabular review of the characters of the suborders.

Free-Living Isopoda

SUBORDER VALVIFERA Sars, 1882

Recent species	Approximately 600.
Size range	5–30 mm.
Carapace	Absent.
Eyes	Sessile, compound, usually well developed.
Antennules	Uniramous.
Antennae	Uniramous; peduncle with 5 segments; sometimes raptorial.
Mandibles	Usually without palp.
Maxillulae	Usually with 2 endites.
Maxillae	Typically with 3 endites.
Maxillipeds	One pair, usually with reduced number of segments.
Thoracic appendages	Thoracopods 3–8 with coxae developed as coxal plates and usually fused to pereon; sometimes ambulatory or prehensile, occasionally reduced in number.
Abdominal appendages	Biramous, branchial; covered by operculate uropods.
Telson	Fused with pleomeres 4–6 (pleotelson).
Tagmata	Cephalon, pereon, and pleon.
Somites	Head with 5 + 1 thoracic (maxilliped); pereon with 7; pleon fused with telson or with 1–3 free somites + pleotelson.
Sexual characters	Gonopores on 6th thoracic somite of female, on 8th of male; female with oostegites; male with appendix masculina and usually stronger spination.
Sexes	Separate.
Larval development	Epimorphic; released at manca stage.

Fossil record	Oligocene to Recent.
Feeding types	Herbivores, scavengers, and predators.
Habitat	Marine; intertidal to abyssal.
Distribution	Worldwide.

SUBORDER ANTHURIDEA Leach, 1814

Recent species	Approximately 100.
Size range	1.5–47.0 mm.
Carapace	Absent.
Eyes	Sessile, compound; usually well developed; occasionally with increased number of ocelli in male.
Antennules	Uniramous.
Antennae	Uniramous, peduncle with 5 segments.
Mandibles	Usually with palp; occasionally reduced or absent.
Maxillulae	Usually absent.
Maxillae	Modified for piercing in some taxa.
Maxillipeds	One pair, highly variable, with 3–7 segments.
Thoracic appendages	Thoracopods 2–8; 1st pair usually strongly subchelate (except in some females), following 2 pairs usually weakly subchelate, and 5–8 simple.
Abdominal appendages	Biramous, elongate; 1st pair operculate in some taxa; 2nd modified in males. Uropods partially overlapping telson.
Telson	True telson in many (not fused with 6th pleomere); often with 1 or 2 statocysts at base.
Tagmata	Cephalon, pereon, and pleon.
Somites	Head with 5 + 1 thoracic (maxilliped); pereon with 7; pleon with 6, excluding telson.
Sexual characters	Gonopores on 6th thoracic somite of female, on 8th of male; female with oostegites; male with appendix masculina and more developed antennules.
Sexes	Separate or protogynic hermaphrodites.
Larval development	Epimorphic; released at manca stage.
Fossil record	Recent.
Feeding types	Little known.
Habitat	Marine, estuarine, and freshwater; benthic.
Distribution	Worldwide.

SUBORDER FLABELLIFERA Sars, 1882

Recent species	Approximately 1400.
Size range	Up to 350 mm.
Carapace	Absent.
Eyes	Sessile, compound, usually well developed; sometimes reduced or absent.
Antennules	Usually uniramous; occasionally vestigial.

Antennae	Uniramous; peduncle with 5 or 6 segments; sometimes reduced.
Mandibles	Usually with palp; molar processes often reduced or modified.
Maxillulae	Usually with 1 endite; sometimes modified for piercing or sucking.
Maxillae	Usually with 2 endites; sometimes modified for piercing or sucking.
Maxillipeds	One pair, variously developed.
Thoracic appendages	Thoracopods 3–8 with coxae developed as plates; ambulatory and/or prehensile.
Abdominal appendages	First 2 or 3 pairs often natatory; posterior pairs branchial. Uropods lateral, with pleotelson forming tailfan.
Telson	Fused with variable number of pleomeres (pleotelson).
Tagmata	Cephalon, pereon, and pleon.
Somites	Head with 5 + 1 thoracic (maxilliped); pereon with 7; pleon variable + pleotelson.
Sexual characters	Gonopores on 6th thoracic somite of female, 8th of male; female with oostegites; male with appendix masculina.
Sexes	Separate or protandrous hermaphrodites.
Larval development	Epimorphic; released at manca stage.
Fossil record	Triassic to Recent.
Feeding types	Omnivorous scavengers and predators; temporary or permanent parasites adapted for piercing and sucking.
Habitat	Marine to freshwater; intertidal to moderate depths, sometimes cave dwellers.
Distribution	Worldwide.

SUBORDER MICROCERBERIDEA Chappuis and Delamare Deboutteville, 1960

Recent species	Approximately 22 in one genus.
Size range	0.5–1.4 mm.
Carapace	Absent.
Eyes	Absent.
Antennules	Uniramous.
Antennae	Uniramous; peduncle with 5 or 6 segments.
Mandibles	Usually without palp.
Maxillulae	With 3 endites.
Maxillae	With 2 reduced endites.
Maxillipeds	One pair; endopod with 5 segments.
Thoracic appendages	Thoracopods 2–8; 1st pair prehensile.
Abdominal appendages	Biramous; uropods with exopods reduced or absent.
Telson	Fused with 4 or fewer pleomeres (pleotelson).
Tagmata	Cephalon, pereon, and pleon.
Somites	Head with 5 + 1 thoracic (maxilliped), pereon with 7; pleon with 2 or more + pleotelson.
Sexual characters	Gonopores on 6th thoracic somite of female, on 8th of male; female with oostegites; male with appendix masculina.

Sexes	Separate.
Larval development	(?) Epimorphic.
Fossil record	Recent.
Feeding types	Unknown.
Habitat	Marine, freshwater; interstitial.
Distribution	Eastern Pacific, South America, Africa, Mediterranean, India.

SUBORDER ASELLOTA Latreille, 1803

Recent species	Approximately 450.
Size range	1–15 mm.
Carapace	Absent.
Eyes	Sessile, compound; frequently absent.
Antennules	Uniramous.
Antennae	Uniramous; peduncle with 6 segments.
Mandibles	Usually with palp.
Maxillulae	Usually with 1 or 2 endites.
Maxillae	Typically with 3 endites.
Maxillipeds	One pair; usually well developed.
Thoracic appendages	Typically thoracopods 2–8, but number sometimes reduced; 1st often subchelate; coxae usually small.
Abdominal appendages	First 1 (female) or 2 (male) pairs operculate, remainder branchial and enclosed. Uropods terminal, uni- or biramous.
Telson	Pleotelson usually of 4–6 pleomeres + telson.
Tagmata	Cephalon, pereon, and pleon.
Somites	Head with 5 + 1 thoracic (maxilliped); pereon typically with 7, often fewer; pleon usually with 1–3 free + pleotelson.
Sexual characters	Gonopores on 6th thoracic somite of female, on 8th of male; female with oostegites; male with appendix masculina.
Sexes	Separate.
Larval development	Epimorphic; released at manca stage.
Fossil record	Recent.
Feeding types	Typically herbivores or detritivores.
Habitat	Marine to freshwater, interstitial to abyssal.
Distribution	Worldwide.

SUBORDER PHREATOICIDEA Stebbing, 1893

Recent species	Approximately 50.
Size range	5–45 mm.
Carapace	Absent.
Eyes	Sessile, compound; sometimes absent.
Antennules	Uniramous.
Antennae	Uniramous; peduncle with 5 segments.
Mandibles	Usually with palp.

Maxillulae	Usually with 2 endites.
Maxillae	Usually with 3 endites.
Maxillipeds	One pair; usually well developed.
Thoracic appendages	Thoracopods 2–8; last 3 pairs directed backwards; coxae not developed as coxal plates.
Abdominal appendages	Usually subequal, natatory as well as branchial. Uropods biramous.
Telson	Fused with 6th pleomere (pleotelson).
Tagmata	Cephalon, pereon, and pleon.
Somites	Head with 5 + 1 thoracic (maxilliped); pereon with 7; pleon with 5 + pleotelson.
Sexual characters	Gonopores on 6th thoracic somite of female, 8th of male; female with oostegites; male with appendix masculina.
Sexes	Separate.
Larval development	Epimorphic; released at manca stage.
Fossil record	Permian to Recent.
Feeding types	Plant detritus feeders.
Habitat	Freshwater; occasionally semiterrestrial.
Distribution	South Africa, India, New Zealand, Australia.

SUBORDER GNATHIIDEA Leach, 1814

Recent species	Approximately 75.
Size range	2–17 mm.
Carapace	Absent.
Eyes	Sessile or rarely on small peduncles, compound; occasionally absent.
Antennules	Uniramous.
Antennae	Uniramous; peduncle with 4 segments.
Mandibles	Absent in adult female; well developed, nonfunctional in male.
Maxillulae	Reduced or vestigial.
Maxillae	Reduced or vestigial.
Maxillipeds	Two pairs, 2nd modified as operculum.
Thoracic appendages	Thoracopods 3–7 (8th absent), not chelate or subchelate.
Abdominal appendages	Usually biramous, unmodified. Uropods well developed, biramous.
Telson	Fused with 6th pleomere (pleotelson).
Tagmata	Cephalon, pereon, and pleon.
Somites	Head with 5 + 2 thoracic (maxillipeds); pereon with 6; pleon with 5 + pleotelson.
Sexual characters	Gonopores on 6th thoracic somite of female, 8th of male as single median cone; female with oostegites; male with mandibles greatly enlarged, modified for burrowing; without specialized copulatory structures.
Sexes	Separate.
Larval development	Metamorphic; segmented larva → praniza (postlarva).

Fossil record	Recent.
Feeding types	Juveniles adapted for piercing and sucking; adults scavengers or nonfeeders.
Habitat	Marine; juvenile, temporary ectoparasites of fish; adult, burrowers (burrows dug by male).
Distribution	Worldwide.

SUBORDER ONISCOIDEA Latreille, 1803

Recent species	Approximately 900.
Size range	Up to 50 mm.
Carapace	Absent.
Eyes	Sessile, compound; usually present.
Antennules	Uniramous, usually reduced.
Antennae	Uniramous; peduncle with 4 or 5 segments.
Mandibles	With palp.
Maxillulae	Typically with 2 endites.
Maxillae	Usually with 3 endites.
Maxillipeds	One pair; terminal segments often reduced.
Thoracic appendages	Thoracopods 3–8 with coxae expanded as coxal plates.
Abdominal appendages	Biramous; at least first 2 pairs, sometimes all 5 pairs with pseudotracheae. Uropods sometimes operculate, with broad exopod and small, ventral endopod.
Telson	Usually fused with 6th pleomere (pleotelson); occasionally not fused with pleomere.
Tagmata	Cephalon, pereon, and pleon.
Somites	Head with 5 + 1 thoracic (maxilliped); pereon with 7; pleon with 5 + pleotelson or 6 + telson.
Sexual characters	Gonopores on 6th thoracic somite of female, on 8th of male; female with oostegites; male with 2nd, sometimes also 1st, pair of pleopods modified for copulation.
Sexes	Separate.
Larval development	Epimorphic; released at manca stage.
Fossil record	Recent.
Feeding types	Usually herbivores or detritivores.
Habitat	Amphibious and terrestrial; several commensal in termite and ant nests.
Distribution	Worldwide.

With few exceptions (Anthuridea, Phreatoicidea), the isopod body is dorsoventrally compressed, a character used as a rule of thumb for distinguishing isopods from amphipods. Another very significant character of isopods is the adaptation of the pleopods for respiration. Examine the intact specimen(s) before beginning a more detailed study of individual appendages, mouthparts, and major organ systems (see Figures 33–35). In dorsal view the typical isopod cephalon is small, usually with small, sessile compound eyes; in *Bathynomus giganteus* A. Milne Edwards, the eyes are very large and occupy much of the anterior part of the cephalon. Uniramous antennules and antennae generally can be observed projecting from the cephalon; however, in species of the Oniscoidea, the antennules are exceptionally short and usually do not protrude beyond the cephalic margin. A

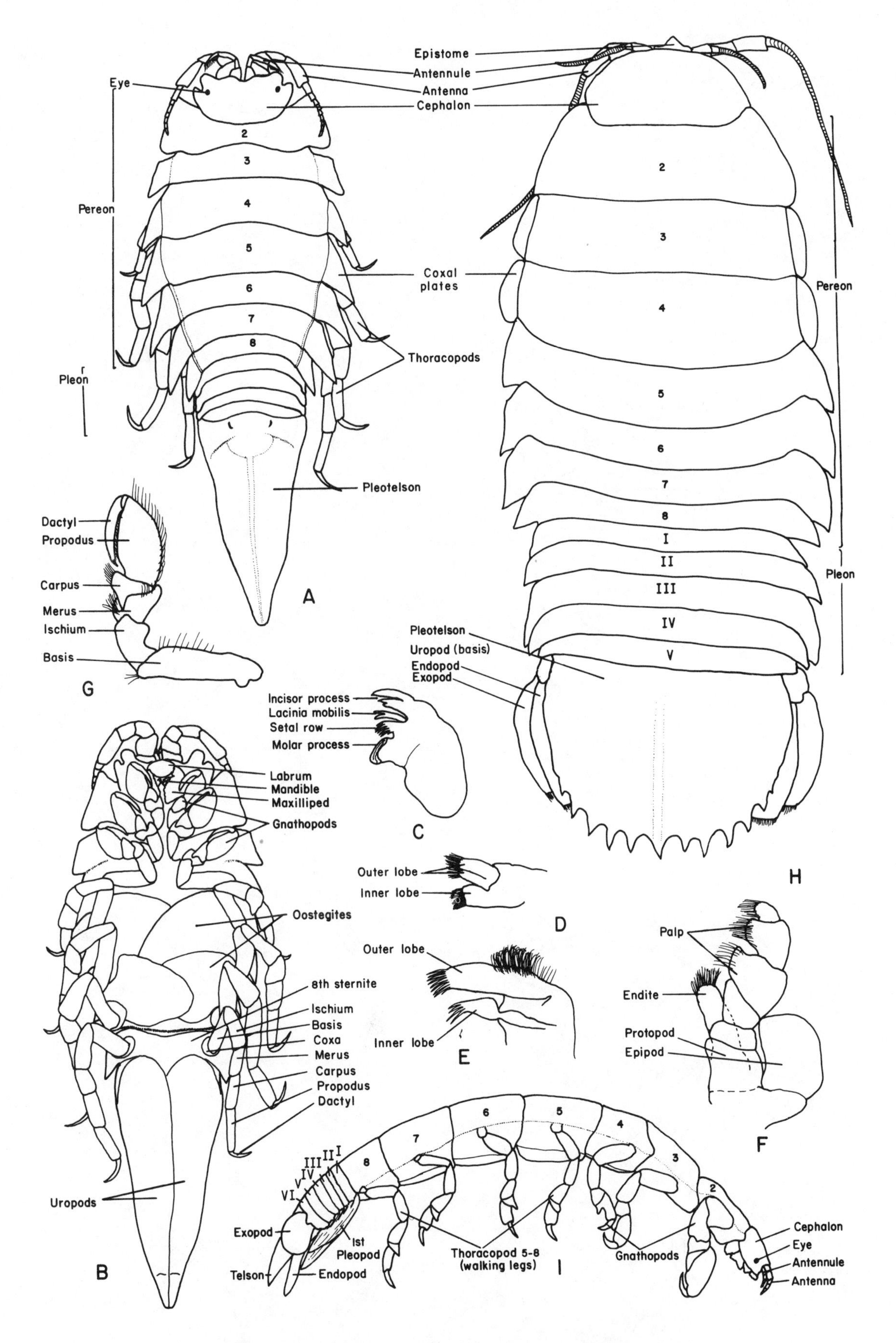

Figure 33 Isopoda: A—G, Valvifera; H. Flabellifera; I, Anthuridea. A. Whole animal (dorsal view); B. Female (ventral view); C. Mandible; D. Maxillule; E. Maxilla; F. Maxilliped; G. Gnathopod; H. Typical flabelliferan (dorsal view); I. Typical anthuridean (lateral view).

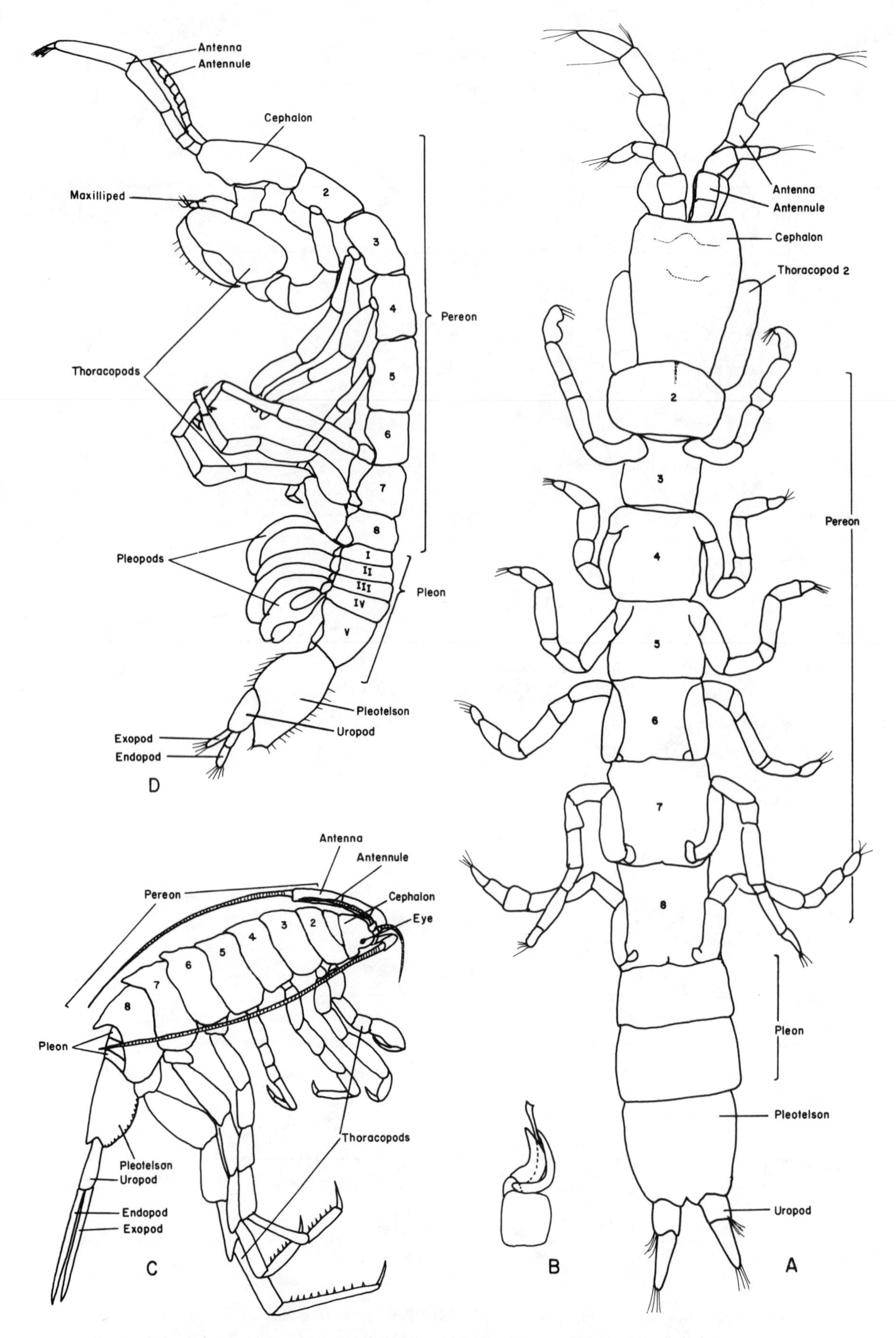

Figure 34 Isopoda: A, B Microcerberidea; C, Asellota; D, Phreatoicidea. A. *Microcerberus* (dorsal view); B. 2nd pleopod of male [A, B after Chappuis and after C. Delamare Deboutteville: *Biologie des eaux souterraines littorales et continentales* (Hermann, Paris, 1960)]; C. Typical asellotan; D. Typical phreatoicidean [after Birstein, 1962].

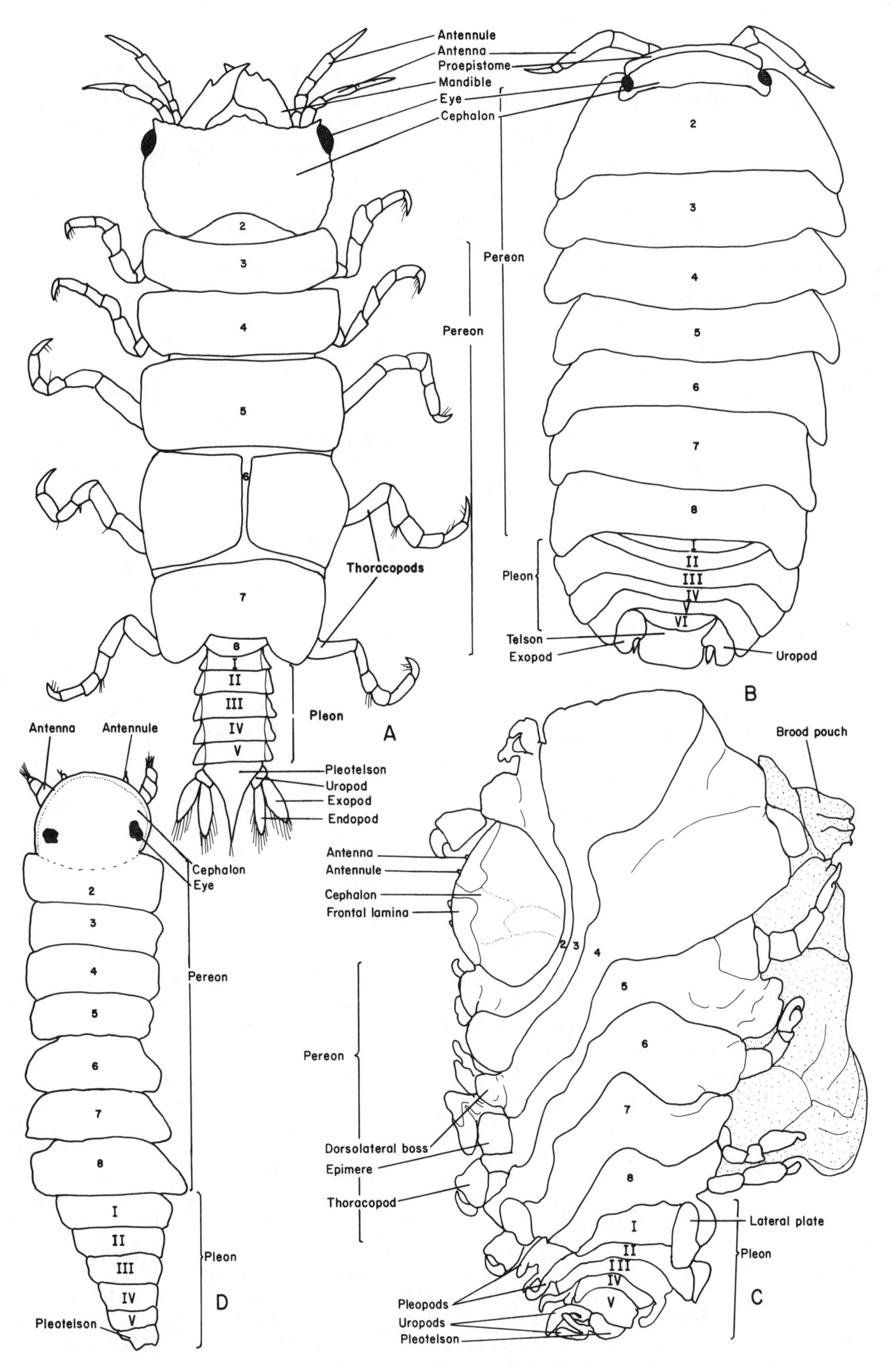

Figure 35 Isopoda: A, Gnathiidea; B, Oniscoidea; C, D, Epicaridea. A. Typical gnathiidean [after Hessler, 1969, from *Treatise on Invertebrate Paleontology,* courtesy of the Geological Society of America and University of Kansas]; B. Typical oniscoidean; C. Female bopyrid (dorsal view); D. Male bopyrid (dorsal view) [C, D after Markham, 1975b].

carapace is lacking but this is a secondarily derived condition. The 1st, rarely also the 2nd, thoracic somite is fused with the cephalon. Although some carcinologists refer to this condition as a cephalothorax, the preferred terminology is cephalon and pereon, as in the tanaidaceans. Seven, or less frequently 6, thoracic somites are free; occasionally 1 or more of these somites may be fused with an adjoining somite. The tergal plates of the somites often are laterally expanded. Very frequently the abdominal somites (pleomeres) are appreciably narrower than those of the pereon; as in tanaidaceans the abdomen is referred to as the pleon. One or more pleomeres usually fuse with the telson, forming the pleotelson. How many thoracic and abdominal somites can you distinguish in your specimen(s)?

In ventral view examine the sternites and appendages. Each pereonite typically carries a pair of appendages modified as walking legs (pereopods); in the Gnathiidea the last pair are absent. The coxal segments of pereopods 2–7 (thoracopods 3–8) often are laterally expanded to form dorsal coxal plates reminiscent of epimeres of other crustaceans and insects. The basal segments usually are directed toward the midline of the body; the remaining segments are directed laterally. A great deal of variation exists in the structure of the pereopods. Frequently the 1st pair is modified either for food gathering or, in males, as an accessory copulatory structure for grasping the female. One to 3 pairs of pereopods frequently are subchelate but never chelate as in the tanaidaceans. In some taxa of every suborder, but particularly in the Anthuridea and Phreatoicidea, 2 groups of pereopods are differentiated: the anterior 3 pairs are directed anteriorly, the posterior 4 pairs posteriorly. In mature females, some thoracic sternites often are covered by large platelike oostegites; typically these are present on the coxae of the 1st to 4th or 5th pereopods, rarely on all 7 pairs. In certain species, the females carry their eggs not in the brood pouch but in paired hypodermal pockets formed by invaginations of the intersternal membranes into the body cavity. These invaginations open to the exterior by narrow slits. Oviducts open through a single median gonopore on the 6th thoracic sternite. The vas deferens of the males usually open through a pair of papillae on the 8th thoracic sternite; in the Gnathiidea and most Oniscoidea the papillae are fused into a median cone.

Five pairs of pleopods usually are present even if the number of pleonal somites has been reduced through fusion. Only rarely are all pleopods lacking, although in some taxa the number may be reduced. Pleopods in isopods serve a respiratory as well as natatory function, and as such, may be modified from the typical pleopodal form. The biramous condition generally persists, and when modifications occur, they usually are modifications of the endopods. In the male, the endopods of the 2nd pleopods bear modified copulatory structures, the appendix masculina. In most Oniscoidea the pleopods are specifically adapted for aerial respiration. If specimens of both aquatic and terrestrial isopods are available, compare the pleopodal adaptations of both groups. The uropods may be uni- or biramous, spatulate or styliform, well developed or reduced, and in a few instances absent. They sometimes are modified for special functions. In valviferans, they are modified as branchial covers. In the Oniscoidea the exopods usually are moderately large and are visible in dorsal view, but the endopods are very small and are ventrally positioned under the telson.

Remove the mouthparts and examine them. The maxilliped frequently is operculate, with a broadened proximal segment; endites frequently are present. The maxilla and maxillule both lack palps. The mandible usually is well developed, with incisor and molar processes, lacinia mobilis, and setal or spine row. The mandibles of male gnathiideans usually are very prominent and modified for burrowing. Occasionally some or all of the mouth appendages are reduced or greatly modified.

The internal anatomy of relatively few isopods has been studied in detail; therefore, the variations among major taxa cannot be predicted. The description of the major organ systems presented is based on *Bathynomus*, one of the more primitive taxa (see Figure 36). Begin your examination by removing the exoskeleton from the dorsal surface and one side of the animal. The tergites of each somite usually will have to be cut and removed individually. Isopods are unique among crustaceans in that the heart is located almost exclusively in the pleon (abdomen). Blood from the venous sinuses is returned to the heart through 2 pairs of ostia, or in the case of most terrestrial isopods, 1 pair. Anteriorly from the heart, trace the unpaired anterior aorta forward; just posterior of the cephalon the walls of the artery are expanded to form a cor frontale or accessory heart that aids in circulation of blood to the cephalon and its appendages. The several branches of the anterior aorta supply blood to the eyes, supraesophageal ganglion, foregut, and cephalic appendages. A ventral branch of the aorta joins the subneural artery anteriorly. Laterally from the heart locate the 5 pairs of lateral arteries. The most anterior pair provides blood to the ventral nerve cord and through large branches to the anterior 4 pairs of thoracic appendages (pereopods 1–4); blood to the following 3 pairs is provided by the 2nd to 4th lateral arteries respectively. Posteroventrally the posterolateral pair of arteries provides blood to the telson and uropods and branches to the abdominal musculature. These blood vessels do not enter the pleopods; blood to the pleopods is supplied by blood sinuses from the gut region and a pair of ventrolateral lacunae. Notice that the lateral arteries leave the heart generally in its anterior half in *Bathynomus*. In some other isopods the arteries are spaced equidistantly along the length of the heart. In terrestrial isopods, not

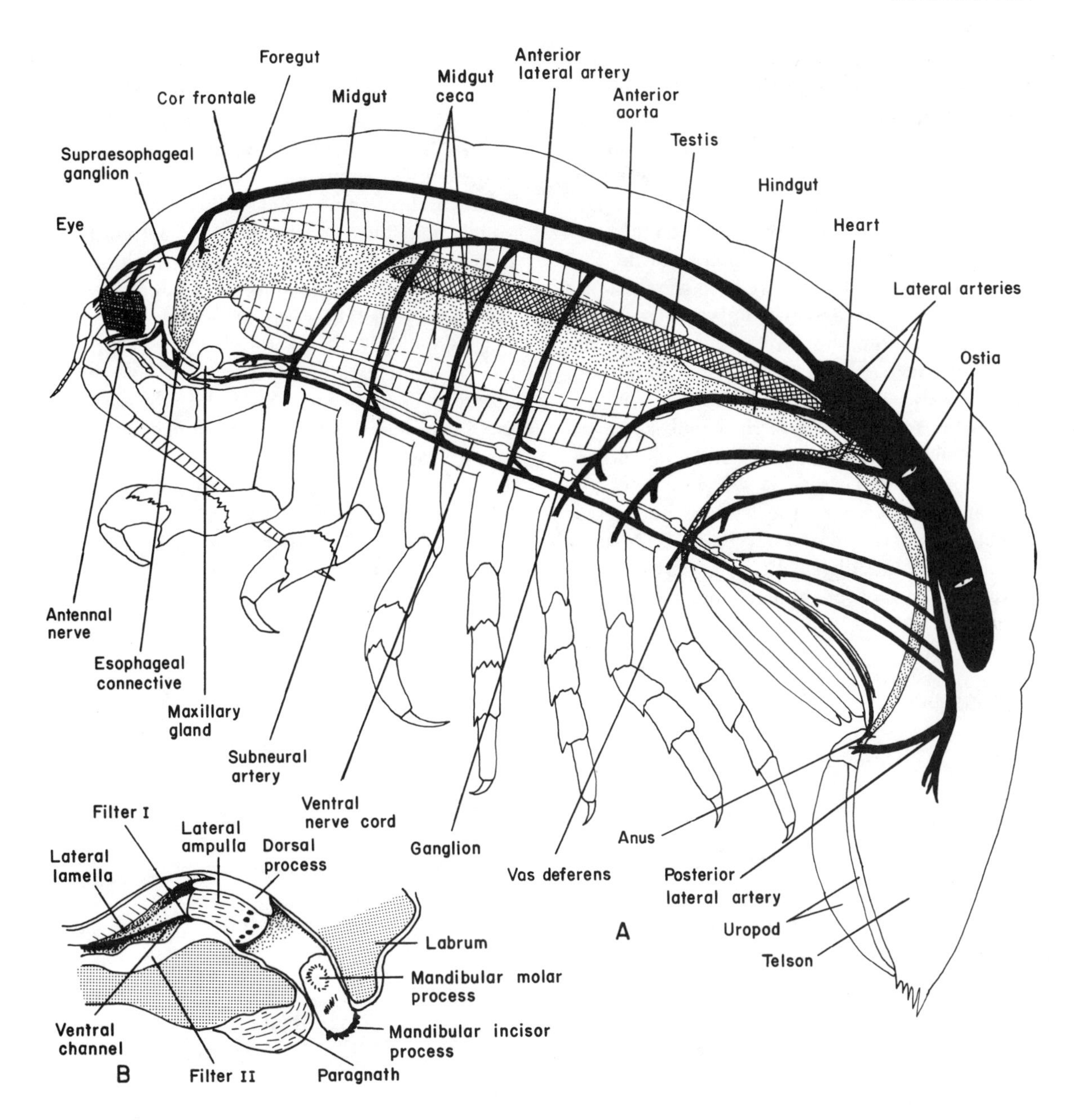

Figure 36 Isopoda: A. Diagrammatic flabelliferan with musculature removed to show major internal organs; B. Diagrammatic valviferan foregut showing internal structure [after Naylor, 1955].

only is there only a single pair of ostia (except in *Ligia*), but the heart tends to be longer and thicker and the arterial system is simpler.

The reproductive system in isopods typically consists of tubular paired ovaries or testes lying dorsolaterally of the gut. Oviducts leave the ovaries posteriorly at the level of the 5th or 6th thoracic somite, turn ventrally, and open individually in a median pore on the sternite between the 6th thoracopods (5th pereopods). Paired vas deferens leave the testes posteriorly and turn ventrally to open through a pair of median papillae or a single median cone on the sternite of the 8th thoracic somite.

The digestive system consists of a foregut, midgut, hindgut, and usually from 1 to 3 pairs (3 in *Bathynomus)* of midgut or hepatic ceca of variable length. In *Bathynomus* the hindgut is the longest part of the digestive tract, commencing at the level of the 4th pereonite and continuing to the anus on the ventroproximal surface of the telson. Near the junction of the midgut and foregut are the ducts of the ceca. The foregut is specialized as a chewing or filtering stomach or, in the cases of isopods adapted for piercing and sucking, as a pumping stomach. As described for *Idotea*, the foregut has a pair of lateral ampullae on each side wall which crush the food material so that fluid can be sifted from the solid matter through paired, ventral, bristled plates into a midventral channel (Filter I). This channel is formed by anterior lamellae (flaplike valves) and divided medially at its posterior end by a second filtering structure (Filter II). On either side of the channel, parallel with its upper margin, is a groove covered with strong bristles. Posteriorly these grooves open into the hepatic or midgut ceca of each side; fluids pass between the setae and into the ceca; a backward projection from the 2nd filter acts as a valve over the opening from the ceca into the midgut. Remove the foregut from your specimen, open it, and compare the structure of the chewing and filtering or pumping mechanisms with that described. Excretion in isopods is via maxillary glands.

With the foregut removed, locate the supraesophageal ganglion and identify the large optic, antennal, and antennular nerves radiating from it. Trace one of the esophageal connectives around the base of the foregut. In *Bathynomus* part of the ventral nerve cord is obscured by a pair of sternal alar plates. Remove part of these plates to expose the subesophageal ganglion, which represents the fusion of the ganglia of the mouthparts. Trace the nerve cord posteriorly, identifying the ganglia of the thoracic appendages and somites. The abdominal ganglia have moved somewhat closer together and no longer appear segmental.

Larval development generally is epimorphic, with the young held for some time after hatching in the brood pouch of the female. At the time of release, the manca stage, the young resemble the adult form a great deal, but usually lack the 7th pereopods. In some isopods (e.g., Gnathiidea, Cymothoidae), development is metamorphic. In Gnathiidea, the larvae are parasitic and quite different in body form from the adults. After release, the larva seeks out a fish host and attaches with its biting mouthparts. Once gorged with blood, the larva drops off the host, sinks to the bottom, and molts to the typical benthic adult. In contrast, the larvae of the Cymothoidae (Flabellifera) are free-living, while the adults are ectoparasites of fishes.

Parasitic Isopoda

SUBORDER EPICARIDEA Latreille, 1831

Recent species	Approximately 375.
Size range	1.5–10.0 mm.
Carapace	Absent.
Eyes	Sessile, compound, frequently present in male; reduced or absent in female.
Antennules	Reduced in male; reduced or absent in female.
Antennae	Often with 3-segmented peduncle in male; reduced or absent in female.
Mandibles	Reduced, absent, or suctorial.
Maxillulae	Reduced, absent, or suctorial.
Maxillae	Reduced, absent, or suctorial.
Maxillipeds	Reduced, absent, or suctorial.
Thoracic appendages	Thoracopods 2–8 in male, often with strong claws; reduced or absent in female.
Abdominal appendages	Usually reduced or absent; if present, branchial. Uropods usually biramous, not well developed in females, variously developed in males.
Telson	Fused with 6th pleomere (pleotelson).
Tagmata	Cephalon, pereon, and pleon.
Somites	Head usually with 5 + 1 thoracomere; pereon usually with 7; pleon usually with 5 + pleotelson.
Sexual characters	Gonopores on 6th thoracic somite of female, on 8th of male; female with oostegites; males extremely reduced in size, without specialized copulatory structures.
Sexes	Separate.

Larval development	Metamorphic; epicaridum or micronicus → cryptoniscus → bopyridum.
Fossil record	(?) U. Jurassic to Recent.
Feeding types	Parasitic, feeding on body fluids of host.
Habitat	Parasitic on other crustaceans, particularly decapods; frequently host specific.
Distribution	Worldwide.

The Epicaridea are ectoparasites of other Crustacea; the best known family is the Bopyridae (see Figure 35). As adults, members of this family are parasitic on decapods, usually lodging in the gill chamber of the host. Juveniles use copepods as intermediate hosts. Sexual dimorphism is extreme. Males retain a general isopodlike appearance, although they are greatly reduced in size. Females are greatly altered. A cephalon usually is present but may be difficult to distinguish. Eyes, when present, are small; cephalic appendages usually are reduced or absent. The thorax generally is segmented and thoracic appendages (pereopods) usually are present, at least on one side, but are considerably reduced in size. The pleon generally also is present, but its segmentation and appendages often are difficult to distinguish. Frequently the greatly enlarged oostegites obscure much of the female's morphology. Examine representatives of several taxa if available, and observe the varying degrees of degeneration of the female.

Larval development is metamorphic. The larvae are released at a stage referred to as the microniscus or epicaridum. At this stage the larva has typical mancalike segmentation, but swims with the antennae and pleopods. It is at this stage that it seeks out the intermediate copepod host. After several molts, sexual differentiation begins, and the maturing juvenile detaches itself from the copepod host. This cryptoniscus is once again planktonic; its pereopods become developed as holdfasts. At this stage the larva seeks its permanent host, usually a decapod. If the host is already infected with a parasite of the same species, the newly arriving larva attaches, molts, and becomes a male. If the host is uninfected, the larva undergoes a series of molts until the adult female form is reached.

References

Barnard, K. H., 1925. Revision of the Anthuridae. *J. Linn. Soc. London, 36:* 109–160.

———, 1927. The freshwater isopodan and amphipodan Crustacea of South Africa. *Trans. Roy. Soc. S. Afr., 14:* 139–215.

Birstein, Ya. A., 1951. *Fauna SSSR*, A. A. Shtakel'berg (ed.), Rakoobranye, 7 (5), Presnovodnye osliki. Izdatel'stvo, Akad. Nauk SSSR. 140 pp. *Fauna of USSR*, Crustacea 7 (5), Freshwater Isopods (Asellota), Translated for Smithson. Inst. and Nat. Sci. Found., 1964, 146 pp.

———, 1962. *Palaeophreatoicus sojanensis* gen. et sp. nov. i nekotorye voprosy filogenii i zoogeografii ravnongikh rakoobrazhykh (Isopoda). *Paleont. J. Akad. Nauk SSSR*, no. 3: 65–80.

———, 1973. *Deep water isopods of the northwestern part of the Pacific Ocean.* [Translation from Russian, U.S. Dept. Documents, TT 67–59075, 316 pp.]

Bonnier, J., 1900. Contribution à l'étude des Epicarides. Les Bopyridae. *Trav. Stat. Zool. Wimereux, 8:* 1–476.

Brusca, G. J., 1966. Studies on the salinity and humidity tolerances of five species of isopods in a transition from marine to terrestrial life. *Bull. So. Calif. Acad. Sci., 65:* 146–154.

Brusca, R. C., 1978. Studies on the cymothoid fish symbionts of the eastern Pacific (Isopoda, Cymothoidae). I. Biology of *Nerocila californica. Crustaceana, 34:* 141–154.

Brusca, R. C., and B. Wallerstein, 1977. The marine isopod crustaceans of the Gulf of California. I. Family Idoteidae. *Amer. Mus. Nov.*, 2634: 1–17.

Burbanck, W. D., 1967. Evolutionary and ecological implications of the zoogeography, physiology and morphology of *Cyathura* (Isopoda). In G. H. Lauff (ed.), Estuaries. *Amer. Assoc. Adv. Sci.*, no. 83: 564–573.

Carvacho, A., 1977. Isopodes intertidaux des côtes du centre et du nord du Chili. 1. Familles des Cirolanidae, Excorallanidae et Corallanidae. *Crustaceana, 32:* 27–44.

Chace, F. A., Jr., J. G. Mackin, L. Hubricht, A. H. Banner, and H. H. Hobbs, Jr., 1959. Malacostraca. In W. T. Edmondson (ed.), *Fresh-water biology,* 2nd ed., pp. 869–901. New York and London: John Wiley and Sons.

Chappuis, P. A., and C. Delamare Deboutteville, 1960. État de nos connaissances sur une famille et une sous-famille: les Microparasellides et les Microcerberines (Isopodes). In C. Delamare Deboutteville, *Biologie des eaux souterraines littorales et continentales,* chap. 15, pp. 293–357. Paris: Hermann.

Chardy, P., 1977. La famille des Haploniscidae (Isopodes, Asellotes): discussion systématique et phylogénique. *Bull. Mus. Natl. Hist. Nat. Zool.*, no. 333: 889–906.

Danforth, C. G., 1970. *Epicaridea (Crustacea: Isopoda) of North America.* Monograph. 191 pp. Ann Arbor, Mich.: Univ. Microfilms.

Dexter, D. M., 1977. Natural history of the Pan-American sand beach isopod *Excirolana brasiliensis* (Crustacea: Malacostraca). *J. Zool., 183:* 103–109.

Elkaim, B., and N. Daguerre de Hureaux, 1976. Contribution a l'étude des Isopodes marins: le genre *Parachiridotea* et la sous-famille nouvelle des Parachiridoteinae (Valvifère, Idoteidae). *Arch. Zool. Exp. Gen., 117:* 275–293.

Ellis, J. P., and R. J. Lincoln, 1975. Catalogue of the types of terrestrial isopods (Oniscoidea) in the collections of the British Museum. II. Oniscoidea, excluding Pseudotracheata. *Bull. Brit. Mus. (Nat. Hist.) Zool., 28:* 63–100.

Fish, S., 1970. The biology of *Eurydice pulchra* (Crustacea: Isopoda). *J. Mar. Biol. Assoc. U.K., 50:* 753–768.

Giard, A., and J. Bonnier, 1887. Contributions à l'étude des bopyriens. *Trav. Sta. Zool. Wimereux, 5:* 1–272.

Hatch, M. H., 1947. The Chelifera and Isopoda of Washington and adjacent regions. *Univ. Wash. Publ. Biol., 10:* 155–274.

Hessler, R. R., 1969. Peracarida. In R. C. Moore, *Treatise on invertebrate paleontology,* Pt. R. Arthropoda 4, *1:* R360–R393. Lawrence, Kans.: Geol. Soc. America and Univ. Kansas.

———, 1970. The Desmosomatidae (Isopoda, Asellota) of the Gay Head-Bermuda Transect. *Bull. Scripps Inst. Oceanogr.,* 15: 1–185.

Hessler, R. R., and D. Thistle, 1975. On the place of origin of deep-sea isopods. *Mar. Biol., 32:* 155–165.

Holdich, D. M., 1968. Reproduction, growth and bionomics of *Dynamene bidentata* (Crustacea: Isopoda). *J. Zool. London, 156:* 137–153.

———, 1973. The midgut/hindgut controversy in isopods. *Crustaceana, 24:* 211–214.

Hurley, D. E., 1961. A checklist and key to the Crustacea Isopoda of New Zealand and the subantarctic islands. *Trans. Roy. Soc. N.Z., 1:* 259–292.

Johnson, W. S., 1976. Biology and population dynamics of the intertidal isopod *Cirolana harfordi. Mar. Biol., 36:* 343–350.

Jones, D. A., 1976. The systematics and ecology of some isopods of the genus *Cirolana* (Cirolanidae) from the Indian Ocean region. *J. Zool. London, 178:* 209–222.

Kensley, B., 1978. *Guide to the marine isopods of southern Africa.* 173 pp. Cape Town: So. Afr. Mus.

Kruczynski, W. L., and G. J. Myers, 1976. Occurrence of *Apanthura magnifica* Menzies and Frankenberg, 1966 (Isopoda: Anthuridae) from the west coast of Florida, with a key to the species of *Apanthura* Stebbing, 1900. *Proc. Biol. Soc. Wash., 89:* 353–360.

McQueen, D. J., 1976. *Porcellio spinicornis* Say (Isopoda) demography. II. A comparison between field and laboratory data. *Canad. J. Zool., 54:* 825–842.

McQueen, D. J., and J. S. Carnio, 1974. A laboratory study of the effects of some climatic factors on the demography of the terrestrial isopod *Porcellio spinicornis* Say. *Canad. J. Zool., 52:* 599–611.

Markham, J. C., 1975a. A review of the bopyrid isopod genus *Munidion* Hansen, 1897, parasitic on galatheid crabs in the Atlantic and Pacific Oceans. *Bull. Mar. Sci., 25:* 422–441.

———, 1975b. Two new species of *Asymmetrione* (Isopoda, Bopyridae) from the western Atlantic. *Crustaceana, 29:* 256–265.

Menzies, R. J., 1950. The taxonomy, ecology and distribution of northern California isopods of the genus *Idothea* with the description of a new species. *Wasmann J. Biol., 8:* 155–195.

———, 1951. New marine isopods, chiefly from northern California, with notes on related forms. *Proc. U.S. Natl. Mus., 101:* 105–156.

———, 1957. The marine borer family Limnoriidae (Crustacea, Isopoda). *Bull. Mar. Sci. Gulf Carib., 7:* 101–200.

———, 1962a. The isopods of abyssal depths in the Atlantic Ocean. In Abyssal Crustacea, *Vema Res. Ser., 1:* 79–206.

———, 1962b. The zoogeography, ecology and systematics of the Chilean marine isopods. Repts. Lunds Univ. Chile Exped. *Lunds Univ. Arsskr.,* 57: 1–162.

Menzies, R. J., and J. L. Barnard, 1959. Marine Isopoda on the coastal shelf bottoms of southern California: systematics and ecology. *Pac. Natur., 1:* 3–35.

Menzies, R. J., and D. Frankenberg, 1966. *Handbook on the common marine isopod Crustacea of Georgia.* 93 pp. Athens, Ga.: Univ. Georgia Press.

Menzies, R. J., and P. Glynn, 1968. The common marine isopod Crustacea of Puerto Rico: A handbook for marine biologists. *Stud. Fauna Curaçao Carib. Isl., 27:* 1–33.

Monod, T., 1926. Les Gnathiidae, essai monographique (morphologie, biologie, systématique). *Mem. Soc. Sci. Nat. Maroc., 13:* 1–667.

Moore, H. F., 1902. Report on Porto Rican Isopoda. *Bull. U.S. Fish Comm., 20:* 163–176.

Mulaik, S. B., 1960. Contribucion al conocimiento de los isopodos terrestres de Mexico (Isopoda, Oniscoidea). *Rev. Soc. Mex. Hist. Nat., 21:* 79–220.

Naylor, E., 1955. The diet and feeding mechanism of *Idotea. J. Mar. Biol. Assoc. U.K., 34:* 347–355.

———, 1972. *British marine isopods.* 86 pp. New York: Academic Press.

Nicolls, G. E., 1943–44. The Phreatoicoidea. *Pap. Proc. Roy. Soc. Tasmania, 1942:* 1–45 (1942); 1–156 (1943).

Nierstrasz, H. F., 1917. Die Isopoden-Sammlung im Naturhistorischen Reichsmuseum zu Leiden. II. Cymothoidae, Spaeromidae, Serolidae, Anthuridae, Idotheidae, Asellidae, Janiridae, Munnopsidae. *Zool. Meded., 3:* 7–120.

Nierstrasz, H. F., and Brender à Brandis, 1926. Isopoda, Epicaridea. In G. Grimpe and E. Wagner (eds.), *Die Tierwelt der Nord- und Ostee, 10:* 1–56. Leipzig: Akad. Verl.

———, 1929. Papers from Dr. Th. Mortensen's Pacific Expedition 1914–16. 48. Epicaridea I. *Vidensk. Medd. Dansk Naturh. Foren., 87:* 1–44.

———, 1931. Papers from Dr. Th. Mortensen's Pacific Expedition 1914–1916. 57. Epicaridea II. *Vidensk. Medd. Dansk Naturh. Foren., 91:* 147–226.

Richardson, H., 1904a. Contributions to the natural history of the Isopoda. *Proc. U.S. Natl. Mus., 27:* 1–89.

———, 1904b. Contributions to the natural history of the Isopoda (second part). *Proc. U.S. Natl. Mus., 27:* 657–681.

———, 1905. Monograph on the isopods of North America. *Bull. U.S. Natl. Mus.,* 54, 727 pp. (reprinted, 1972, Antiquariaat Junk).

———, 1909. Isopods collected in the northwest Pacific by the U.S. Bureau of Fisheries steamer "Albatross" in 1906. *Proc. U.S. Natl. Mus., 37:* 75–129.

Sars, G. O., 1898. *An account of the Crustacea of Norway, 2. Isopoda*, 270 pp. Bergen: Bergen Museum.

Schultz, G. A., 1969. *How to know the marine isopod crustaceans.* 359 pp. Dubuque, Iowa: Wm. Brown Co.

———, 1970. A review of the species of the genus *Tylos* Latreille from the New World (Isopoda, Oniscoidea). *Crustaceana, 19:* 297–305.

Sheader, M., 1977. The breeding biology of *Idotea pelagica* (Isopoda: Valvifera) with notes on the occurrence and biology of its parasite *Clypeoniscus hanseni* (Isopoda: Epicaridea). *J. Mar. Biol. Assoc. U.K., 57:* 659–674.

Shiino, S. M., 1942. On the parasitic isopods of the family Entoniscidae, especially those found in the vicinity of Seto. *Mem. Coll. Sci., Kyoto Imper. U.,* (B) *17:* 37–76.

———, 1965. Phylogeny of the genera within the family Bopyridae. *Bull. Mus. Natl. Hist. Nat.,* (2) *37:* 462–465.

Vandel, A., 1960, 1962. Isopoda Terrestres. In *Faune de France. 64:* 1–416; *66:* 417–932. Paris: Lechevalier.

Van Name, W. G., 1936. American land and freshwater isopod Crustacea. *Bull. Amer. Mus. Nat. Hist., 71:* 1–535.

Williams, W. D., 1970. A revision of North American epigean species of *Asellus* (Crustacea: Isopoda). *Smiths. Contr. Zool.,* 49: 1–79.

Wolff, T., 1962. The systematics and biology of bathyal and abyssal Isopoda Asellota. *Galathea Rept., 6:* 7–320.

ORDER AMPHIPODA Latreille, 1816

The nearly 5000 amphipod species are divided into four major taxa, the Gammaridea, Hyperiidea, Caprellidea, and Ingolfiellidea. The gammarideans are the most successful and, while primarily members of the benthos, have adapted to a variety of habitats. The morphological specializations exhibited by the suborders are of sufficient interest that they are presented separately.

Gammaridea and Hyperiidea

SUBORDER GAMMARIDEA Latreille, 1803

Recent species	Approximately 4500.
Size range	1–150 mm.
Carapace	Absent.
Eyes	Sessile, compound; usually not very large.
Antennules	Usually biramous; 2nd flagellum sometimes reduced or absent.
Antennae	Uniramous.
Mandibles	With palp.
Maxillulae	With endopodal palp and basal endite.
Maxilla	Usually reduced.
Maxillipeds	One pair; usually with 7 segments.
Thoracic appendages	Uniramous; thoracopods 2–8, each often with large coxal plate; 2nd and 3rd usually subchelate.
Abdominal appendages	Pleopods and uropods usually biramous, easily differentiated.
Telson	Entire or bilobed.
Tagmata	Cephalon, pereon, and pleon.
Somites	Head with 5 + 1 thoracic (maxilliped); pereon with 7; pleon with 6 excluding telson.
Sexual characters	Gonopores on 6th thoracic somite of female, on 8th of male; female with oostegites; male with paired penes. Eyes, antennae, and 2nd gnathopod sexually dimorphic.
Sexes	Separate.

Larval development	Epimorphic; without postlarval stage.
Fossil record	U. Eocene to Recent.
Feeding types	Scavengers, filter feeders, often microphagous and predators; sometimes parasitic, adapted for piercing and sucking.
Habitat	Marine, brackish, and freshwater; primarily benthic; few semi-terrestrial.
Distribution	Worldwide.

SUBORDER HYPERIIDEA Latreille, 1831

Recent species	Approximately 450.
Size range	1–90 mm.
Carapace	Absent.
Eyes	Sessile, compound; often very large.
Antennules	Uniramous; usually reduced or vestigial in female.
Antennae	Uniramous; often vestigial in female.
Mandibles	With or without palp.
Maxillulae	Usually with palp; may be reduced or absent.
Maxillae	Often reduced, occasionally absent.
Maxillipeds	One pair, with 1 to 4 segments; usually without palp.
Thoracic appendages	Uniramous; thoracopods 2–8 with coxal plates small or fused to body; 6th sometimes raptorial.
Abdominal appendages	Pleopods and uropods usually well developed.
Telson	Entire.
Tagmata	Cephalon, pereon, and pleon.
Somites	Head with 5 + 1 thoracic (maxilliped); pereon with 7; pleon with 5, excluding telson (urosomites 2 and 3 fused).
Sexual characters	Gonopores on 6th thoracic somite of female, on 8th of male; female with oostegites; male with small paired penes. Antennae of female reduced or vestigial.
Sexes	Separate.
Larval development	Epimorphic; usually without postlarval stage.
Fossil record	Recent.
Feeding types	Predators and parasites; usually associates of gelatinous plankton.
Habitat	Marine, pelagic; usually commensal or parasitic on tunicates, coelenterates, ctenophores and colonial radiolarians.
Distribution	Worldwide.

Of the four suborders of the Amphipoda, the two largest, both in size and in number of species, the Gammaridea and Hyperiidea, generally are typical of the conceptual amphipod. The two smaller suborders, the Caprellidea and Ingolfiellidea, will be discussed separately. Undoubtedly the most abundant amphipods available for dissection and study are members of the Gammaridea. The following instructions and discussion have been sufficiently generalized so that they are applicable to most taxa. Basic differences between gammaridean and hyperiidean morphology will be pointed out (see Figure 37).

In contrast to the Isopoda, the amphipod body usually is laterally compressed; however, a few gammarideans are more or less dorsoventrally compressed. Usually the 1st thoracic somite is fused to the cephalon and its pair of appendages modified as maxillipeds, which often are

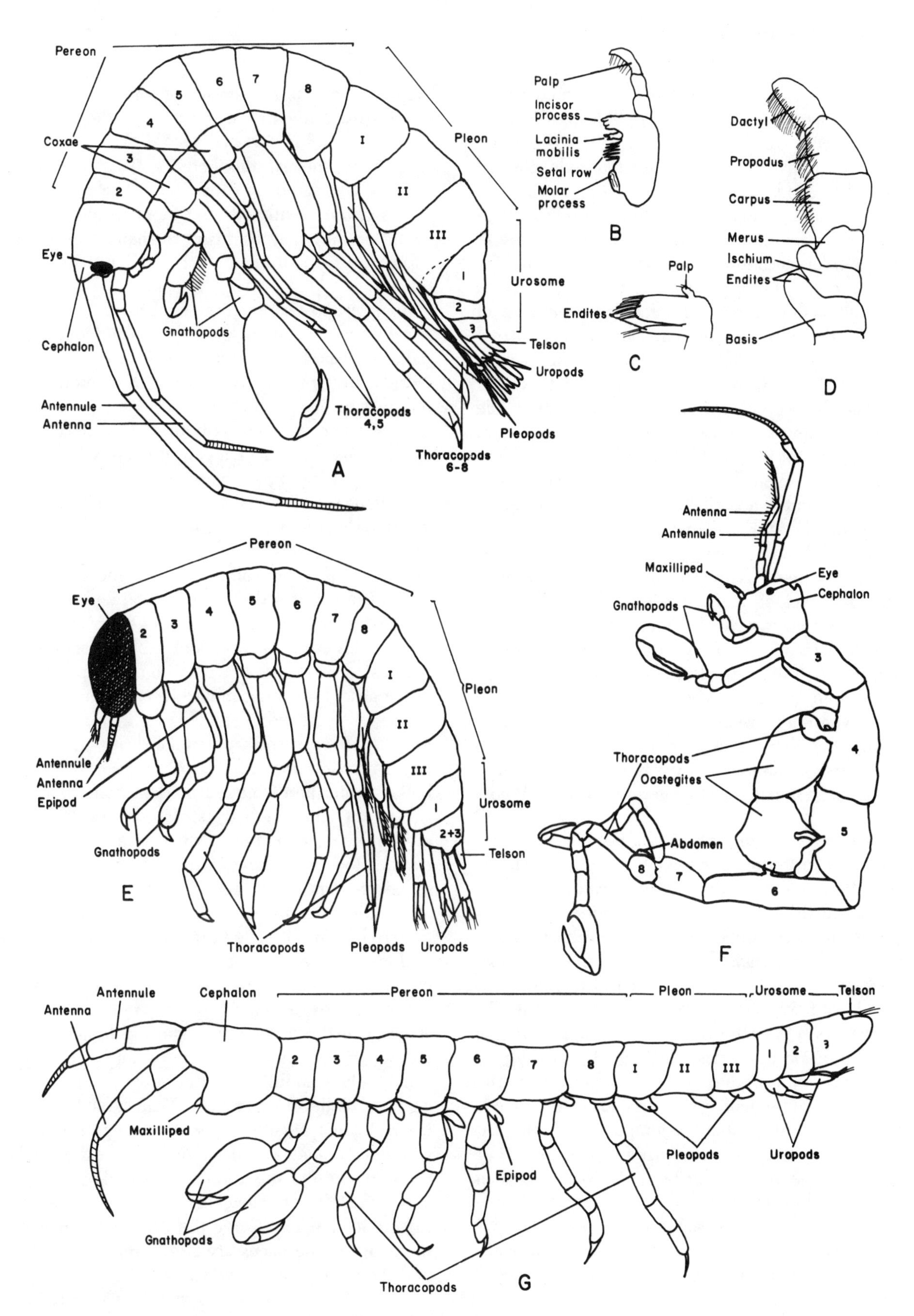

Figure 37 Amphipoda: A—D, Gammaridea; E, Hyperiidea; F, Caprellidea; G, Ingolfiellidea. A. Typical gammaridean (lateral view); B. Mandible; C. Maxillule; D. Maxilliped; E. Typical hyperiidean; F. Typical caprellidean; G. *Ingolfiella* [after C. Delamare Deboutteville: *Biologie des eaux souterraines littorales et continentales* (Hermann, Paris, 1960)].

fused basally. Eyes usually are present and moderately small in gammarideans, but in hyperiideans the eyes frequently occupy the major portion of the cephalon. The antennules have a 1- to 3-segmented peduncle and flagellum. A second, accessory flagellum also usually is present in living specimens, but often is broken off in preserved material. Inspect your specimen(s) closely to see if an accessory flagellum is, or has been, present. The antennae are uniramous, each consisting of a 1- to 5-segmented peduncle and a flagellum. The flagella of both antennules and antenna may carry sensory structures such as calceoli. The 7 thoracic somites collectively, as in isopods and tanaidaceans, are referred to as the pereon. The thoracopods are uniramous, with the coxae ventrally expanded to form coxal plates that cover and protect the gills, which are thoracic rather than abdominal as in isopods. Typically the first 2 pairs of pereopods (thoracopods 2 and 3) are subchelate and are referred to as gnathopods; the 2nd gnathopod, in particular, may be sexually dimorphic in mature males. The shape of this gnathopod has been used extensively as a taxonomic character; however, in many species it is polymorphic, thus it is not always a valid diagnostic character. The following 5 pairs of pereopods (thoracopods 4–8) are commonly called walking legs, although amphipods rarely use them for this purpose. Most obtain their mobility by swimming, usually on one side or the other. Frequently these appendages are differentiated into two types; the first 2 pairs are relatively short, often quite setose; the last 3 pairs are longer and directed posteriorly.

The abdomen or pleon is typically 6-segmented, not exhibiting the fusion of the pleomeres seen in isopods; however, there is a major distinction between the first 3 pleomeres and the last 3. The first 3 somites are generally normal in size and carry paired, multiarticulate, setose pleopods, well adapted for swimming. The posterior 3 somites usually are smaller and the paired appendages have fewer setae and only 1 or 2 segments. In hyperiideans only 2 posterior somites are apparent as a result of the fusion of the 2nd and 3rd. By convention, the last 2 or 3 somites are collectively referred to as the urosome, and their appendages as uropods. The uropods are often adapted for pushing or jumping. Only the last pair actually is homologous with uropods of other peracaridans. The telson may be entire or bilobed in gammarideans but is always entire in hyperiideans. Remove and examine 1 of each pair of appendages. If several stages of males of the same species are available, study the polymorphism exhibited by the 2nd gnathopods.

Not only are the 2nd gnathopods sexually dimorphic; the eyes and antennae of males frequently are larger than those of females. Males also possess a pair of penes, sometimes quite small and rather inconspicuous, on the sternite of the 8th thoracomere. In the female oostegites are attached to the coxae of the 2nd through 5th pereopods; gonopores are present on the sternite of the 6th thoracomere.

In hyperiideans, the body, at least anteriorly, frequently is inflated or subcircular. The number of segments of the maxillipeds and thoracic appendages is reduced or they may be fused to their somites, and the coxal plates usually are much smaller than in gammarideans.

Remove and examine the mouthparts. The maxillipeds usually have 7 segments, a peduncle and four-segmented palp, and moderately well-developed endites in gammarids. In hyperiids the number of segments usually is reduced to 3 or 4 and the palp is absent. Quite frequently the maxillipeds in both groups are fused basally. The maxillae each have a pair of endites, but often they are reduced and occasionally may be absent. The maxillulae generally are well developed. A palp is present in gammarids and at least in some hyperiids (mouthparts have not been described for many species in the latter group). A pair of well-developed endites usually are present in gammarids, but the basal endite may be absent in hyperiids. The mandibles usually are well developed, with strong incisor and molar processes, lacinia mobilis, and setal row. A mandibular palp may be present or absent in both gammarideans and hyperiideans.

The following discussion of major amphipod organ systems has been based on gammarideans (see Figure 38). The gammarid exoskeleton adheres very closely to the body somites and is not easily detached. To avoid damage to the internal body structures use a pair of microscissors or a sharpened dissecting needle to cut the exoskeleton. Beginning at the 8th thoracic somite and proceeding anteriorly, cut the cuticle of each somite just above the coxal plate and carefully tease it free of the underlying muscles. Be particularly careful in the dorsal midline as the heart and pericardium attach to the epidermis directly beneath the exoskeleton. You need remove the exoskeleton from only one side of the body. To remove the cuticle from the head, make a shallow incision in the midline dorsally, anteriorly, and ventrolaterally. Carefully free the integument from the underlying tissue and remove it. Damage to the sessile eye is usually unavoidable; however, enough ocelli probably will remain to mark its position. Remove the exoskeleton from one side of the abdomen by making a lateral cut approximately at the level of the thoracic coxal plates and teasing it free. To observe the ventral nerve cord in the thorax it may be necessary also to remove carefully several of the thoracic appendages.

In the dorsal midline observe the elongate heart that extends from the 2nd to the 7th thoracic somite. Pairs of ostia are located in the 3rd to 5th somites respectively. Three pairs of lateral arteries lead from the heart to the stomach and midgut ceca. The heart is enclosed in a pericaridal sinus, which in gammarideans often extends

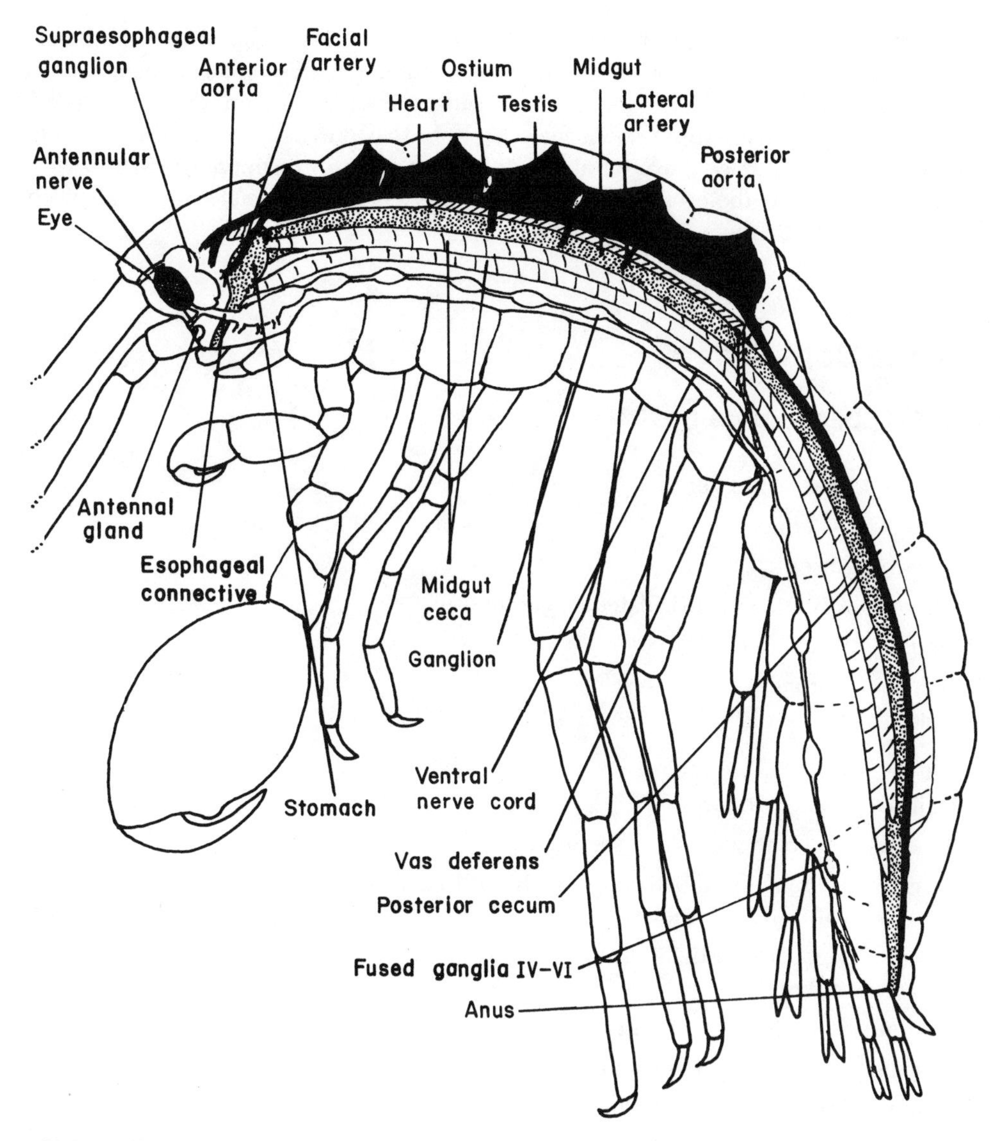

Figure 38 Amphipoda: Diagrammatic gammaridean with musculature removed to show major organ systems.

to the telson. Anteriorly identify the anterior aorta that provides blood to the cephalic region and an anterior lateral or facial artery that ramifies beneath the cephalic integument. Blood from the head enters the ventral or sternal sinus that bathes the ventral nerve cord and the hepatic ceca. From the sinus blood is provided to the appendages by subdivided channels (podopericardial channels); blood is returned from the appendages to the pericardium. Posteriorly identify the posterior aorta; it may be paired or a single vessel that extends the length of the abdomen. At the level of the telson it branches, turns downward, and empties into the ventral sinus. There is no definitive ventral or subneural artery in amphipods.

Beneath the heart and lateral to the midgut above the hepatic or midgut ceca observe the paired, elongate, tubular gonads. If your specimen is a male, locate one of the vas deferens at the posterior end of the testis and trace it to the ejaculatory duct at the tip of one of the penial papillae on the 8th thoracic sternite. If your specimen is a female locate the oviduct and trace it from an ovary to its opening on the 6th thoracic sternite.

Food enters the body through a small, often slitlike mouth and passes through a short wide esophagus into the stomach. The stomach is thin walled, but armed internally with a pair of vertical plates anteriorly, cuticular toothed ridges, and a ventral groove covered in part by horizontal overlapping plates. The stomach opens posteriorly into the elongate midgut. Anteriorly from the midgut arise a forwardedly directed short cecum and 2 to 4 posteriorly directed, elongate tubular ceca. Trace these ceca to their terminations. In the 1st urosomal somite, the midgut gives off a forwardly directed posterior pair of ceca. The hindgut may be distinguished posteriorly to the ceca. The digestive tract terminates in a ventrally directed anus at the base of the telson. Remove the stomach, being careful to leave the esophagus intact, and examine the internal structure. The excretory organ in amphipods is the antennal gland.

Anteriorly in the head distinguish the supraesophageal ganglion. It may be possible also to trace the antennular, antennal, and optic nerves. Follow the esophageal connective around the esophagus to the ventral ganglia. Note that the ganglia of the mouthparts have fused to form a large subesophageal ganglion. Trace the ventral nerve cord and segmental ganglia through the body. Note that the ganglia are fused in each somite, but between somites the nerve cords are widely separated. The ganglia of the urosomal somites are fused.

Ingolfiellidea and Caprellidea

SUBORDER INGOLFIELLIDEA Hansen, 1903

Recent species	Several in 1 genus.
Size range	1–14 mm.
Carapace	Absent.
Eyes	Absent; with articulated ocular lobes.
Antennules	Uniramous.
Antennae	Uniramous.
Mandibles	?
Maxillulae	?
Maxillae	Usually reduced.
Maxillipeds	One pair; with 6 or 7 segments.
Thoracic appendages	Uniramous; thoracopods 2–8 with small coxae; 2nd and 3rd with large subchelae.
Abdominal appendages	Pleopods vestigial; uropods 1 and 2 biramous, 3rd reduced or absent.
Telson	(?) Reduced or absent.
Tagmata	Cephalon, pereon, and pleon.
Somites	Head with 5 + 1 thoracic (maxilliped); pereon with 7; pleon with 6 excluding telson.
Sexual characters	Gonopores on 6th thoracic somite of female, on 8th of male; female with small oostegites; male probably with small paired penes.
Sexes	Separate.
Larval development	Epimorphic; without postlarval stage.
Fossil record	Recent.
Feeding types	Unknown.
Habitat	Marine, interstitial; abyssal to shallow; also cave dwellers.
Distribution	Worldwide.

SUBORDER CAPRELLIDEA Leach, 1814

Recent species	Approximately 300.
Size range	1–32 mm.
Carapace	Absent.
Eyes	Sessile, compound.
Antennules	Uniramous.

Antennae	Uniramous.
Mandibles	With or without palp and molar process.
Maxillulae	With endopodal palp; without basal endite.
Maxillae	Usually reduced.
Maxillipeds	One pair; usually with 7 segments, occasionally reduced.
Thoracic appendages	Uniramous; thoracopods 2–8, with 4th and 5th pairs reduced or absent; 6th pair occasionally reduced or absent.
Abdominal appendages	Usually absent in female; male with 1–3 pairs, often vestigial.
Telson	Vestigial or absent.
Tagmata	Cephalon and pereon; pleon reduced or vestigial.
Somites	Head with 5 + 2 thoracic (maxilliped); pereon with 4–6; pleon reduced or vestigial.
Sexual characters	Gonopores of female on 6th thoracic somite, of male on 8th; female with oostegites on 4th and 5th thoracic appendages only; male with paired penes.
Sexes	Separate.
Larval development	Epimorphic; without postlarval stage.
Fossil record	Recent.
Feeding types	Predators; sometimes parasitic.
Habitat	Marine, benthic, usually clinging to plants; 1 family parasitic on whales.
Distribution	Worldwide.

The Ingolfiellidea are interstitial inhabitants. The body is elongate and cylindrical (see Figure 37). The cephalon, which is fused with the 1st thoracic somite, has a small pair of ocular lobes but lacks eyes. A pair of maxillipeds is present; the thoracic appendages consist of 2 pairs of gnathopods and 5 pairs of pereopods. The coxal segments are small, as are the females' oostegites. The pleon and urosome are nearly equal in size; the pleopods are reduced but the first 2 pairs are biramous. The first 2 pairs of uropods also are reduced, but not as much as the pleopods; the last pair of uropods is vestigial.

The Caprellidea are the most structurally modified of all the amphipods. Despite their small size, they are easily recognized. The cephalon and first 2 thoracic somites are fused. Free-living caprellids have elongate bodies, while the bodies of parasitic species are dorsoventrally flattened. The pleon and telson are extremely reduced or vestigial in all but a few taxa. Although the first 2 thoracic somites are fused with the cephalon, only the 1st pair of thoracopods is modified as maxillipeds. The following 2 pairs usually are gnathopods. Of the remaining 5 pairs of pereopods, occasionally all are well developed, but most frequently the first 2 pairs are vestigial or absent. Occasionally even the 3rd pereopod (6th thoracopod) is reduced. The last 3 pairs (or as indicated above, 2 pairs) are well developed and adapted for holding the animal on the substrate. The cephalon carries small or moderately small compound sessile eyes, elongate antennules, short antennae, and relatively well-developed mouthparts. Examine several specimens, if possible, from different genera. Very often the only structures apparent on the 4th and 5th thoracic somites are small gills, or in females, oostegites. The segments of the pereopods may be difficult to distinguish because of fusion, but see if you can determine the number of segments present.

Remove and examine the mouthparts. These structures are important in caprellidean systematics. Typically the maxilliped, including palp, consists of 7 segments, although this number may be reduced. The palp is present in most free-living taxa; occasionally it is reduced or more rarely absent. As in other amphipod taxa, the maxillae are reduced. In all species for which the maxillulae have been described, a palp is present but the basal endite is absent. The mandible is an especially valuable diagnostic character. A mandibular palp may be present or absent, as may the molar process. Although not as asymmetrical as seen in tanaidaceans, the right and left lacinia mobilis often differ. Not infrequently the right is 5-toothed, either serrate or smooth, while the left is 5-toothed apically. In some taxa accessory plates have been developed in addition to the lacinia mobilis. Do you find differences in the mandibular structure of your specimen(s)? In parasitic forms all the mouthparts are modified as piercing structures.

Larval development in all amphipods is epimorphic

and only in a few of the hyperiideans do the young hatch as postlarvae. In these cases, the appendages of the pleon are only buds and develop through 4 or 5 juvenile stages.

References

Barnard, J. L., 1958. Index to families, genera and species of gammaridean Amphipoda. *Occas. Pap. Allan Hancock Found., Publ., 19:* 1–145.

———, 1960. New bathyal and sublittoral ampeliscid amphipods from California, with an illustrated key to *Ampelisca. Pac. Natur., 1*(16): 3–36.

———, 1967. Bathyal and abyssal gammaridean Amphipoda of Cedros Trench, Baja California. *Bull. U.S. Natl. Mus.*, no. 260: 1–205.

———, 1969a. Gammaridean Amphipoda of the rocky intertidal of California: Monterey Bay to La Jolla. *Bull. U.S. Natl. Mus.*, no. 258: 1–230.

———, 1969b. The families and genera of marine gammaridean Amphipoda. *Bull. U.S. Natl. Mus.*, no. 271: 1–536.

———, 1970. Sublittoral Gammaridea (Amphipoda) of the Hawaiian Islands. *Smiths. Contr. Zool.*, no. 34: 1–286.

———, 1971. Keys to the Hawaiian marine Gammaridea, 0–30 meters. *Smiths. Contr. Zool.*, no. 58: 1–135.

———, 1972. Gammaridean Amphipoda of Australia, Part I. *Smiths. Contr. Zool.*, no. 103: 1–333.

———, 1974. Gammaridean Amphipoda of Australia, Part II. *Smiths. Contr. Zool.*, no. 139: 1–148.

Barnard, J. L., and R. R. Given, 1960. Common pleustid amphipods of Southern California, with a projected revision of the family. *Pac. Natur., 1*(17): 1–18.

Barnard, K. H., 1930. Crustacea. Part XI. Amphipoda. *British Antarctic ("Terra Nova") Expedition 1910, Nat. Hist. Rep. Zool., 8:* 307–454.

———, 1932. Amphipoda. *Discovery Reports, 5:* 1–326.

Bary, B. M., 1959. Ecology and distribution of some pelagic Hyperiidea (Crustacea, Amphipoda) from New Zealand waters. *Pac. Sci., 13:* 317–334.

Behning, A., 1939. Die Amphipoda-Hyperiidea der den Fernen Osten der Ud. SSR umgrenzenden Meere. *Intern. Rev. Ges. Hydrobiol. Hydrogr., 38:* 353–367.

Bousfield, E. L., 1951. Pelagic Amphipoda of the Belle Isle Strait region. *J. Fish. Red. Bd. Canad., 8:* 134–163.

———, 1973. *Shallow-water gammaridean Amphipoda of New England.* xii + 312 pp. Ithaca, N.Y.: Comstock Publ. Assoc., Cornell University Press.

Bowman, T. E., 1960. The pelagic amphipod genus *Parathemisto* (Hyperiidea: Hyperiidae) in the North Pacific and adjacent Arctic Ocean. *Proc. U.S. Natl. Mus., 112:* 343–392.

———, 1973. Pelagic amphipods of the genus *Hyperia* and closely related genera (Hyperiidea: Hyperiidae). *Smiths. Contr. Zool.*, 136: 1–76.

Bowman, T. E., and H-E. Gruner, 1973. The families and genera of Hyperiidea (Crustacea: Amphipoda). *Smiths. Contr. Zool.*, 146: 1–64.

Chevreux, E., and L. Fage, 1925. Amphipodes. *Faune de France, 9:* 1–488. Paris: Paul Lechevalier.

Dahl, E., 1977. The amphipod functional model and its bearing upon systematics and phylogeny. *Zool. Scr., 6:* 221–228.

Delamare Deboutteville, C., 1960. *Biologie des eaux souterraines littorales et continentales.* 740 pp. Paris: Hermann.

Fox, R. S., and K. H. Bynum, 1975. The amphipod crustaceans of North Carolina estuarine waters. *Chesapeake Sci., 16:* 223–237.

Hurley, D. E., 1955. Pelagic amphipods of the sub-order Hyperiidea in New Zealand waters. I. Systematics. *Trans. Roy. Soc. N.Z., 33:* 119–194.

———, 1956. Bathypelagic and other Hyperiidea from California waters. *Occ. Pap. Allan Hancock Found.*, no. 18: 1–25.

Ivester, M. S., and B. C. Coull, 1975. Comparative study of ultrastructural morphology of some mouthparts of four haustoriid amphipods. *Can. J. Zool., 53:* 408–417.

Karaman, G. S., 1975a. The family Ampeliscidae of the Adriatic Sea. (64. Contribution to the knowledge of the Amphipoda). *Acta Adriat., 17:* 1–67.

———, 1975b. The higher classification in amphipods. *Crustaceana, 28:* 304–310.

Karaman, G. S., and S. Pinkster, 1977. Freshwater *Gammarus* species from Europe, North Africa and adjacent regions of Asia (Crustacea—Amphipoda). Part I. *Gammarus pulex*—group and related species. *Bijdr. Dierk., 47:* 1–97.

Keith, D. E., 1971. Substrate selection in caprellid amphipods of southern California, with emphasis on *Caprella californica* Stimpson and *Caprella equilibra* Say (Amphipoda). *Pac. Sci., 25:* 387–395.

Krapp-Schickel, G., 1975. Revision of Mediterranean *Leucothoe* species (Crustacea, Amphipoda). *Boll. Mus. Civ. Stor. Natur. Verona, 2:* 91–118.

———, 1976. Die Gattung *Stenothoe* (Crustacea, Amphipoda) im Mittelmeer. *Bijdr. Dierk., 46:* 1–34.

Laubitz, D. R., 1976. On the taxonomic status of the family Caprogammaridae Kudrjaschov & Vassilenko (Amphipoda). *Crustaceana, 31:* 145–150.

Lowry, J. K., and S. Bullock, 1976. Catalogue of the marine gammaridean Amphipoda of the Southern Ocean. *Bull. Roy. Soc. N.Z.*, no. 16, vi + 187 pp.

McCain, J. C., 1968. The Caprellidae (Crustacea: Amphipoda) of the western North Atlantic. *Bull. U.S. Natl. Mus.*, no. 278: 1–147.

Mayer, P., 1882. Die Caprelliden des Golfes von Neapel und der angrenzenden Meeres-Abschnitte. Eine Monographie. *Fauna Flora Golfe Neapel, 6:* 1–201.

———, 1903. Die Caprellidae der "Siboga"-Expedition. *Siboga Exped., 34:* 1–160.

Mills, E. L., 1965. The zoogeography of North Atlantic and North Pacific ampeliscid amphipod crustaceans. *Syst. Zool., 14:* 119–130.

———, 1967. Deep-sea amphipods from the western North Atlantic ocean; Ingolfiellidea and Pardaliscidae. *Canad. J. Zool., 45:* 347–355.

Schellenberg, A., 1942. Flohkrebse oder Amphipoda. In F.

Dahl, *Die Tierwelt Deutschlands, 40:* 1–252. Jena: G. Fischer.

Schiecke, U., 1976. Eine marine *Ingolfiella* (Amphipoda: Ingolfiellidae) im Golf von Neapel: *Ingolfiella ischitana* n. sp. *Boll. Mus. Civ. Stor. Natur. Verona, 3:* 413–420.

Siewing, R., 1963. Zur Morphologie der aberranten Amphipoden Gruppe Ingolfiellidae. *Zool. Anz., 171:* 76–91.

Stock, J. H., 1976. A new member of the crustacean suborder Ingolfiellidea from Bonaire, with a review of the entire suborder. *Stud. Fauna Curaçao Carib. Isl., 50:* 56–75.

———, 1977a. The taxonomy and zoogeography of the hadziid Amphipoda with emphasis on the West Indian taxa. *Stud. Fauna Curaçao Carib. Isl.*, no. 92: 1–130.

———, 1977b. The zoogeography of the crustacean suborder Ingolfiellidea, with descriptions of new West Indian taxa. *Stud. Fauna Curaçao Carib. Isl.*, no. 92: 131–146.

Tsvetkova, N. L., 1975. *Pribrezhnye gammaridy severnykh i dalnevostochnykh morei SSSR i sopredelnykh vod.* 256 pp. Leningrad: Akad. Nauk SSSR, Zool. Inst., Izdatelstvo Nauka.

Vasilenko, S. V., 1974. *Kaprellidy (Morskie kozochki) morei SSSR i sopredelnykh vod.*, 107. 287 pp. Leningrad: Akad. Nauk SSSR, Izdatelstvo Nauka.

Superorder EUCARIDA Calman, 1904

The eucarids are highly evolved and developed eumalacostracans. The carapace is fused to all the thoracomeres to form a cephalothorax; the eyes typically are carried on movable stalks; the mandible lacks a lacinia mobilis; and larval development is typically metamorphic.

This taxon includes the orders Euphausiacea, Amphionidacea, and Decapoda. The latter, representing the familiar shrimps, lobsters, and crabs, is so large and diverse that discussions of the different morphological groups are provided at the levels of infraorder, superfamily, and family.

ORDER EUPHAUSIACEA Dana, 1852

Recent species	Approximately 85, in eleven genera.
Size range	Up to 80 mm.
Carapace	Well developed; fused to all thoracic somites; gills not covered.
Eyes	Stalked, compound, usually well developed; sometimes subdivided by constriction.
Antennules	Biramous; sometimes sexually modified in males.
Antennae	Biramous; exopod scalelike.
Mandibles	Usually with palp; without lacinia mobilis in adult.
Maxillulae	Usually with 2 endites and unsegmented endopodal palp; palp 3-segmented *Bentheuphausia*, 2-segmented in some larvae.
Maxillae	Typically with 4 segments; exopod, endopodal palp, and 2 endites, often bi- or trilobed.
Maxillipeds	None.
Thoracic appendages	Usually 8 biramous thoracopods; often 8th, sometimes 7th and 8th reduced or vestigial; sometimes 1st and 2nd or 2nd and 3rd elongate, occasionally chelate-like.
Abdominal appendages	Pleopods biramous, usually with appendix interna; 1st and 2nd pairs of males modified for copulation. Uropods biramous.
Telson	Well developed, usually with movable subterminal spines; together with uropods forms tailfan.
Tagmata	Cephalothorax and abdomen.
Somites	Head with 5; thorax with 8; abdomen with 6, excluding telson.
Sexual characters	Gonopores on coxae of 6th thoracopods of female or on separate coxal plate; on 8th thoracomere of male; copulatory structures developed on 1st and 2nd pleopods of male. Antennules sometimes sexually dimorphic.

Sexes	Separate; female sometimes with ovisac(s).
Larval development	Anamorphic: nauplius → calyptopis → furcilia
Fossil record	Recent.
Feeding types	Usually planktonic filter feeders; occasionally predators.
Habitat	Marine, planktonic; often occurring in dense swarms; usually inhabiting open ocean waters.
Distribution	Worldwide.

The euphausiaceans have a relatively straightforward body plan; therefore, they are a comparatively simple group to study. The euphausiid depicted for external morphology is a specimen of the genus *Thysanopoda* in the family Euphausiidae (see Figure 39). The illustrated internal structures represent a composite. Your attention will be directed to the major differences among taxa. (Also see Table 1 on generic characters.) The number of thoracopods is variable. In several genera either the 8th or the 7th and 8th thoracopods are reduced or vestigial. Because of the development of filamentous gills on the coxae of the thoracopods, particularly the posteriormost pairs, reduced or vestigial appendages may be difficult to distinguish. A good rule of thumb is to start to count the number of appendages at the posterior end of the cephalothorax. A prominent luminescent organ is located on the base of the 7th thoracopod, except in *Bentheuphausia*.

Begin your study with an examination of the intact animal. The compound eyes are either simple or divided into anterodorsal and lateral parts. The antennular peduncle is either simple or the 1st segment has a flap-like lappet. The antennular flagella always are paired. The antenna carries a single flagellum; the exopod is developed into a broad scale or squama. Note that the carapace, which may be produced into a prominent rostrum, covers the thorax but does not enclose the gills. Frequently the lateral margins of the carapace have an interior or posterior spine or occasionally both. The gills are developed from the coxae of the 2nd through 8th thoracopods; those of the anterior appendages consist of a single branch. The posterior gills have several branches. Your attention already has been directed to the prominent luminescent or light organ at the base of the 7th thoracopod; similar luminescent organs generally also are present on each eyestalk and on the bases of each of the 2nd through 6th thoracopods. Additional luminescent structures can be observed on the ventral surfaces of each of the first 4 abdominal somites.

Generally all the thoracopods are similar in structure; however, in certain genera either the 2nd or the 3rd pair, or less frequently both the 1st and 2nd pairs, is extremely elongate. Most often these appendages terminate in a series of bristles; however, in the genus *Stylocheiron* the ultimate and penultimate segments may form a chela-like structure.

The abdomen bears 5 pairs of well-developed pleopods used for swimming. In males, the endopods of the 1st and 2nd are modified as copulatory organs. The development of the 1st pair is more elaborate and their structure is an important taxonomic character. The appendages of the 6th abdominal somite are the uropods, which, together with the telson, form a tailfan. In most species the telson bears a pair of subapical movable spines. At the posteroventral margin of the 6th somite, a preanal spine frequently is present. Dorsally the abdominal somites may be armed with spines or produced to form prominent keels.

Once you have identified all of the major morphological structures of the euphausiid, remove one of the first 2 thoracic appendages. Generally the basis, ischium, and merus bear a series of closely set, plumose setae that join with the opposing members to form a food basket. If your specimen(s) belongs to one of the genera with specialized 1st, 2nd, or 3rd thoracopods, remove them and compare their modifications with the structure of the more typical thoracopods. What function would you ascribe to these specialized appendages?

Among the Eucarida, the euphausiaceans are the only group in which maxillipeds are not developed. The mouthparts consist of a labrum, labium, and paired mandibles, maxillulae, and maxillae. Examine each and identify the individual structural parts. How do the euphausiid mouthparts, particularly the mandibles, compare with those of the Peracarida you have studied?

Remove and make a temporary slide mount preparation of the male copulatory structure from the 1st pleopod. The processes vary, depending upon the species under study. Identify the structural elements of your particular specimen. A sketch may be helpful for comparison with the copulatory structure illustrated in Figure 39D.

Well-preserved, relatively large specimens should be used for a study of the major internal organ systems (see Figure 40). With a fine scalpel or sharpened microneedle sever the lateral muscle attachments of the carapace. Dorsally the carapace is held by a pair of prominent muscles and must be teased free very carefully, as the

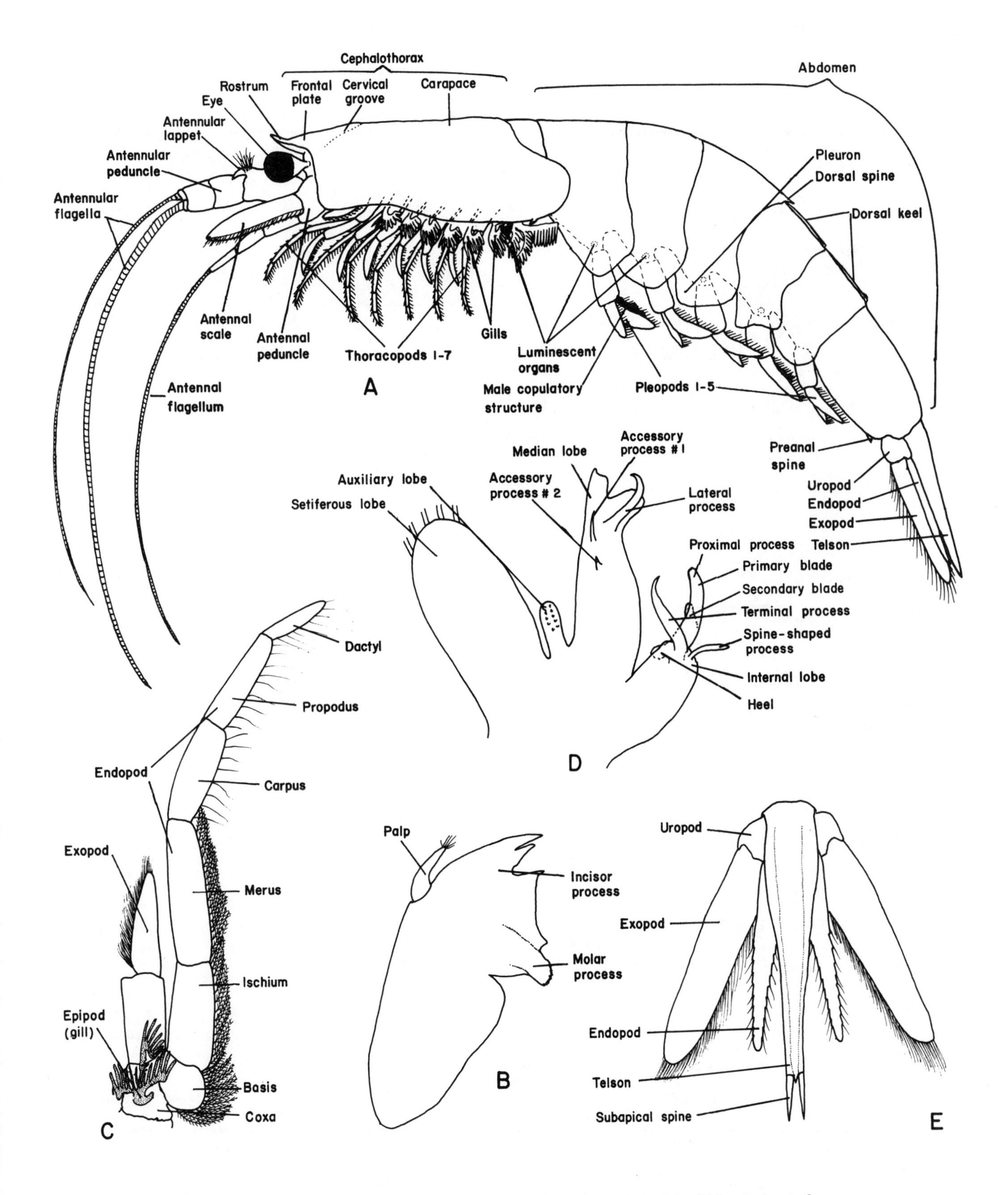

Figure 39 Euphausiacea: A. Male *Thysanopoda* (lateral view); B. Mandible; C. Typical thoracopod; D. Male copulatory structure; E. Telson and uropods.

TABLE 1 GENERIC CHARACTERS OF ADULT EUPHAUSIACEA

Genus	Thoracopods	Eyes	Rostrum	Ovisacs	Luminescent Organs
I *Bentheuphausia*	8 pairs, well developed	Imperfect	Triangular, weakly produced	Unknown	None
II *Euphausia*	7th and 8th rudimentary, 5th and 6th similar		Variable	Absent	Typical distribution
Meganyctiphanes	8th rudimentary		Obsolete or weakly produced	Absent	Typical distribution
Nyctiphanes	8th rudimentary; exopod 6th and 7th in ♀, terminal segments 7th in ♂, all absent		Variable	Double	Typical distribution
Pseudoeuphausia	7th and 8th rudimentary; 6th distal segments reduced		Absent; frontal plate truncate	Double, fused	Typical distribution
Thysanopoda	8th endopod reduced, exopod normal; 7th short		Variable	Absent	Typical distribution
III *Nematobrachion*	8th rudimentary; 7th short; 3rd greatly elongate		Variable, or rostral process	Unknown	Typical distribution
Nematoscelis	8th vestigial; 7th endopod biarticulate in ♀, exopod only in ♂ ; 2nd elongate		Variable	Single	Typical distribution
Stylocheiron	1st and 2nd feeble; 8th rudimentary; 3rd greatly produced		Variable	Single	Present on 7th thoracopod and 1st abdominal somite
Tessarabrachion	8th vestigial; 7th endopod absent in ♂ ; 2nd and 3rd elongate		Absent; frontal plate short, triangular	Unknown	Typical distribution
*Thysanoessa**	8th vestigial; 7th endopod absent in ♂ ; 2nd elongate		Well developed, long to very long	Absent	Typical distribution

*Included in group III because, although variable, the majority of species exhibit characters of this group.

delicate blood vessels and heart lie directly beneath and posterior to this pair of muscles. The carapace in the rostral region often will have to be cut to free it. Remove the carapace, but before beginning your examination of the cephalothoracic region cut open the abdominal exoskeleton. Returning to the cephalothorax, observe the antennal and mandibular muscles as well as the very prominent lateral and dorsal muscles. Anteriorly be-

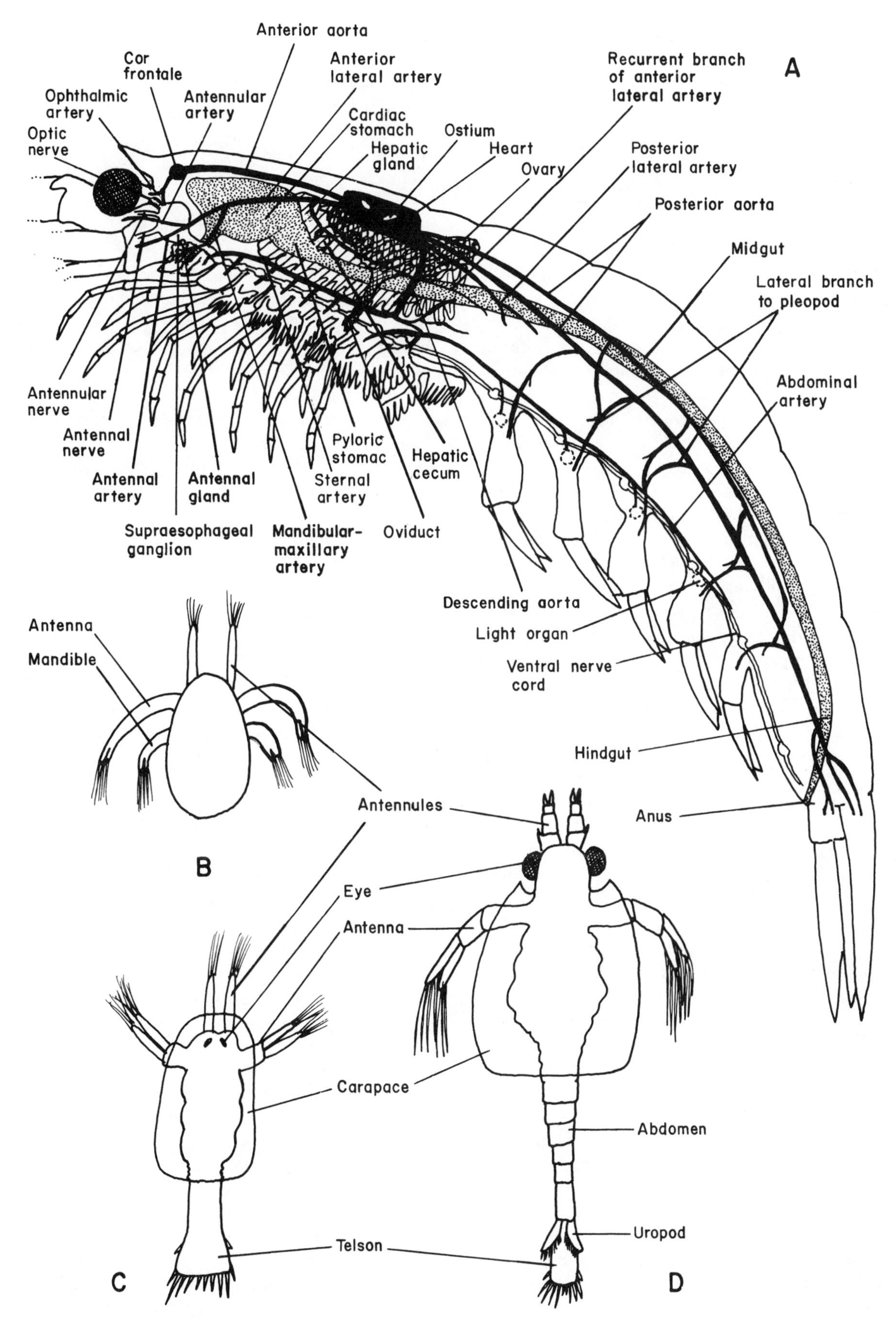

Figure 40 Euphausiacea: A. Diagrammatic euphausiacean with musculature removed to show major organ systems; B. Nauplius; C. Calyptopis; D. Furcilia.

tween the separated dorsal muscles note the dorsal surface of the cardiac stomach. The anterior aorta overlies the stomach, but will be difficult to detect until it can be traced from the heart. Carefully separate the dorsal muscles posteriorly to determine the position of the heart and the numerous blood vessels leaving it. Remove the dorsal and lateral muscles from one side of the body with caution, being careful in male specimens not to destroy the vas deferens, which crosses the musculature exteriorly on its course to the 8th thoracic somite. Removal of the musculature often will destroy the terminal portions of the anterior lateral artery, but its main path from the heart can be followed for some distance. Trace the anterior aorta over the cardiac stomach; anteriorly an enlargement of the vessel, a cor frontale, provides an additional pumping mechanism to improve circulation to the head. The anterior aorta provides blood to the eyes and anterior head region, while the anterior lateral artery serves the cephalic appendages. A lateral recurrent branch of the lateral artery and a small hepatic artery from the heart may be difficult to distinguish. The heart itself is subrectangular in shape, relatively small, and opens to the pericardium through 2 pairs of ostia. Observe the blood vessels leaving the heart posteriorly. The posterior aorta, which can be traced into the abdomen when the pair of dorsal abdominal muscles are removed, may be single or paired. A pair of posterior lateral arteries leads into the abdomen beside the midgut ceca, where they give off numerous small branches. Ventrally from the heart observe the descending aorta. This artery also may be paired, but if so, the second artery is much smaller and probably will not be discernable. The descending aorta provides blood to the sternal artery anteriorly and the abdominal artery posteriorly; however, neither of these blood vessels will be visible until you have removed the midgut and ceca. Blood from the sternal artery is supplied to the maxillae and first 6 pairs of thoracopods through lateral branches; branches of the abdominal artery supply the last 2 pairs of thoracopods and the luminescent organs of the abdomen. Follow the posterior aorta posteriorly. In each abdominal somite branches are given off to the appendages, musculature, and body tissues.

Return to the cephalothorax and examine the reproductive system. In the male the testes may be paired tubular structures lying to either side of the ceca and midgut, or a single testis may be horseshoe-shaped with lateral diverticula. The vas deferens may be coiled or straight. Trace this duct from the testis to its opening on the 8th thoracic somite. In the female, the ovaries lie posteroventrally of the heart and extend anteriorly over the anterior midgut, midgut ceca, and pyloric stomach. The posterior arteries from the heart often pass between the pair posteriorly. Trace an oviduct from the ovary to its opening on the coxa of the 6th thoracopod.

The euphausiid digestive system consists of an esophagus, cardiac and pyloric sections of the stomach, midgut, midgut ceca, and hindgut. The short esophagus leads from the mouth into the enlarged cardiac stomach. The structure of the cardiac stomach differs between filter feeding and predatory euphausiaceans. The cardiac stomach opens into the pyloric stomach through a cardiopyloric valve. At the entrance of the midgut a pair of hepatic ceca arise, which are highly convoluted and occupy much of the thoracic body cavity. Follow the midgut posteriorly. The beginning of the hindgut varies among taxa, but usually is at the level of the last abdominal somite. In some taxa a short blind diverticulum arises from the hindgut. The digestive tract terminates in a muscular anus. Excretion in euphausiaceans is by means of antennal glands.

The cardiac stomach frequently need not be removed to examine the nervous system, but the hepatic ceca should be, as should the lateral abdominal muscles from one side of the abdomen. Anteroventral to the cardiac stomach, observe the supraesophageal ganglion and the larger cephalic nerves. Follow the esophageal connective around one side of the esophagus to the fused ventral ganglia of the 3 pairs of mouthparts; then follow the ventral nerve cord the length of the body, distinguishing the individual segmental ganglia. In close proximity to the ventral nerve cord identify the sternal artery in the thorax and the abdominal artery in the abdomen.

Euphausiaceans pass through several stages and many instars during their larval development. Young hatch from eggs as free-swimming nauplii. This stage is followed by a metanauplius stage and then by a calyptopis stage with several instars or substages. The 4th stage and its instars is referred to as the furcilia stage. The nauplius is characterized by the presence of 3 pairs of appendages: the antennules, antennae, and mandibles; no eyes are present and the body lacks segmentation. One or 2 instars or substages frequently occur. The metanauplius stage may be a single instar in some taxa or several substages. The metanauplius is characterized by the development of rudiments of the other cephalic appendages and the 1st thoracopods; eyes begin development and the mandibles become reduced. Neither the nauplius nor the metanauplius feeds. The calyptopis stage is characterized by the presence of the carapace, the elongation of the trunk, and, in progressive instars or substages, the differentiation of somites. The appendages already present become more fully developed and subsequent limb buds develop. The furcilia stage with its several substages is characterized by the eyes being exposed and movable, the development of the thoracic and abdominal appendages, including the uropods, and by the development of the antennules toward the adult form. In the older literature a final cyrtopia stage, characterized by the adaptation of the pleopods for swimming, was described; however, this stage now is included in the furcilliae as the 3rd of 4 phases. With the 4th phase the telson exhibits a reduction in the number of terminal spines. Some euphausiaceans have an abbreviated larval development in which one or more naupliar stages are passed through in the egg.

References

Banner, A. H., 1950. A taxonomic study of the Mysidacea and Euphausiacea (Crustacea) of the North Pacific, Pt. 3 Euphausiacea. *Trans. Roy. Soc. Canad. Inst.*, *28:* 2–49.

Boden, B. P., 1954. The euphausiid crustaceans of southern African waters. *Trans. Roy. Soc. S. Afr.*, *34:* 181–243.

Boden, B. P., M. W. Johnson, and E. Brinton, 1955. The Euphausiacea (Crustacea) of the North Pacific. *Bull. Scripps Inst. Oceanogr.*, *6:* 287–400.

Brinton, E., 1953. *Thysanopoda spinicaudata*, a new bathypelagic giant euphausiid crustacean, with comparative notes on *T. cornuta* and *T. egregia*. *J. Wash. Acad. Sci.*, *43:* 408–412.

———, 1962. Variable factors affecting the apparent range and estimated concentrations of euphausiids in the North Pacific. *Pac. Sci.*, *16:* 374–408.

Einarsson, H., 1942. Notes on Euphausiacea I–III. *Vidensk. Medd. Dansk Natur.*, *106:* 263–286.

———, 1945. Euphausiacea. 1. Northern Atlantic species. *Dana Rept.*, no. 27: 1–185.

Hansen, H. J., 1911. The genera and species of the order Euphausiacea, with an account of remarkable variation. *Bull. Inst. Oceanogr. Monaco*, *210:* 1–54.

Holt, E. W. L., and W. M. Tattersall, 1905. Schizopodous Crustacea from the northeast Atlantic slope. *Fish. Ireland, Sci. Invest.*, 1902–03, *4:* 99–152.

Lomakina, N. B., 1978. Eufauziidy mirovogo okeana (Euphausiacea). *AN SSSR*, *118*, "Nauka," Leningrad, 222 pp.

MacDonald, R., 1927. Food and habits of *Meganyctiphanes norvegica*. *J. Mar. Biol. Assoc., U.K.*, *14:* 785–794.

McLaughlin, P. A., 1965. A redescription of the euphausiid crustacean, *Nematoscelis difficilis* Hansen, 1911. *Crustaceana*, *9:* 41–44.

Mauchline, J., and L. R. Fisher, 1969. The biology of euphausiids. In F. S. Russell and M. Younge, *Advances in Marine Biology*, *7:* 1–454. New York and London: Academic Press.

Rustadt, D., 1930. Euphausiacea, with notes on their biogeography and development. *Norsk. Vidensk. Acad., Oslo*, *1:* 1–83.

Ruud, J. T., 1932. On the biology of southern euphausiids. *Norsk. Vid.-Akad. Oslo Hvalradetskr.*, no. 2: 5–105.

———, 1936. Euphausiacea. *Rept. Danish Ocean. Exped. 1908–1910 Med. Adj. Seas*, 2 (Biology): 1–86.

Sheard, K., 1953. Taxonomy, distribution and development of the Euphausiacea (Crustacea). *B.A.N.Z.A.R.E. Repts.* (B) *8:* 1–72.

Tattersall, W. M., 1939. The Euphausiacea and Mysidacea of the John Murray Exped. to the Indian Ocean. *John Murray Exped. 1933–34, Sci. Rept.*, *8:* 203–256.

Zimmer, C., 1909. Die Nordische Schizopoden. *Nordisches Plankton*, *12:* 1–178.

———, 1956. Euphausiacea. In H. G. Bronn, *Klassen und Ordnungen des Tierreichs*, 5 Arthropoda, 1. Crustacea, 6 (3), pts. 1 and 2. 286 pp. Leipzig: Akad. Verl.

ORDER AMPHIONIDACEA Sars, 1867

Recent species	One, *Amphionides reynaudii* (H. Milne Edwards, 1832).
Size range	2–3 cm.
Carapace	Cylindrical; extremely thin.
Eyes	Stalked, compound; with large inner tubercle.
Antennules	Biramous; without statocyst; with terminal dorsal papilla.
Antennae	Presumably with 5-segmented peduncle and with broad scale.
Mandibles	Vestigial.
Maxillulae	Vestigial.
Maxillae	Biramous; endites vestigial.
Maxillipeds	One pair.
Thoracic appendages	Six biramous pairs, with exopods reduced, in females; 7 pairs, 2–7 biramous with exopods reduced, 8th uniramous in males.
Abdominal appendages	Five biramous pairs of pleopods in males; 1st uniramous, 2nd–5th biramous in females. Uropods biramous.
Telson	Together with uropods forms tailfan.

Tagmata	Cephalothorax and abdomen.
Somites	Head with 5 + 1 thoracic (maxilliped); thorax with 7; abdomen with 6, excluding telson.
Sexual characters	Gonopores of female on coxae of 6th thoracopods, presumably 8th of male. Female with enlarged uniramous 1st pleopods.
Sexes	Separate.
Larval development	Metamorphic; zoea → megalopa.
Fossil record	Recent.
Feeding types	Adults presumably do not feed.
Habitat	Marine, pelagic.
Distribution	Worldwide.

Although *Amphionides reynaudii* has a broad distribution, relatively few specimens are present in collections and thus available for study. The species is best known from its larvae. There are 9 to 12 larval substages, all bearing a general resemblance to the zoeal larvae of caridean shrimp, and for this reason *Amphionides* often has been classified among the Caridea. However, both larvae and adults have a number of characters that differ so greatly, not only from the carideans but from decapods in general, that the taxon is now classified as a distinct order of the Eucarida. Of particular import is the very specialized brood chamber of the female, formed by the elongate broad 1st pleopods and the single pair of maxillipeds. *Amphionides* is distinguished from the Caridea by the fact that the 2nd abdominal pleuron never overlaps the 1st. In this character the species agrees with the Dendrobranchiata, but can be distinguished from members of this suborder by the presence of an appendix interna, lack of a petasma in the male and thelycum in the female, in particular.

Adult females have an extremely thin carapace with well-developed postrostral spine and frequently also a rostral spine (see Figure 41). The eyes are small and a large tubercle, presumably a photophore, is present on the inner side. The antennules lack statocysts but do each have a terminal dorsal papilla. The scaphocerite of the antenna is very large and almost circular.

The thoracopods (2–7) are quite distant from the mouthparts; the 2nd are much shorter than the following pairs. Although the thoracopods are biramous, the exopods are small and apparently functionless; the endopods are segmented. The 5th pair is elongate. The 8th thoracomere lacks appendages in the female.

The 1st pair of pleopods in the female is elongate, reaching to the bases of the 2nd thoracopods, and uniramous; the fringe of setae and the shape, proximally concave and distally flat, suggest its function as a brood chamber. The remaining 4 pairs of pleopods are biramous swimming appendages; the endopods are each provided with an appendix interna.

The mandible and maxillule are vestigial; the maxilla and maxilliped are biramous but small. The degenerate condition of the digestive system suggests that the adult animals do not feed, or obtain food through absorption.

Adult males have been captured only rarely; therefore, descriptions of their morphology are limited. The 8th thoracopods are developed; the 1st pair of pleopods is biramous and adapted for swimming.

References

Boas, J. E. V., 1879. *Amphion* und *Polycheles* (Willemoesia). *Zool. Anz.*, *11:* 256–259.

Gordon, I., 1955. Importance of larval characters in classification. *Nature*, Lond., *176:* 911–912.

Gurney, R., 1936. Larvae of decapod Crustacea, 2. Amphionidae. *Discovery Rept.*, *12:* 392–399.

———, 1942. Larvae of decapod Crustacea. *Monogr. Ray Soc. London*, *129:* i–viii, 1–309.

Heegaard, P., 1969. Larvae of decapod Crustacea. The Amphionidae. *Dana Rept.*, *77:* 1–82.

Koeppel, E., 1902. Beiträge zur Kenntnis der Gattung *Amphion*. *Arch. Naturgesch.*, *68:* 262–298.

Milne Edwards, H., 1832. Note sur un nouveau genre de Crustacés de l'ordre des Stomapodes. *Ann. Soc. Ent. France*, *1:* 336–340.

———, 1837. *Histoire naturelle des Crustacés, comprenant l'anatomie, la physiologie et la classification de ces animaux*. *2:* 1–532.

Ortmann, A. E., 1893. Decapoden und Schizopoden. *Ergebn. Plankton Exped. Humboldt Stiftung*, *2*(Gb): 1–120.

Willemöes-Suhm, R. von, 1876. Preliminary remarks on the development of some pelagic decapods. *Proc. Roy. Soc. London*, *24:* 132–134.

Williamson, D. I., 1973. *Amphionides reynaudii* (H. Milne Edwards), representative of a proposed new order of eucaridan Malacostraca. *Crustaceana*, *25:* 35–50.

Zimmer, C., 1904. *Amphionides valdiviae* n. g., n. sp. *Zool. Anz.*, *28:* 225–228.

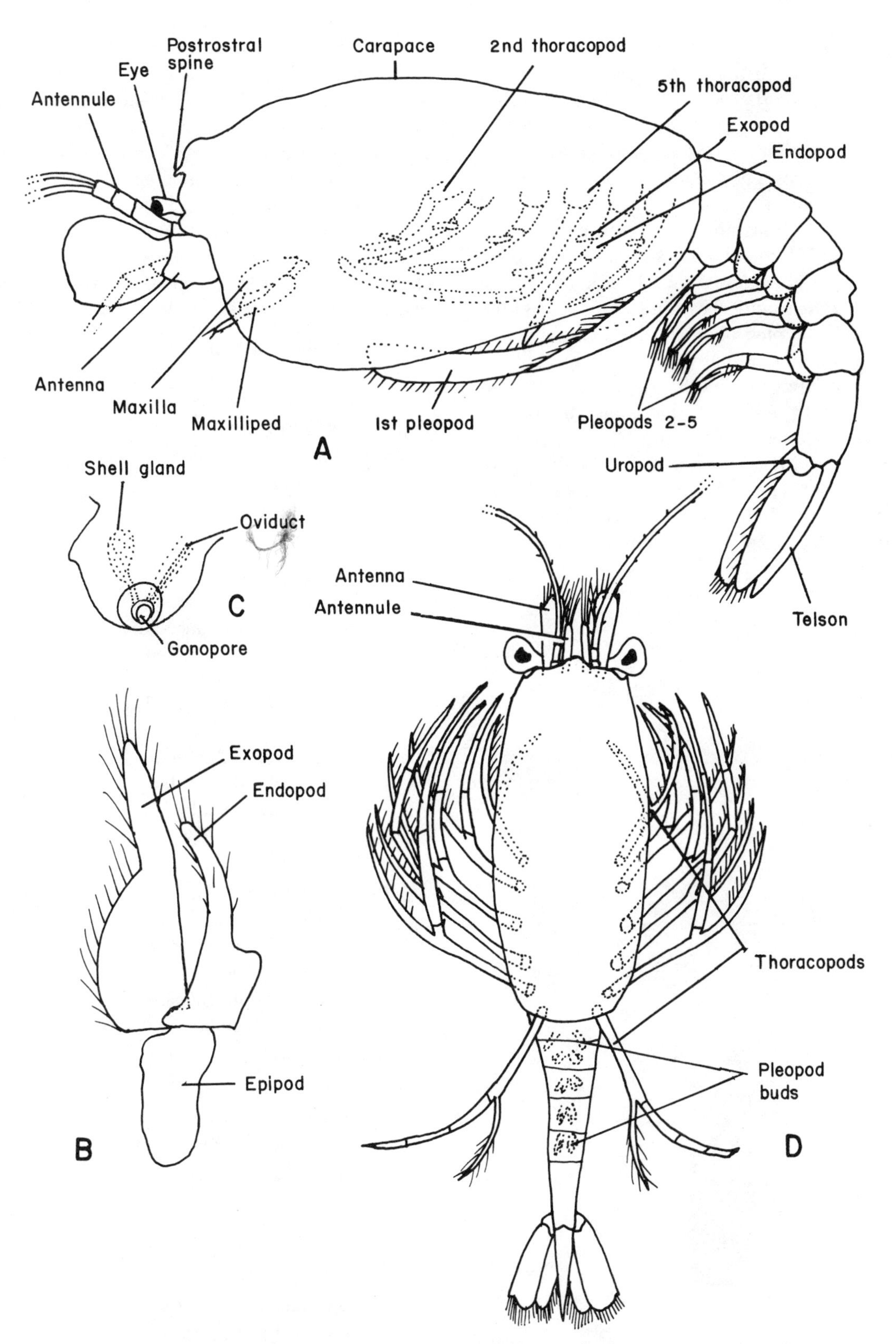

Figure 41 Amphionidacea: A. Diagrammatic *Amphionides*, adult female; B. Maxilliped; C. Base of 6th thoracopod showing gonopore, shell gland, and oviduct [A–C after Williamson, 1973]; D. Stage 2 larva [after Gurney, 1936].

ORDER DECAPODA Latreille, 1803

The Decapoda includes a great diversity of marine, freshwater, and semiterrestrial taxa characterized by the presence of a carapace developed to enclose the branchial chambers (branchiostegites) and by the modification of the 1st 3 pairs of thoracopods as maxillipeds; the remaining 5 pairs are typically walking legs or pereopods.

Within the Decapoda in general there is a basic gill pattern: a series of 4 gills are attached to each somite and/or appendage of the thorax; the gills have been named according to their positions. The gill attached to the lateral wall of the somite above the articulation of the appendage is called a pleurobranch; the 2 gills attached to the articular membrane between the body wall and the coxa are arthrobranchs; and the gill (a differentiation of the distal lobe of the epipod) that is inserted on the coxa is a podobranch. The proximal lobe of the epipod, when retained, often is referred to as a mastigobranch. There are 3 basic types of gills: dendrobranchiate, trichobranchiate, and phyllobranchiate. The phyllobranchiate gill (see Figure 42A) is the simplest in structure, consisting of an axis with blood vessels and a pair of flattened branches. The trichobranchiate gill (see Figure 42B) consists of a series of filamentous branches arranged in several series around the axis. In dendrobranchiate gills (see Figure 42C) the primary biserial branches are divided into arborescent bunches.

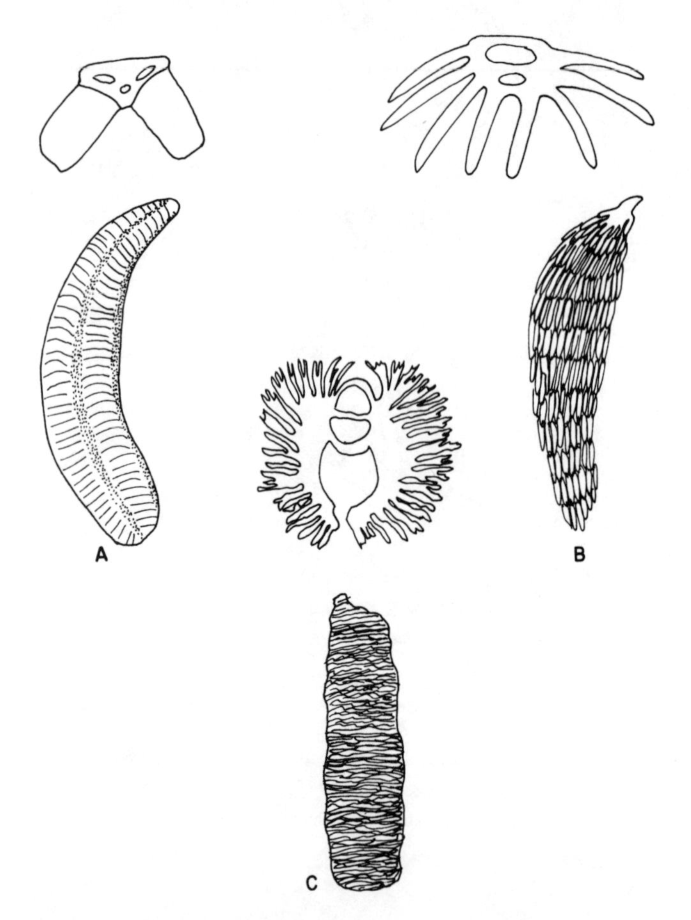

Figure 42 Gill structure of decapods: A. Phyllobranchiate, in transverse section (upper) and entire (lower); B. Trichobranchiate, in transverse section (upper) and entire (lower); C. Dendrobranchiate, in transverse section (upper) and entire (lower) [after Calman, 1909].

As previously indicated, discussions of the major decapod groups will be at unequal levels within the taxonomic hierarchy. For reference and orientation the classification, as used in the text, is given below; taxa marked with an asterisk are discussed in detail.

Suborder Dendrobranchiata
- Superfamily Penaeoidea
  - Family Penaeidae*
- Superfamily Sergestoidea
  - Family Sergestidae

Suborder Pleocyemata
- Infraorder Stenopodidea
  - Family Stenopodidae*
- Infraorder Caridea*
  - Superfamily Procaridoidea
  - Superfamily Oplophoroidea
  - Superfamily Bresilioidea
  - Superfamily Stylodactyloidea
  - Superfamily Pasiphaeoidea
  - Superfamily Palaemonoidea
  - Superfamily Psalidopodoidea
  - Superfamily Alpheoidea
  - Superfamily Pandaloidea
  - Superfamily Crangonoidea
  - Superfamily Physetocaridoidea
- Infraorder Astacidea
  - Superfamily Astacoidea*
  - Superfamily Parastacoidea
  - Superfamily Nephropoidea
    - Family Nephropidae
    - Family Thaumastochelidae
- Infraorder Austroastacidea
- Infraorder Palinura
  - Superfamily Glypheoidea
  - Superfamily Eryonoidea
  - Superfamily Palinuroidea
- Infraorder Anomura
  - Superfamily Thalassinoidea
  - Superfamily Coenobitoidea
  - Superfamily Paguroidea*
  - Superfamily Galatheoidea
  - Superfamily Hippoidea
- Infraorder Brachyura
  - Section Dromiacea
  - Superfamily Dromioidea
  - Superfamily Homoloidea

Section Oxystomata
- Superfamily Dorippoidea
- Superfamily Calappoidea
- Superfamily Raninoidea

Section Oxyrhyncha
- Superfamily Majoidea
- Superfamily Mimilambroidea
- Superfamily Parthenopoidea

Section Cancridea
- Superfamily Corystoidea

Section Brachyrhyncha
- Superfamily Portunoidea*
- Superfamily Xanthoidea
- Superfamily Bellioidea
- Superfamily Grapsidoidea
- Superfamily Pinnotheroidea
- Superfamily Potamoidea
- Superfamily Ocypodoidea
- Superfamily Hapalocarcinoidea

Classification from Bowman and Abele (personal communication); incomplete at family level.

SUBORDER DENDROBRANCHIATA Bate, 1888

Recent species	Approximately 350.
Size range	Up to 31 cm.
Carapace	Laterally compressed or cylindrical.
Eyes	Stalked, compound; rarely reduced.
Antennules	Biramous; with stylocerite.
Antennae	With 5-segmented peduncle and scaphocerite.
Mandibles	With palp.
Maxillulae	With segmented endopodal palp.
Maxillae	Biramous; with 2 bilobed endites.
Maxillipeds	Endopods of 1st with 5 segments; without crista dentata.
Thoracic appendages	First 3 pairs chelate; sometimes without exopods.
Abdominal appendages	Pleopods biramous; usually with petasma, sometimes also with appendix masculina; without appendix interna. Uropods biramous.
Telson	Together with broad uropods forms tailfan.
Tagmata	Cephalothorax and abdomen.
Somites	Head with 5 + 3 thoracic (maxillipeds); thorax with 5; abdomen with 6, excluding telson.
Sexual characters	Gonopores on coxae of 3rd pereopods (6th thoracopods) of female, 5th of males (8th thoracopods); male with petasma, female with thelycum; eggs shed into sea.
Sexes	Presumably separate.
Larval development	Metamorphic; nauplius → protozoea → zoea → postlarva.
Fossil record	Permotriassic to Recent.
Feeding types	Various.
Habitat	Marine, brackish, or freshwater; often benthic, sometimes burrowing in mud.
Distribution	Worldwide.

Although differences are present between the penaeoids (superfamily Penaeoidea) and sergestoids (Superfamily Sergestoidea), the instructions given for the study of the Dendrobranchiata have been made sufficiently general so they will be applicable to most of the taxa. The illustrations presented have been based on species of *Penaeus*.

Begin your study with an examination of the general morphology of the animal (see Figure 43). The carapace completely covers the cephalothorax and encloses the

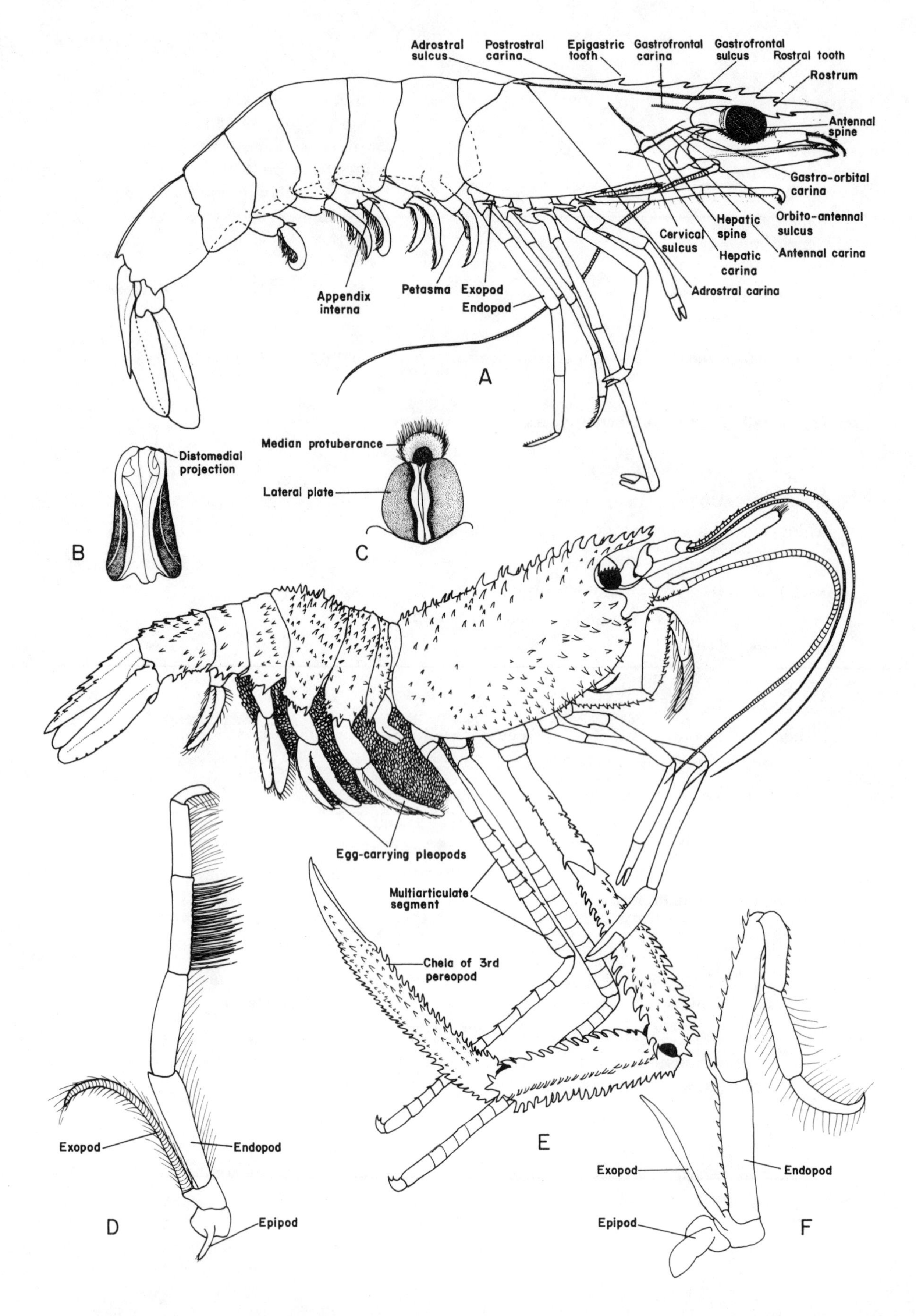

Figure 43 Decapoda: A—D, Dendrobranchiata; E—F, Stenopodidea. A. Typical penaeoidean (lateral view), showing carapace armature and ornamentation; B. Representative petasma of male; C. Representative thelycum of female; D. 3rd maxilliped; E. Typical stenopodidean (lateral view); F. 3rd maxilliped.

gills; it is fused dorsally to all the thoracic somites. The carapace usually is produced anteriorly to form a well-developed, elongate rostrum that is armed with dorsal and sometimes also ventral marginal spines. Posterior to the rostrum, in the dorsal midline, a postrostral carina or median ridge very frequently separates a pair of rostrolateral grooves. The carapace usually is provided with a hepatic spine and frequently with 1 or more lateral carapace grooves. The compound eyes are set on a pair of movable peduncles and usually are directed laterally during life.

The antennules each consist of a basal peduncle or protopod of 3 segments and a pair of flagella. Very frequently the anterolateral angle of the 1st peduncular segment is produced into a small acute spine, the stylocerite. Examine the dorsal surface of the 1st segment and locate the opening to the statocyst. This segment also contains a depression for the eye accompanied by a dense row of setae often called an eye brush or prosartema. The antennular flagella appear to serve a chemoreceptory function; frequently in males the median or inner flagellum is sexually modified through a dorsoventral flattening of the flagellum and the addition of a number of stout processes on the dorsal margin. It is presumed that these modifications aid the males in locating mature females. Each antenna is represented by a basal 2- or 3-segmented endopod, a scalelike exopod, and a long flagellum. It is customary to refer to the segments of the protopod and endopod together as the antennal peduncle and to the exopod as the scaphocerite or antennal acicle or scale. The antennal peduncle usually consists of a coxa, basis, and 3 endopodal segments. The flagellum is long, often 2 or 3 times the length of the body, straight in penaeoids, but with a bend in sergestoids.

The 1st pair of pereopods usually are short and often are used more for food manipulation than for locomotion. The first 3 pairs are chelate; the 4th and 5th may be subchelate or simple, well developed or, as in sergestoids, reduced. The pereopods usually consist of a 2-segmented protopod (coxa and basis) and 5-segmented endopod (usually ischium, merus, carpus, propodus, and dactyl). Exopods may be present or absent, but when present usually are quite small. The gills are dendrobranchiate; the gill formula typically is 19 + 6 mastigobranchs (7 pleurobranchs, 11 arthrobranchs, and 1 podobranch). Examine your particular specimen and list the number and types of gills that you find associated with each appendage.

As in other eucarideans the abdomen is well developed and elongate. Each tergite is laterally expanded as a latero-tergal plate or pleuron. Each successive pleuron is overlapped by the preceding one. The abdomen consists of 6 somites, each usually with a pair of biramous appendages, and a well-developed telson. The first 5 pairs of appendages are developed as pleopods, the primary swimming structures. In males, the endopods of the 1st pair are strongly modified to form a copulatory structure, the petasma. The medial margin of each half of the petasma is joined to its counterpart by a series of minute knobs. The structural details of the petasma frequently are used as diagnostic characters of species. In some genera the 2nd pair of pleopods also are modified. In these the endopods form the appendix masculina. Be sure to examine male specimens and study the details of these appendages. In females the sternal plates of the 7th and 8th thoracic somites develop a series of grooves and protuberances to form a sperm receptacle called the thelycum. The shape and configuration of the thelycum also is an important diagnostic character for species determination. The appendages of the 6th abdominal somite are the biramous uropods, which are well developed, and together with the telson form the tailfan.

As previously mentioned, the mouthparts of decapods include 3 pairs of maxillipeds. Remove and examine each of the maxillipeds from one side of the body. Identify the segments of the endopod. How does the 1st maxilliped compare with the 1st maxillipeds of the peracaridans you have examined? Be sure that you can distinguish endites from exites. Internal to the 1st maxilliped is the maxilla. Care should be taken when removing the maxilla, as the scaphognathite lays considerably laterad, in the incurrent water channel of the gills; it acts as a water pump for the gills. Removal of the maxilla will expose the maxillule. Between the maxillule and the mandible observe the fleshy paragnath. Remove the maxillule and then the mandible. Identify the incisor and molar processes, which often are separated only by a narrow groove. In some, however, a setal row also may be present. The mandibular palp is well developed.

The description of the major organ systems of the Dendrobranchiata is based on species of the Penaeidae (see Figure 44). Begin your examination by removing the carapace and the abdominal pleura from one side of the animal. To remove the carapace, which in decapods is fused to all the thoracic somites, make a shallow incision along the adrostral sulcus and gently free the carapace from the underlying membrane. Care is required in removing the carapace because the heart and pericardium are very close to the dorsal surface.

The penaeid body is quite muscular, and while we will not attempt a detailed description of the musculature, certain prominent muscles will have to be removed before the organ systems can be observed. Dorsolateral and lateral to the heart are the lateral thoracoabdominal muscles. Carefully remove enough of the muscle fibers to clearly observe the heart and its arteries. The heart receives blood from the pericardium through 3 pairs of ostia, 2 dorsal pairs and 1 lateral pair. From the anterior apical end of the heart a small unpaired artery extends a short distance; this may be the vestige of the anterior aorta of lower malacostracans. Anteriorly from the heart

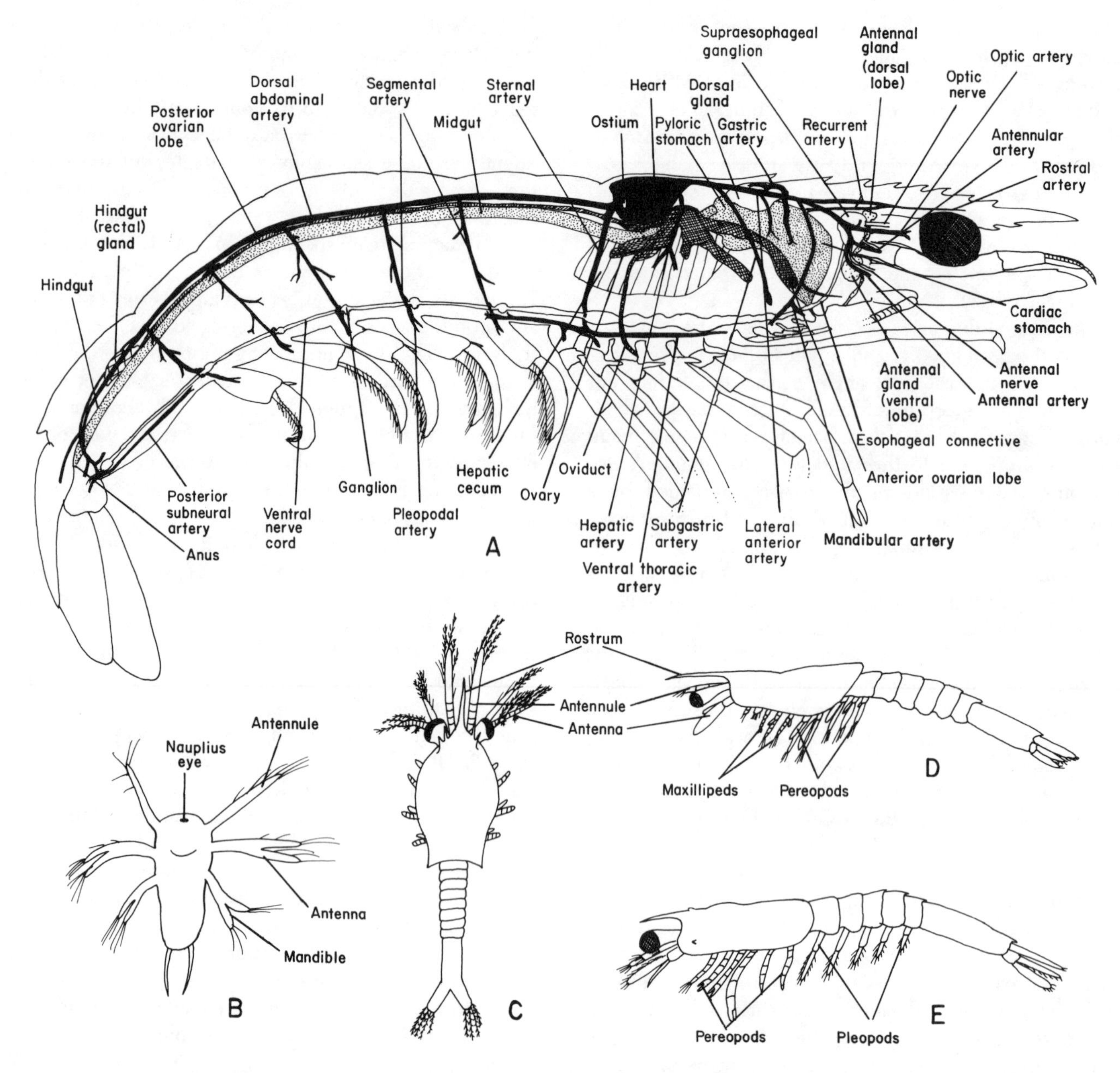

Figure 44 Decapoda, Dendrobranchiata: A. Diagrammatic penaeoidean with musculature removed to show major organ systems. B. Nauplius I; C. Zoea I; D. Mysis I; E. Postlarva I.

trace one of the pair of lateral anterior arteries and its several branches. The first to branch off is the subgastric artery that provides blood to the glandlike organ beneath the cardiac stomach. A short distance anteriorly locate the gastric artery, which serves the dorsal and ventral regions of the cardiac stomach, and the dorsal gland (a structure of unknown function). Further rostrad the lateral anterior artery turns toward the midline and joins its counterpart from the opposite side to form a median longitudinal vessel. At the point of fusion, or slightly before, a pair of optic arteries proceed to the ocular peduncles, giving off the cephalic arteries to the supraesophageal ganglion enroute. The anterior part of the lateral anterior artery continues into the rostrum as the rostral artery and the posterior part turns posteriorly (as the recurrent artery) to the cardiac stomach. Other branches of the lateral anterior arteries are the mandibular and antennal arteries that supply blood to the muscles of the mandibles, antennules, and antennae. Returning to the heart, identify the pair of hepatic arteries that also leave the heart anteriorly. These vessels provide the blood supply to the hepatic cecum or hepatopancreas.

From the posterior end of the heart locate the large, unpaired posterior abdominal artery. Almost immediately the sternal artery is given off; in some malacostracans this artery arises from the heart itself.

Follow the sternal artery ventrally; it passes through the ventral nerve cord between the ganglia of the 3rd and 4th pereopods. Beneath the nerve cord the sternal artery divides into the two branches of the ventral thoracic artery. The anterior branch provides blood to the anterior thoracic appendages and to the nerve cord; the posterior branch provides large vessels to the 4th and 5th pereopods and then proceeds into the abdomen. Return to the posterior abdominal artery and trace its course posteriorly between the dorsal abdominal muscles. At the level of each somite it gives off a pair of segmental arteries that supply blood to the musculature, pleopods, and ventral nerve cord. Posteriorly the dorsal abdominal artery bifurcates around the hindgut or rectal gland and proceeds along the hindgut as a paired vessel, fusing again near the end of the hindgut. Branches are provided to the telson, uropods, and muscles of the region before the fused artery turns anteriorly beneath the ventral nerve cord as the posterior subneural artery. This latter artery appears to extend only into the 6th abdominal somite. Blood is returned to the pericardium and heart through a series of sinuses in the appendages and hemocoel.

The reproductive system of the male consists of a pair of testes lying on the dorsal surface of the hepatic cecum (hepatopancreas) ventral to the heart. Each testis is comprised of several lobes, frequently extending over the surface of the cecum. From the posterior end of the testis the vas deferens extends ventrally to a terminal ampoule situated above the male gonopore on the sternum of the 8th thoracic somite. This ampoule is primarily a glandular structure that secretes the spermatophore. In the female the ovaries also are paired. Usually an anterior projection of the ovary can be traced along the cardiac stomach, and several lobes can be distinguished on the dorsal surface of the hepatic cecum. An additional lobe, the dorsal ovarian lobe, should be located posterior to the heart; trace this lobe into the abdomen dorsolateral to the midgut. Locate the oviduct and follow it to the female gonopore on the coxa of the 3rd pereopod.

The penaeid digestive system consists of a muscular esophagus, prominent stomach with well-defined cardiac and pyloric sections, midgut, hindgut, and digestive gland. The cardiac stomach contains the grinding part of the stomach, the gastric mill, and this structure should be examined when the stomach is removed. The primary digestive gland is the hepatic cecum, also referred to by some as the hepatopancreas. It is a large gland situated in the posterior region of the thorax beneath and somewhat anterior to the heart, consisting of a mass of closely packed tubules. Juices from the gland enter the pyloric stomach ventrally near its junction with the midgut. In penaeids the midgut is elongate, extending nearly the entire length of the abdomen. Near the beginning of the hindgut, a small hindgut or rectal gland arises. The function of this gland is not known, but a role in osmotic regulation has been proposed for it. The hindgut terminates in a muscular anus. Returning to the stomach, remove it to reveal the nervous system and ventral blood vessels. Be sure to examine the gastric mill, which in penaeids is relatively simple; it consists of a strong median tooth and series of lateral denticles. Excretion is via antennal glands, but the glands themselves are diffused and not easily recognized. Externally the excretory pore can be identified.

Anteriorly in the cephalon beneath the rostrum locate the supraesophageal ganglion. Identify the large optic, antennal, and antennular nerves leaving the ganglion. The smaller oculomotor and tegumental nerves may not be as easily distinguished. Trace the esophageal connectives around the esophagus to the ventral nerve cord. The ganglia of the mouthparts are conspicuously fused; those of the first 3 pereopods are indicated by segmental swellings. The nerve cord is divided into 2 longitudinal tracts between the 3rd and the coalesced 4th and 5th ganglia to allow for passage of the sternal artery. Trace the nerve cord into the abdomen; each of the anterior 5 ganglia gives off a pair of nerves, one to the musculature and one to the pleopod on each side of the body.

Larval development in the Dendrobranchiata is metamorphic and has been described best for commercial species of *Penaeus*. All penaeid species have 3 distinct larval stages and numerous substages. The principal stages are the naupliar, zoeal, and mysis stages followed by metamorphosis to postlarvae. The naupliar stage usually consists of 5 or 6 substages during which the antennules, antennae, and mandibles are the only well-developed appendages. During the 3 zoeal substages the maxillulae and maxillae become functional, stalked eyes develop, and abdominal segmentation appears. During the 3 mysis substages the body takes on a shrimplike appearance, the pereopods become well developed and functional, and toward the conclusion to this stage the pleopods develop. Postlarvae may be recognized by the appearance of chelate first 3 pairs of pereopods; locomotion is a function of the pleopods.

References

Alcock, A., 1901. *A descriptive catalogue of the Indian deep-sea Crustacea Decapoda Macrura and Anomala, in the Indian Museum.* 286 pp. Calcutta: Indian Museum.

Anderson, W. W., and M. J. Linder, 1945. A provisional key to the shrimps of the family Penaeidae with especial reference to American forms. *Trans. Am. Fish. Soc.*, *73:* 284–319.

Balss, H., 1925. Macrura der Deutschen Tiefsee Expedition, 2. Natantia. *Wiss. Ergebn. Deutsch. Tiefsee Exped. Valdivia*, *20:* 215–315.

Bate, C. S., 1881. On the Penaeidae (Crustacea Decapoda). *Ann. Mag. Nat. Hist.*, (5) *8:* 169–196.

Bate, C. S., 1888. Report on the Crustacea Macrura collected by H.M.S. "Challenger" during 1873–1876. *Rep. Voy. Challenger, Zool.*, *24:* i–xc, 1–942.

Boschi, E. E., 1963. Los camarones comerciales de la familia Penaeidae de la costa atlántica de América del Sur. Clave para el reconocimiento de las especies y datos bioecológicos. *Bol. Inst. Biol. Mar., Mar del Plata*, *3:* 1–39.

Boschi, E. E., and V. Angelescu, 1962. Descripción de la morfología externa e interna del langostino con algunas aplicaciones de índole taxonómica y biológica. *Hymenopenaeus mulleri* (Bate) Crustacea, fam. Penaeidae. *Bol. Inst. Biol. Mar., Mar del Plata*, *1:* 1–73.

Bouvier, E. L., 1908. Crustacés décapodes (Pénéidés) provenant des campagnes de l'"Hirondelle" et de la "Princesse-Alice" (1886–1907). *Résult. Camp. Sci. Monaco*, 33: 1–122.

Burkenroad, M. D., 1934. The Penaeidea of Louisiana with a discussion of their world relationships. *Bull. Am. Mus. Nat. Hist.*, *68:* 61–143.

———, 1936. The Aristaeinae, Solenocerinae and pelagic Penaeinae of the Bingham Oceanographic collection. *Bull. Bingham Oceanogr. Collect.*, *5:* 1–151.

———, 1939. Further observations on Penaeidae of the northern Gulf of Mexico. *Bull. Bingham Oceanogr. Collect.*, *6:* 1–62.

———, 1963. The evolution of the Eucarida (Crustacea, Eumalacostraca), in relation to the fossil record. *Tulane Stud. Geol.*, *2:* 3–16.

Eldred, B., and R. F. Hutton, 1960. On the grading and identification of domestic commercial shrimps (family Penaeidae) with a tentative world list of commercial penaeids. *Quart. J. Fla. Acad. Sci.*, *23:* 89–118.

Glaessner, M. F., 1969. Decapoda. In R. C. Moore (ed.), *Treatise on invertebrate paleontology,* Part R Arthropoda 4, *2:* R399–R566. Lawrence, Kans.: Geol. Soc. America and Univ. Kansas.

Ivanov, B. G., and A. M. Hassan, 1976a. Penaeid shrimps (Decapoda, Penaeidae) collected off east Africa by the fishing vessel "Van Gogh," 1. *Solenocera ramadani* sp. nov., and commercial species of the genera *Penaeus* and *Metapenaeus*. *Crustaceana*, *30:* 241–251.

———, 1976b. Penaeid shrimps (Decapoda, Penaeidae) collected off east Africa by the fishing vessel "Van Gogh", 2. Deep-water shrimps of the genera *Penaeopsis* and *Parapenaeus* with description of *Penaeopsis balssi* sp. nov. *Crustaceana*, *31:* 1–10.

Kensley, B. F., 1968. Deep sea decapod Crustacea from west of Cape Point, South Africa. *Ann. S. Afr. Mus.*, *50:* 283–323.

Kubo, I., 1949. Studies on penaeids of Japan and its adjacent waters. *J. Tokyo Coll. Fish.*, *36:* 1–467.

Omori, M., 1975. The systematics, biogeography and fishery of epipelagic shrimps of the genus *Acetes* (Decapoda, Sergestidae). *Bull. Ocean Res. Inst., Univ. Tokyo*, no. 7: 1–91.

Pequegnat, W. E., and T. W. Roberts, 1971. Decapod shrimps of the family Penaeidae. In W. E. Pequegnat, L. H. Pequegnat, R. W. Firth, Jr., B. M. James, and T. W. Roberts, Gulf of Mexico deep-sea fauna Decapoda and Euphasuiacea, pp. 8–9. *Ser. Atlas Mar. Environ. Am. Geogr. Soc.*, Folio 20.

Pérez Farfante, I., 1969. Western Atlantic shrimps of the genus *Penaeus*. *Fish. Bull.*, *67:* 461–591.

———, 1971. Western Atlantic shrimps of the genus *Metapenaeopsis* (Crustacea, Decapoda, Penaeidae), with descriptions of three new species. *Smiths. Contr. Zool.*, 79: 1–37.

———, 1975. Spermatophores and thelyca of the American white shrimps, genus *Penaeus* subgenus *Litopenaeus*. *Fish. Bull.*, *73:* 463–468.

———, 1977. American solenocerid shrimps of the genera *Hymenopenaeus*, *Haliporoides*, *Pleoticus*, *Hadropenaeus* new genus and *Mesopenaeus* new genus. *Fish. Bull.*, *75:* 261–364.

Roberts, T. W., and W. E. Pequegnat, 1970. Deep-water decapod shrimps of the family Penaeidae. In W. E. Pequegnat and F. A. Chace, Jr. (eds.), *Contributions on the biology of the Gulf of Mexico*. Texas A. M. Univ. Oceanogr. Stud., *1:* 21–57. Houston, Texas: Gulf Publ. Co.

Voss, G. L., 1955. A key to the commercial and potentially commercial shrimp of the family Penaeidae of the western North Atlantic and the Gulf of Mexico. *Fla. St. Bd. Conserv. Tech.* Ser. 14: 1–22.

Williams, A. B., 1965. Marine decapod crustaceans of the Carolinas. *Fish. Bull.*, *65:* 1–298.

Young, J. H., 1959. Morphology of the white shrimp *Penaeus setiferus* (Linnaeus, 1758). *Fish. Bull.*, *59:* 1–168.

SUBORDER PLEOCYEMATA Burkenroad, 1963

INFRAORDER STENOPODIDEA Huxley, 1879

Recent species	Approximately 30.
Size	2–7 or 8 cm.
Carapace	Usually cylindrical.
Eyes	Stalked, compound.
Antennules	Biramous.
Antennae	With 5-segmented peduncle and scaphocerite.
Mandibles	With palp.
Maxillulae	With endopodal palp.

Maxillae	Biramous; with 2 bilobed endites.
Maxillipeds	Endopod of 1st with 2 segments; without crista dentata.
Thoracic appendages	First 3 pairs chelate; without exopods.
Abdominal appendages	Pleopods biramous; usually with appendix masculina; without appendix interna. Uropods biramous.
Telson	Together with broad uropods forming tailfan.
Tagmata	Cephalothorax and abdomen.
Somites	Head with 5 + 3 thoracic (maxillipeds); thorax with 5; abdomen with 6, excluding telson.
Sexual characters	Gonopores on coxae of 3rd pereopods of female; on 5th of male; male usually with appendix masculina; eggs carried on female pleopods.
Sexes	Presumably separate.
Larval development	Metamorphic; protozoea → zoea → postlarva.
Fossil record	Recent.
Feeding types	Usually particulate.
Habitat	Marine; frequently commensal.
Distribution	Tropical or subtropical.

Of the infraorders of the Pleocyemata, the Stenopodidea is by far the smallest, with only a little more than three dozen species described. It is probably best known by one of its more conspicuous members, *Stenopus hispidus* Olivier, the largest of the cleaner shrimp. Stenopodideans share a number of characters in common with the Dendrobranchiata, but are separated immediately by their trichobranchiate gills and by the massive, elongate 3rd pereopods (see Figure 43). If specimens are available, compare the general morphology of the stenopodideans with that of the dendrobranchs. Note particularly the development of the rostrum and scaphocerite and the armature of the cephalothorax and abdomen. Multiarticulate carpi and propodi of the 4th and 5th pereopods are characteristic of the genus *Stenopus*.

Larvae hatch as inert pre- or protozoeae, but within hours molt into the 1st zoeal substage. At this stage the rostrum is well developed; the carapace lacks spines but several abdominal spines are present. All cephalic appendages are present, as are the 3 pairs of maxillipeds and 1st pereopods. Six zoeal substages or instars are followed by metamorphosis to the megalopa or postlarva, in which the pleopods are used for locomotion.

References

Bruce, A. J., and K. Baba, 1973. *Spongiocaris*, a new genus of stenopodidean shrimp from New Zealand and South African waters, with a description of two new species (Decapoda Natantia, Stenopodidea). *Crustaceana, 25:* 153–170.

Gurney, R., 1936. Larvae of decapod Crustacea. I. Stenopidea. *Discovery Rept., 12:* 379–392.

Haan, W., de, 1833–1850. Crustacea. In P. F. von Siebold, *Fauna Japonica:* i–xvii, i–xxx, 1–244.

Hansen, H. J., 1908. Crustacea Malacostraca, 1. *Danish Ingolf-Exped., 3:* 1–120.

Holthuis, L. B., 1946. The Stenopodidae, Nephropsidae, Scyllaridae and Palinuridae. Pt. 1. In The Decapoda Macrura of the "Snellius" Expedition. *Temminckia, 7:* 1–178.

———, 1955. The Recent genera of the caridean and stenopodidean shrimps (Class Crustacea: Order Decapoda: Supersection Natantia) with keys for their determination. *Zool. Verhandl.*, no. 26: 1–157.

Schmitt, W. L., 1965. *Crustaceans.* 204 pp. Ann Arbor: University of Michigan Press.

INFRAORDER CARIDEA Dana, 1852

Recent species	Approximately 1600.
Size range	Up to 32 cm.
Carapace	Cylindrical, laterally compressed, or slightly compressed dorsoventrally.

Eyes	Stalked, compound; occasionally reduced or absent.
Antennules	Bi- or triramous; with stylocerite.
Antennae	With 5-segmented peduncle and scaphocerite.
Mandibles	With or without palp.
Maxillulae	With endopodal palp.
Maxillae	Biramous; usually with bilobed distal and entire proximal endites.
Maxillipeds	Endopod of 1st with 2 or fewer segments; without crista dentata.
Thoracic appendages	First 2 pairs usually chelate; sometimes without exopods.
Abdominal appendages	Pleopods usually biramous; often with appendix masculina; usually also with appendix interna. Uropods biramous.
Telson	Together with broad uropods forms tailfan.
Tagmata	Cephalothorax and abdomen.
Somites	Head with 5 + 3 thoracic (maxillipeds); thorax with 5; abdomen with 6, excluding telson.
Sexual characters	Gonopores on coxae of 3rd pereopods of female, 5th of male; male often with appendix masculina; eggs carried on pleopods of female.
Sexes	Separate; protandry and protogyny occur.
Larval development	Metamorphic; protozoea → zoea → megalopa; sometimes abbreviated.
Fossil record	M. Jurassic to Recent.
Feeding types	Various.
Habitat	Marine, brackish, or freshwater; planktonic or benthic; sometimes commensal; rarely semiterrestrial.
Distribution	Worldwide.

Although the Caridea resemble the Penaeoidea, Sergestoidea, and Stenopodidea in general body form, two important characters serve immediately to distinguish caridean taxa (see Figure 45). Examine the pleura of the first 3 abdominal somites. In contrast to the condition observed in penaeoids, sergestoids, and stenopodids, the 2nd abdominal pleura overlap the pleura of both the 1st and the 3rd somites in all carideans. The other important difference is the lack of terminal chelae on the 3rd pereopods.

The carapace of carideans is subdivided into regions (see Figure 45B), and for ease in understanding and working with descriptions of the numerous and diverse caridean taxa, you should commit these regions to memory. The basic subdivisions of the cephalothorax can be recognized through the number of changes in body shape that occur within the Decapoda. Note the number of carapace spines, not present in penaeoids, that occur in the various caridean taxa. The presence or absence of particular spines is of generic and specific import. The caridean rostrum is extremely variable in both length and armature, and in some taxa it even is articulated with, rather than fused to, the carapace. In the Alpheidae the carapace forms a broad platelike covering over the eyes, thus restricting their movement; the rostrum, when present, is a small spine between the orbital hoods.

As in dendrobranchs and stenopodids, the antennules consist of a 3-segmented peduncle and a pair of flagella; however, in some caridean taxa the lateral (outer) flagellum is subdivided distally for part of its length, causing the impression that there are 3 flagella. A stylocerite usually is present. The antennal peduncle consists of a 2-segmented protopod and a 3-segmented endopod. The scaphocerite (exopod) usually is made up of a sharp spinelike process and a broad blade; however, one or the

Figure 45 Decapoda, Caridea: A. Typical caridean (lateral view); B. Cephalothorax (lateral view) showing carapace regions and armature; C. Mandible; D. Maxillule; E. Maxilla; F. 1st maxilliped; G. 2nd maxilliped; H. 3rd maxilliped; I. Chela with pectinate fingers; J. 3rd pereopod.

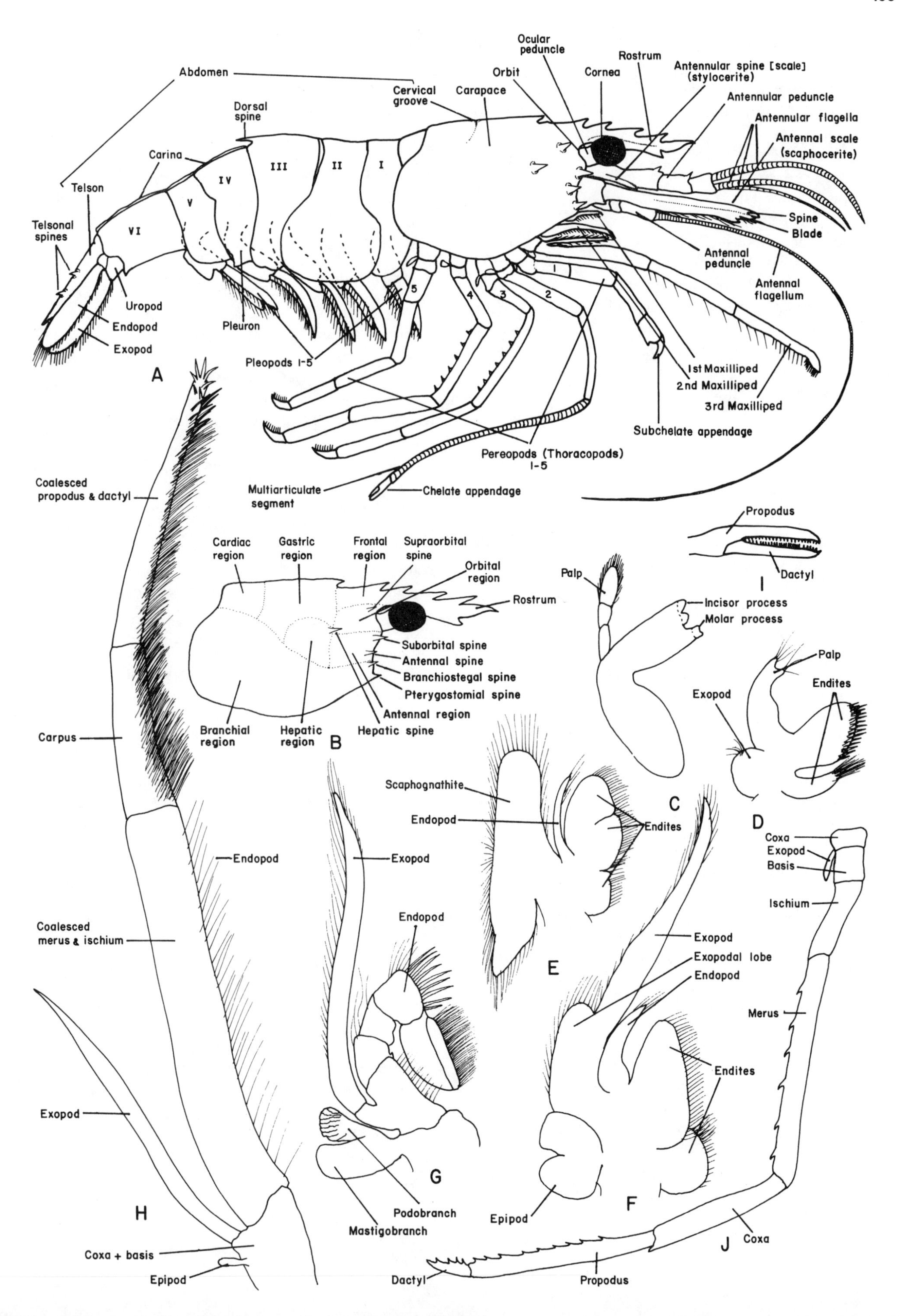
Abdomen
Cervical groove
Carapace
Ocular peduncle
Orbit
Cornea
Rostrum
Antennular spine [scale] (stylocerite)
Antennular peduncle
Antennular flagella
Antennal scale (scaphocerite)
Dorsal spine
Carina
Telson
Telsonal spines
VI
V
IV
III
II
I
Spine
Blade
Antennal peduncle
Antennal flagellum
Uropod
Endopod
Exopod
Pleuron
Pleopods 1-5
1st Maxilliped
2nd Maxilliped
3rd Maxilliped
Subchelate appendage
Pereopods (Thoracopods) 1-5
Multiarticulate segment
Chelate appendage
A
Coalesced propodus & dactyl
Cardiac region
Gastric region
Frontal region
Supraorbital spine
Orbital region
Rostrum
Suborbital spine
Antennal spine
Branchiostegal spine
Pterygostomial spine
Antennal region
Hepatic spine
Branchial region
Hepatic region
B
Propodus
Dactyl
I
Palp
Incisor process
Molar process
C
Palp
Endites
Exopod
D
Carpus
Scaphognathite
Endopod
Endites
E
Endopod
Exopod
Coalesced merus & ischium
Endopod
Exopod
Exopodal lobe
Endopod
Endites
F
Epipod
Coxa
Exopod
Basis
Ischium
Merus
Exopod
G
Podobranch
Mastigobranch
H
Coxa + basis
Epipod
J
Coxa
Dactyl
Propodus

other, or occasionally both, may be absent. The antennal flagellum usually is very long.

The 5 pairs of pereopods generally consist of 2 chelate pairs (1st and 2nd) and 3 simple; however, there are many variations. The Crangonoidea, for example, all have subchelate 1st pereopods; in the Pandalidae, if chelae are present on the 1st pereopods, they are microscopic in size. In contrast, the first 2 pairs in the Alpheidae are considerably larger than the other pereopods, and one usually is massive. Examine the first 2 pairs of pereopods of your specimens and identify the individual segments. Depending upon the taxon, the carpus of the 2nd pereopod consists of a single article or may be subdivided into 2 to many articles. The cutting edges of the dactyl and propodal finger of the chela also are diagnostic in certain taxa. In your specimens are the edges pectinate? Exopods may be present or absent, depending upon the taxon. The gill formulae among carideans is extremely variable. In some species of *Pandalus*, for example, the formula is 12 + 7 mastigobranchs, whereas in *Hippolyte* spp. it is 5 + 2 mastigobranchs. Determine the gill formulae for your specimens.

The abdomens of carideans frequently are armed with spines or sometimes developed into carinae, and the telsons frequently carry several small spines. Observe the abdominal armature and ornamentation of your specimens. Examine the first 2 pairs of pleopods of a male specimen. A petasma, such as occurs in penaeoids, is absent; however, an appendix masculina often is developed on the 2nd pair. In both sexes the pleopods of each pair usually are coupled by an appendix interna (a pair of outgrowths from the endopods, provided with a series of small coupling hooks, or retinacula) enabling them to stroke simultaneously.

Remove the mouthparts, examine each, and compare them with the mouthparts of the penaeoids examined previously. Frequently the number of segments of the maxillipeds is reduced by the coalescence of 2 or more segments. In some taxa the terminal segment of the 2nd maxilliped articulates with the external margin of the penultimate segment rather than with the distal margin. The endopod of the 1st maxilliped is short; the exopod frequently has an epipod. The maxilla has a well-developed scaphognathite and endites. The maxillule has a pair of endites and an endopodal palp. The mandible may or may not have a palp.

Internally the organ systems are similar to those described for the Penaeidae; however, your attention is called to certain differences. In some caridean hearts 5 pairs of ostia are present. A median ophthalmic artery (or anterior aorta) also may be present and may be provided with a cor frontale. The lengths of the midgut and hindgut vary among taxa. The diverticula of the hepatic cecum usually are much longer in carideans than in penaeids. Excretion in carideans is via the antennal glands; however, the length of the excretory canal varies considerably.

Larval development in the Caridea also is metamorphic, but more abbreviated than in the Penaeoidea. The naupliar stage is passed through in the egg and the young hatch as zoeae. One to several zoeal substages usually are followed by the mysid stage and then metamorphosis to postlarval form. The degree of development in each of the stages is comparable with that of the Penaeidae.

References

Abele, L. G., 1970. Semi-terrestrial shrimp *(Merguia rhizophorae). Nature, London, 226:* 661–662.

Allen, J. A., 1959. On the biology of *Pandalus borealis* Krøyer, with reference to a population of the Northumberland coast. *J. Mar. Biol. Assoc. U.K., 38:* 189–220.

Banner, A. H., 1953. The Crangonidae, or snapping shrimp, of Hawaii. *Pac. Sci., 7:* 1–144.

Banner, A. H., and D. M. Banner, 1966. The alpheid shrimp of Thailand: The alpheid shrimp of the Gulf of Thailand and adjacent waters. *Siam Soc. Mongr. Ser., 3:* 1–168.

Bouvier, E. L., 1940. Décapodes marcheurs. In *Faune de France, 37:* 1–399. Paris: Paul Lechevalier.

Bowman, T. E., and J. C. McCain, 1967. Distribution of the planktonic shrimp, *Lucifer,* in the western Atlantic. *Bull. Mar. Sci., 17:* 660–671.

Chace, F. A., Jr., 1937. Caridean decapod Crustacea from the Gulf of California and the west coast of Lower California. Pt. 7, in: The Templeton Crocker Expedition. *Zoologica, 22:* 109–138.

———, 1940. The bathypelagic caridean Crustacea. Pt. 9, in: Plankton of the Bermuda oceanographic expeditions. *Zoologica, 25:* 117–209.

———, 1972. The shrimps of the Smithsonian-Bredin Caribbean expeditions with a summary of the West Indian shallow-water species (Crustacea: Decapoda: Natantia). *Smiths. Contr. Zool.*, no. 98: 1–170.

———, 1976. Shrimps of the pasiphaeid genus *Leptochela* with descriptions of three new species (Crustacea: Decapoda: Caridea). *Smiths. Contr. Zool.*, no. 222: 1–51.

Chace, F. A., Jr., and H. H. Hobbs, Jr., 1969. The freshwater and terrestrial decapod crustaceans of the West Indies with special reference to Dominica. *Bull. U.S. Natl. Mus.*, 292: 1–258.

Chace, F. A., Jr., and R. B. Manning, 1972. Two new caridean shrimps, one representing a new family, from marine pools on Ascension Island (Crustacea: Decapoda: Natantia). *Smiths. Contr. Zool.*, no. 131: 1–18.

Forster, G. R., 1951. Biology of the common prawn *Leander serratus. J. Mar. Biol. Assoc. U.K., 30:* 333–360.

Glaessner, M. F., 1969. Decapoda. In R. C. Moore (ed.), *Treatise on invertebrate paleontology,* Part R Arthropoda 4, 2: R399–R566. Lawrence, Kans.: Geol. Soc. America and Univ. Kansas.

Gurney, R., 1942. *Larvae of decapod Crustacea. 129:* 1–306. London: Ray Soc.

Hayashi, K., 1975. The Indo-West Pacific Processidae (Crustacea, Decapoda, Caridea). *J. Shimonoseki Univ. Fish.*, *24:* 47–145.

Hobbs, H. H., Jr., H. H. Hobbs III, and M. A. Daniel, 1977. A review of the troglobitic decapod crustaceans of the Americas. *Smiths. Contr. Zool.*, no. 244: 1–183.

Hobbs, H. H. III, and H. H. Hobbs, Jr., 1976. On the troglobitic shrimps of the Yucatan Peninsula, Mexico (Decapoda: Atyidae and Palaemonidae). *Smiths. Contr. Zool.*, no. 240: 1–23.

Holthuis, L. B., 1946. The Stenopodidae, Nephropsidae, Scyllaridae and Palinuridae. Pt. 1, in: The Decapoda Macrura of the "Snellius" Expedition. *Temminckia*, *7:* 1–178.

———, 1947. The Hippolytidae and Rhynchocinetidae collected by the "Siboga" and "Snellius" Expeditions with remarks on other species. Pt. 9, in: The Decapoda of the "Siboga" Expedition. *Siboga-Expeditie*, *39a:* 1–100.

———, 1951a. The caridean Crustacea of tropical West Africa. *Atlantide Rept.*, *2:* 7–187.

———, 1951b. The subfamilies Euryrhynchinae and Pontoninae. Pt. 1, in: A general revision of the Palaemonidae (Crustacea Decapoda Natantia) of the Americas. *Allan Hancock Found. Occ. Pap.*, no. 11: 1–332.

———, 1952. The subfamily Palaemoninae. Pt. 2. A general revision of the Palaeomonidae (Crustacea Decapoda Natantia) in the Americas. *Allan Hancock Found. Occ. Pap.*, no. 12: 1–396.

———, 1955. The Recent genera of the caridean and stenopodidean shrimps (Class Crustacea: Order Decapoda: Supersection Natantia) with keys for their determination. *Zool. Verhandl.*, no. 26: 1–157.

———, 1959. The Crustacea Decapoda of Suriname (Dutch Guiana). *Zool. Verhandl.*, no. 44: 1–296.

Limbaugh, C., H. Pederson and F. A. Chace, Jr., 1961. Shrimps that clean fishes. *Bull. Mar. Sci. Gulf Carib.*, *11:* 237–257.

Manning, R. B., and F. A. Chace, Jr., 1971. Shrimps of the family Processidae (Crustacea, Decapoda, Caridea) from the northwestern Atlantic. *Smiths. Contr. Zool.*, no. 89: 1–41.

Makarov, R. R., 1975. Larvae of decapod Crustacea of the International Indian Ocean Expedition collections. Family Crangonidae (Decapoda: Caridea). *Indian J. Mar. Sci.*, *4:* 68–76.

Schmitt, W. L., 1921. *The marine decapod Crustacea of California*. 470 pp. Berkeley: Univ. Calif. Press.

———, 1935. Crustacea Macrura and Anomura of Porto Rico and the Virgin Islands. *Scientific Survey of Porto Rico and the Virgin Islands*, Pt. 2, *15:* 125–227.

———, 1936. Macruran and anomuran Crustacea from Bonaire, Curaçao and Aruba. *Zool. Jahrb.*, *67:* 363–378.

Williams, A. B., 1965. Marine decapod crustaceans of the Carolinas. *Fish. Bull.*, *65:* 1–298.

INFRAORDER ASTACIDEA Latreille, 1803

Recent species	Approximately 750.
Size range	Up to 64.3 cm.
Carapace	Usually subcylindrical or subovoid.
Eyes	Stalked, compound; occasionally reduced or absent.
Antennules	Biramous; peduncle with 3 segments.
Antennae	Peduncle with 5 segments; with or without scaphocerite.
Mandibles	Usually with palp.
Maxillulae	With endopodal palp.
Maxillae	Biramous; with 2 bilobed endites.
Maxillipeds	Endopod of 1st with 2 or fewer segments; usually with crista dentata.
Thoracic appendages	Uniramous; first 3 pairs chelate, 5th often chelate.
Abdominal appendages	Pleopods uni- or biramous; often with appendix masculina or other copulatory structures in males. Uropods biramous.
Telson	Together with broad uropods forms tailfan.
Tagmata	Cephalothorax and abdomen.
Somites	Head with 5 + 3 thoracic (maxillipeds); thorax with 5; abdomen with 6, excluding telson.
Sexual characters	Gonopores on coxae of 3rd pereopods of female, on 5th of male; male often with appendix masculina or other copulatory structures; female often with annulus ventralis.
Sexes	Presumably separate.

Larval development	Metamorphic or epimorphic.
Fossil record	Permotriassic to Recent.
Feeding types	Various; often on plant material in freshwater species.
Habitat	Marine, brackish, or freshwater; usually benthic.
Distribution	Worldwide, except for Africa and parts of Asia.

This infraorder is subdivided into three superfamilies; the Astacoidea, containing the freshwater crayfishes of the northern hemisphere; the Parastacoidea, the freshwater crayfishes of the southern hemisphere; and the Nephropoidea, the marine lobsters. Several species of North American crayfishes are classical laboratory study animals, so you probably have examined specimens in general invertebrate courses. Your present study of the crayfish and marine lobsters, if specimens are available, should include not only the examination of the morphology of this group but also a comparison with similar structures of the previously studied decapods.

Most crayfish are relatively unspecialized morphologically, although their successful adaptation to fresh water has been accompanied by some physiological specializations and adaptations. The cephalothorax typically is divided into anterior and posterior regions by the cervical groove (see Figure 46). The rostrum is a large, broad, and dorsoventrally flattened structure that is flanked posterolaterally by a pair of postorbital ridges. The branchiostegites are delineated laterally. The anterior carapace frequently is armed with small orbital or hepatic spines. The posterior carapace is marked dorsally by a pair of branchiocardiac grooves; the area between the grooves usually is referred to as the areola. If specimens of nephropoideans are available, compare the carapaces of the two taxa. Characteristically the nephropoidean carapace is marked by numerous grooves, ridges, spines, teeth, and tubercles, all of which are of considerable diagnostic importance (see Figure 47B). The carapace lines and grooves also are especially significant in paleontological studies.

Of the 5 pairs of pereopods, the first 3 always are chelate in this infraorder; the 1st pair usually is much more massive than the others. The pleopods generally are not as well developed as in the Dendrobranchiata and other infraorders of the Pleocyemata studied previously. In some taxa the 1st pair of pleopods is absent. The males frequently have the 1st and 2nd pairs modified as copulatory structures, or an appendix masculina is developed on the 2nd pair (Nephropidae). In females, a median sternal sclerite between the last 2 thoracic somites often is modified to form a sperm receptacle, the annulus ventralis (Astacoidea and Parastacoidea) or a thelycum is developed on the sternite of the 4th pereopods (Nephropoidea). This structure is functionally equivalent to the thelycum of the female penaeoidean.

Before removing the mouthparts, remove the branchiostegite from one side and expose the gills. What type of gills are present and what are the gill formulae for your specimens? Remove the maxillipeds and other mouthparts. The crista dentata on the ischium of the 3rd maxilliped is well developed in most crayfishes and frequently in nephropoideans as well. The mandible usually is well developed and usually has a palp. Note the absence of a lacinia mobilis. Are the incisor and molar processes distinguishable?

The major organ systems are described for species of Astacoidea (see Figure 47). Before beginning your dissection, carefully remove the carapace and proximal region of the rostrum. At this time it would be advisable also to remove the pleura dorsally and laterally from one side of your specimen. Although we will not discuss the musculature in detail, 2 pairs of prominent muscles immediately are apparent dorsally in the anterior part of the cephalothorax when the carapace is removed. The anteriormost are the muscles of the proventriculus or stomach; the posterior pair are the adductor muscles of the mandibles. These muscles will have to be removed to expose the underlying organs.

Posteriorly in the cephalothorax just beneath the hypodermis locate the heart. Blood enters the heart from the pericardial sinus or pericardium through 3 pairs of ostia, 1 dorsal, 1 lateral, and 1 ventral. From the anterior end of the heart follow the unpaired median ophthalmic artery forward over the stomach. A swelling in the artery (the cor frontale or accessory heart) aids in circulating blood to the cephalic region. Beyond the cor frontale the artery branches to provide blood to the eyes, antennules, and supraesophageal ganglion. A pair of large antennary arteries leaving the heart anteriorly parallel the ophthalmic artery for a short distance before turning ventrolaterally. Branches from the antennary arteries supply blood to the gonads, branchiostegites, hepatic cecum, stomach, and musculature of the cephalic appendages. Trace as many of these branches as possible. Anteroventrally from the heart locate one of the pair of hepatic arteries that carry blood to the hepatic cecum (hepatopancreas). From the posterior margin of the heart locate the large dorsal abdominal artery. As this artery leaves the heart a pair of asymmetrical arteries are given off. Both of these arteries provide blood to the gonads; the larger gives rise to the sternal artery. Trace the sternal artery ventrally as it passes to one side of the

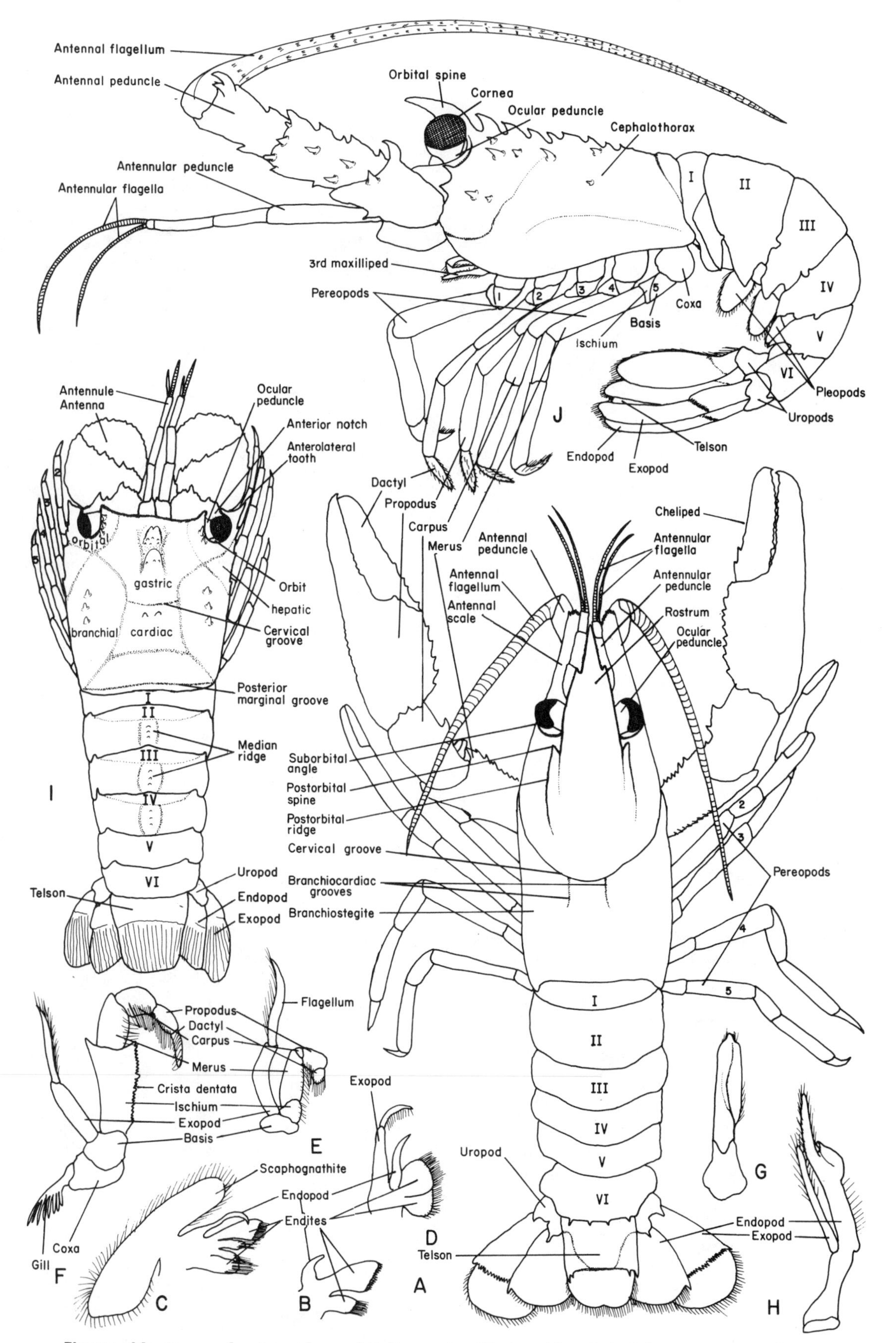

Figure 46 Decapoda, Astacidae and Palinura: A—H, Astacidea; I, J, Palinura. A. Typical astacidean (crayfish); B. Maxillule; C. Maxilla; D. 1st maxilliped; E. 2nd maxilliped; F. 3rd maxilliped; G. 1st pleopod of male; H. 2nd pleopod of male; I. Typical scyllaridean (slipper lobster); J. Typical palinuridean (spiny lobster).

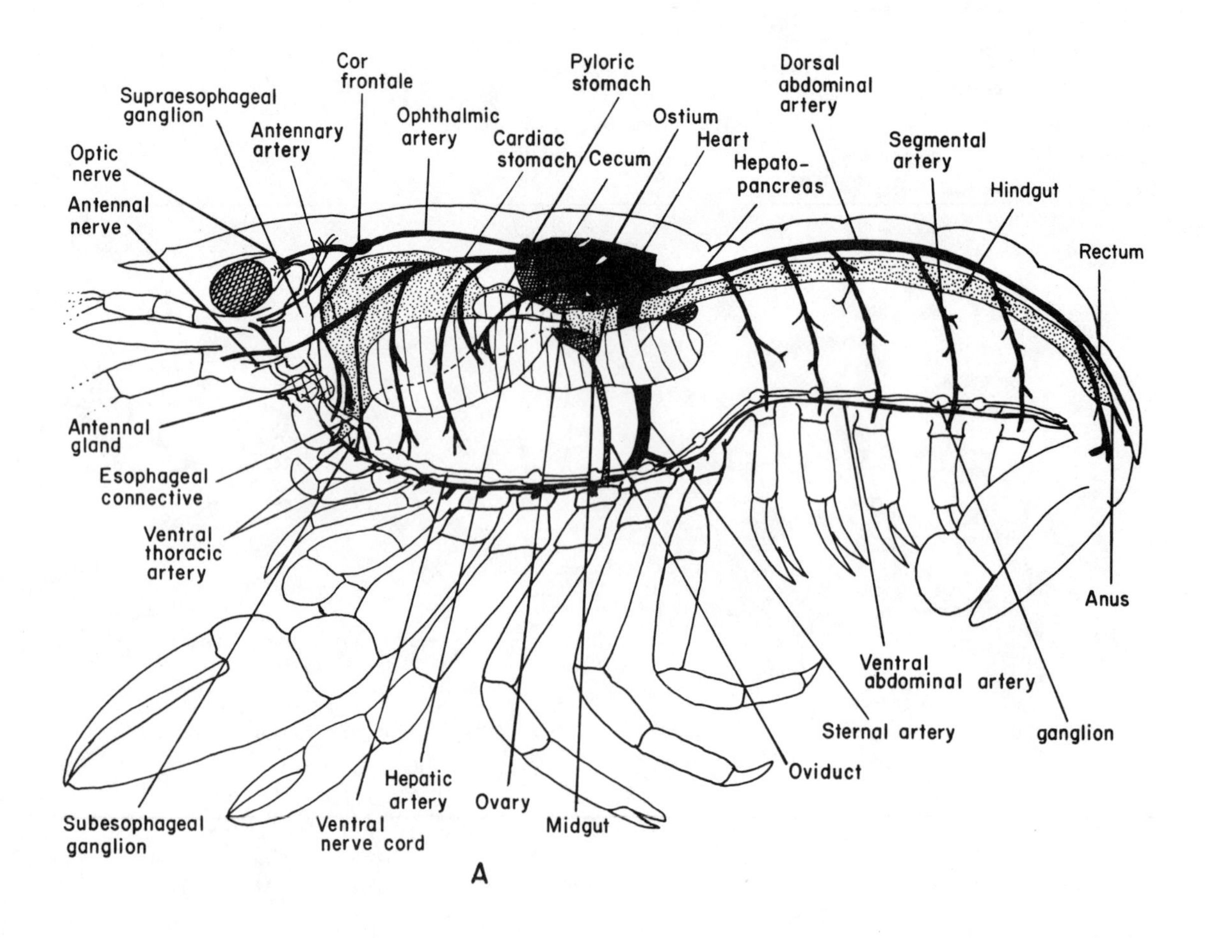

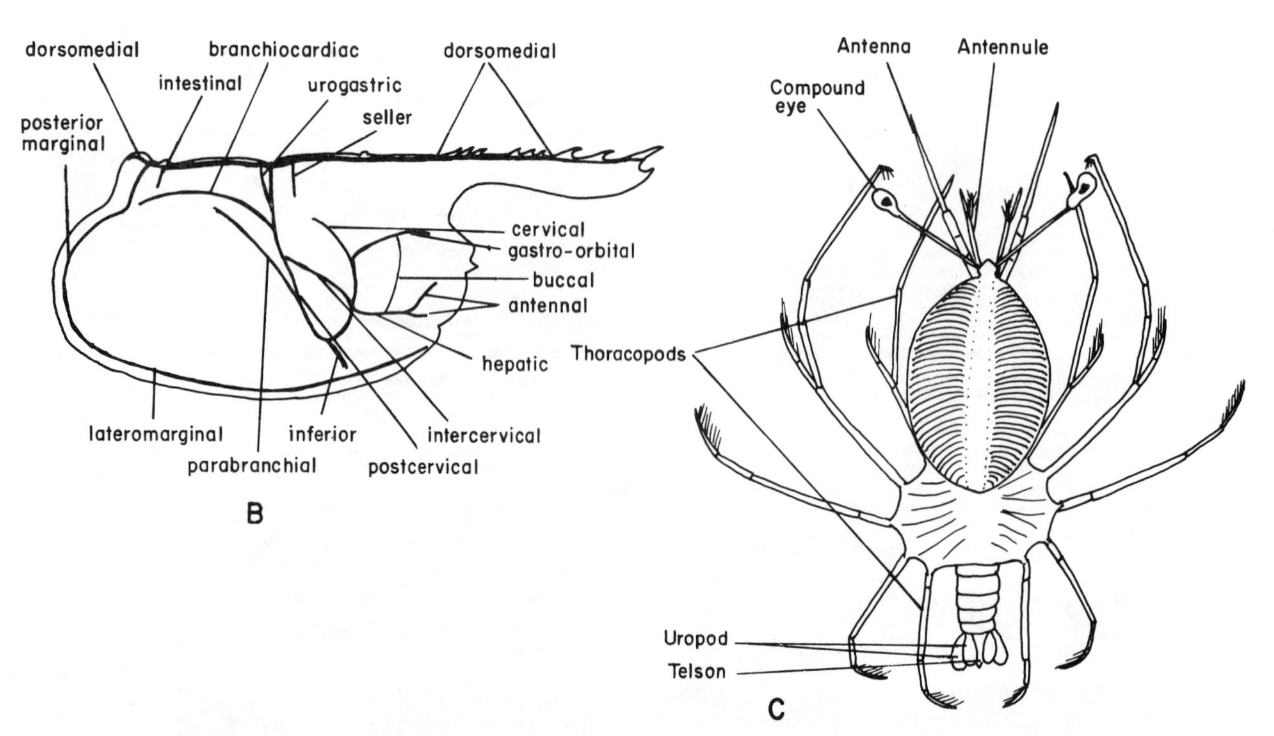

Figure 47 Decapoda: A. Diagrammatic astacidean with gills and musculature removed to show major organ systems; B. Diagrammatic nephropoidean carapace illustrating carapace grooves [after Holthuis, 1974]; C. Phyllosoma larva.

midgut or the other, pierces the ventral nerve cord, and then branches anteriorly and posteriorly. The anterior branch, the ventral thoracic artery, supplies blood to the mouthparts, nerve cord, and 1st 3 pairs of pereopods. The course of this artery cannot be traced until the stomach and hepatic cecum have been removed. The posterior branch, the ventral abdominal artery, which also should be traced later, provides blood to the 4th and 5th pairs of pereopods, nerve cord, and parts of the ventral abdomen. Returning to the heart, follow the dorsal abdominal artery posteriorly through the abdomen. In each somite a pair of segmental arteries are given off that supply blood to the midgut, abdominal musculature, and pleopods. The posterior pair supplies the uropods and telson.

Remove the heart and underlying pericardial membrane to expose the gonad. In females the gonad typically is Y-shaped, with lobes extending anteriorly on either side of the stomach. The male testis is similarly positioned. If possible trace the oviducts or vas deferens from their origins on the gonad ventrally to the gonopores on the coxae of the pereopods (3rd in females, 5th in males).

If the anterior cephalothoracic muscles have not yet been removed, remove them at this time to expose the esophagus and stomach. The esophagus is short and opens into the cardiac portion of the stomach. A constriction marks the boundary between the cardiac and pyloric parts of the stomach. At the junction of the pyloric stomach and midgut, a pair of ducts from the hepatopancreas open into the midgut. In addition to the prominent bilobed hepatopancreas, locate the small cecum that projects anteriorly adjacent to the pyloric stomach. Just posterior to this cecum a slight ridge indicates the junction of the midgut and hindgut; in most crayfish the latter extends nearly the entire length of the abdomen. In the 6th abdominal somite it usually enlarges to form a rectum before terminating at the anus on the ventral surface of the telson. Remove the stomach and examine the structure of the gastric mill in the cardiac portion of the stomach. The pyloric stomach is made up of plates, bars, and channels. How does this stomach compare with that of the penaeid? Excretion in astacideans is via antennal glands. In contrast to the condition found in penaeoideans, the antennal gland of astacideans is distinct. Locate this gland and trace its duct to the excretory pore at the base of the antenna.

The supraesophageal ganglion lies in the midline just ventral to the ocular peduncles. Locate this ganglion and identify the 4 large pairs of nerves radiating from it, the antennular, optic, antennal, and tegumental nerves. Ventrally trace the esophageal connective of one side around the esophagus; it passes beneath the endophragmal shelf before joining the subesophageal ganglion. Portions of this shelf must be removed to expose the ventral thoracic ganglia. The subesophageal ganglion is indistinct from the fused ganglia of the mandibles, maxillulae, maxillae, and first 2 pairs of maxillipeds. The ganglia of the first 3 pairs of pereopods are segmental; the ganglia of the 4th and 5th pairs lie very close together. Follow the ventral nerve cord into the abdomen and identify the abdominal ganglia.

Larval development is direct (epimorphic) in all freshwater taxa; in marine taxa early developmental stages are passed through in the egg and hatching usually occurs at the mysis stage. Development at this stage corresponds to similar stages in other decapods.

References

Balss, H., 1940–1957. Decapoda. In H. G. Bronn, *Klassen und Ordnungen des Tierreichs,* 5(1) *7:* 321–480. Leipzig: Akad. Verl.

Bott, R., 1950. Die Flusskrebse Europas (Decapoda, Astacidae). *Abhandl. Senckenberg. Naturf. Ges., 483:* 1–36.

Budd, T. W., J. C. Lewis, and M. L. Tracey, 1978. The filter-feeding apparatus in crayfish. *Canad. J. Zool.,* 56: 695–707.

Calman, W. T., 1909. *Treatise on zoology.* Pt. 7, Appendiculata, Fasc. 3, Crustacea. 346 pp. London: Ray Soc.

Creaser, E. P., 1933. Descriptions of some new and poorly known species of North American crayfishes. *Occ. Pap. Mus. Zool., Univ. Mich.,* 275: 1–21.

Crocker, D. W., and D. W. Barr, 1968. *Handbook of the crayfishes of Ontario.* 158 pp. Toronto: Roy. Ontario Mus., Univ. Toronto Press.

Glaessner, M. F., 1969. Decapoda. In R. C. Moore (ed.), *Treatise on invertebrate paleontology,* Pt. R, Arthropoda 4, *2:* R399–R566. Lawrence, Kans.: Geol. Soc. America and Univ. Kansas.

Hobbs, H. H., Jr., 1942a. A generic revision of the crayfishes of the subfamily Cambarinae (Decapoda, Astacidae) with the description of a new genus and species. *Am. Midl. Natur., 28:* 334–357.

———, 1942b. The crayfishes of Florida. *Univ. Florida Publ. Biol. Sci. Ser., 3:* 1–179.

———, 1967. The current status of the crayfishes listed by Girard (1852) in his "A revision of the North American Astaci . . ." *Crustaceana 12:* 124–132.

———, 1968. Crustacea: Malacostraca. In F. K. Parrish, *Keys to water quality indicative organisms (Southeastern United States).* Fed. Water Poll. Contr. Admin. Dept. Interior: K1–K36. Reissued 1975.

———, 1969. On the distribution and phylogeny of the crayfish genus *Cambarus.* In P. C. Holt (ed.), The distributional history of the biota of the Southern Appalachians, Part I: Invertebrates. *Virginia Polytechnic Inst. Res. Div. Monogr. 1:* 93–178.

Hobbs, H. H., Jr., and A. Villalobos, 1964. Los Cambarinos de Cuba. *An. Inst. Biol. Univ. Nal. A. de México 84:* 307–366.

Huxley, T. H., 1880. *The crayfish.* Intern. Sci. Series 28. New York: D. Appleton and Co. 371 pp. (Reprinted 1973, MIT Press).

Kingsley, J. S., 1899. Astacoid and thallassinoid Crustacea. Synopses of North American invertebrates. IV. *Am. Natur.*, *33:* 819–824.

Meredith, W. G., and F. J. Schwartz, 1960. Maryland crayfishes. *Maryland Dept. Res. Ed. Educational Series* 46, 32 pp.

Ortmann, A. E., 1906a. Mexican, Central American and Cuban cambari. *Proc. Wash. Acad. Sci.*, *8:* 1–24.

———, 1906b. The crawfishes of the state of Pennsylvania. *Mem. Carnegie Mus.* *2:* 343–523.

———, 1907. Grabende Krebse in Nordamerika. *Aus der Natur, 1906/07:* 705–716.

———, 1931. Crawfishes of the southern Appalachians and the Cumberland Plateau. *Ann. Carnegie Mus.* *20:* 61–160.

Rhoades, R., 1944. The crayfishes of Kentucky, with notes on variation, distribution and descriptions of new species and subspecies. *Am. Midl. Natur.* *31:* 111–149.

Rick, E. F., 1969. The Australian freshwater crayfish (Crustacea: Decapoda: Parastacidae), with descriptions of new species. *Aust. J. Zool.*, *17:* 855–918.

———, 1971. The freshwater crayfishes of South America. *Proc. Biol. Soc. Wash.*, *84:* 129–136.

Schwartz, F. J., and W. C. Meredith, 1960. Crayfishes of the Cheat River watershed in West Virginia and Pennsylvania. Part I. Species and localities. *Ohio J. Sci.*, *60:* 1–46.

———, 1962. Crayfishes of the Cheat River watershed in West Virginia and Pennsylvania. Part II. Observations upon ecological factors relating to distribution. *Ohio J. Sci.*, *62:* 260–273.

INFRAORDER AUSTROASTACIDEA Clark, 1936

Recent species	Few in one genus, *Austroastacus*.
Size range	10 to ~ 80 mm.
Carapace	Strongly vaulted posteriorly.
Eyes	Stalked, compound, well developed.
Antennules	With 2nd flagellum small or absent.
Antennae	Inserted beneath antennules; scaphocerite long or short, slender, sometimes with long slender spine.
Mandibles	(?) With palp.
Maxillulae	(?) With endopodal palp.
Maxillae	(?) Biramous.
Maxillipeds	Epipods of 1st without branchial filaments; 3rd without exopods.
Thoracic appendages	Uniramous; first 3 pairs chelate; 5th pereopods without gills.
Abdominal appendages	Pleopods biramous; 1st pair absent in both sexes. Uropods biramous, without transverse sutures, well calcified.
Telson	Entirely calcareous.
Tagmata	Cephalothorax and abdomen.
Somites	Head with 5 + 3 thoracic (maxillipeds); thorax with 5; abdomen with 6, excluding telson.
Sexual characters	Gonopores on coxae of 3rd pereopods of female; on short, simple papillae on coxae of 5th pereopods of male.
Sexes	Presumably separate.
Larval development	Presumably epimorphic.
Fossil record	Recent.
Feeding types	Presumably herbivores.
Habitat	Burrows in swampy and dry areas.
Distribution	Australia.

The type species, *Austroastacus hemicirratulus* (Smith and Schuster), of this small group of semiterrestrial Australian crayfishes originally was described in the astacidean genus *Engaeus*. Clark (1936) removed *A. hemicirratulus* from its former genus and assigned it to his newly established *Austroastacus*, and at the same time described for it the family Austroastacidae. Bowman and Abele (in press) have since elevated the family to infraorder rank.

Several characters distinguish the austroastacideans from other crayfish. The telson and uropods are completely calcified. The first pair of pleopods are absent in both sexes, and the abdomen is relatively small, although larger in females than in males. The carapace is higher than broad and vaulted posteriorly; the cervical and branchiocardiac grooves are strongly impressed and the areola narrow. The inner flagella of the antennules, when present, are much shorter than the outer.

These crayfish live in family groups and dig deep burrows both in swampy patches and on dry hillsides in heavily timbered areas. The entrances to the burrows dug in swampy ground usually have cones a few inches high with an opening at the top. Burrows dug on hillsides lack cones; the entrances are indicated only by small round holes, usually beneath fallen logs or stones.

References

Bowman, T. E., and L. G. Abele, in press. Classification. In L. G. Abele (ed.) *The biology of Crustacea,* vol. 1, Systematics, Genetics, Morphology and the Fossil Record. New York: Academic Press.

Clark, E., 1936. The freshwater and land crayfishes of Australia. *Mem. Natl. Mus. Melbourne, 10:* 5–58.

Glaessner, M. F., 1969. Decapoda. In R. C. Moore (ed.), *Treatise on invertebrate paleontology,* Part R Arthropoda, *2:* R399–R533. Lawrence, Kans.: Geol. Soc. America and Univ. Kansas.

Healy, A., and J. Yaldwyn, 1970. *Australian crustaceans in colour.* 112 pp. Sydney: A. H. and A. W. Reed.

INFRAORDER PALINURA *Latreille, 1803*

Recent species	Approximately 130.
Size range	Up to 61 cm.
Carapace	Cylindrical, subovoid, or dorsoventrally compressed.
Eyes	Stalked, compound, generally well developed.
Antennules	Biramous; peduncle with 3 segments.
Antennae	Peduncle with 5 or fewer segments; exopod reduced; flagella sometimes modified.
Mandibles	Usually with palp; molar and incisor processes usually not distinct.
Maxillulae	With endopodal palp.
Maxillae	Biramous; with 2 bilobed endites.
Maxillipeds	Endopod of 1st with 2 or fewer segments; usually with crista dentata.
Thoracic appendages	Uniramous; 1st pair sometimes chelate; other pairs often subchelate.
Abdominal appendages	Pleopods usually biramous; 1st pair frequently absent; male copulatory structures frequently present. Uropods biramous.
Telson	Together with broad uropods forms tailfan.
Tagmata	Cephalothorax and abdomen.
Somites	Head with 5 + 3 thoracic (maxillipeds); thorax with 5; abdomen with 6, excluding telson.
Sexual characters	Gonopores on coxae of 3rd pereopods of female, on 5th of male; male frequently with copulatory structure.
Sexes	Presumably separate.
Larval development	Metamorphic; phyllosoma → puerulus (postlarva) [Palinuroidea].
Fossil record	(?) L. Triassic, M. Triassic to Recent.
Feeding types	Various.
Habitat	Marine.
Distribution	Worldwide.

This infraorder contains three superfamilies; however, specimens from only one, the Palinuroidea, can be obtained relatively easily. It contains, among others, the Palinuridae (spiny lobsters) and the Scyllaridae (slipper or shovel-nosed lobsters). If specimens are available, representatives of these taxa should be compared with the crayfishes and true lobsters (see Figure 46). In scyllarids, the great modification of the antenna is the most striking character. The antennal flagella have been transformed into spinose or tuberculate platelike structures; the first segments of the antennular peduncles are immovable. In the Palinura the rostrum is not prominent, and the carapace is fused to the epistome. In the Palinuridea, the 5th pereopods of females are the only chelate appendages; however, chelate appendages usually are present in the other superfamilies.

Larval development is typically metamorphic. Early larval stages are passed through in the egg; the first postembryonic stage in the Palinuroidea is the phyllosoma. This is a very distinctive larval form, with an extremely flattened, disk-shaped body and elongate appendages; it is equivalent to the mysis stage. The postlarval stage in this superfamily is referred to as the puerulus, nisto, or pseudibacus stage, a benthic form more closely resembling the adult.

References

Alcock, A., 1901. *A descriptive catalogue of the Indian deep-sea Crustacea Decapoda Macrura and Anomala in the Indian Museum.* 286 pp. Calcutta: Indian Museum.

Balss, H., 1940–1957. Decapoda. In H. G. Bronn, *Klassen und Ordnungen des Tierreichs* 5(1) 7: 321–480. Leipzig: Akad. Verl.

Bonde, C. von, and J. M. Marchand, 1935. The natural history and utilization of the Cape crawfish, Kreef, or spiny lobster, *Janus (Palinurus) lalandii* (Milne Edwards) Ortmann. *Fish. Bull. Fish Mar. Biol. Sur. S. Afr., 1:* 1–55.

Bouvier, E. L., 1925. Les Macroures Marcheurs. Reports of the results of dredging in the Gulf of Mexico (1877–78), in the Caribbean Sea (1878–79), and along the Atlantic coast of the United States (1880) by the U.S. Coast Survey Steamer "Blake" . . . *Mem. Mus. Comp. Zool. Harvard, 47:* 397–472.

———, 1940. Décapodes marcheurs. In *Faune de France, 37:* 1–399. Paris: Paul Lechevalier.

Calman, W. T., 1909. *Treatise on zoology.* Pt. 7, Appendiculata, Fasc. 3, Crustacea, 346 pp. London: Ray Soc.

Corrivault, G. W. and J. L. Tremblay, 1948. Contribution à la biologie du homard *(Homarus americanus* Milne-Edwards) dans la Baie-des-Chaleurs et le golfe Saint-Laurent. *Contr. Stn Biol. St. Laurent, 19:* 1–222.

Dawson, C. E., 1954. *A bibliography of the lobster and the spiny lobster, families Homaridae and Palinuridae.* 86 pp. Florida State Board of Conservation.

Firth, R. W., and W. E. Pequegnat, 1971. *Deep-sea lobsters of the Families Polychelidae and Nephropidae (Crustacea Decapoda) in the Gulf of Mexico and Caribbean Sea.* 103 pp.

Glaessner, M. F., 1969. Decapoda. In R. C. Moore (ed.), *Treatise on invertebrate paleontology,* Pt. R, Arthropoda 4, *2:* R399–R566. Lawrence, Kans.: Geol. Soc. America and Univ. Kansas.

Herrick, F. H., 1895. The American lobster: a study of its habits and development. *Bull. U.S. Fish. Comm., 15:* 1–252.

Holthuis, L. B., 1946. The Stenopodidae, Nephropsidae, Scyllaridae and Palinuridae. The Decapoda Macrura of the "Snellius" Expedition. *Temminckia, 7:* 1–178.

———, 1974. The lobsters of the superfamily Nephropidea of the Atlantic Ocean (Crustacea: Decapoda). *Bull. Mar. Sci., 24:* 723–884.

Lewis, R. D., 1970. A bibliography of the lobsters, genus *Homarus. U.S. Fish Wildl. Serv. Spec. Sci. Rept. Fish.,* 591: 1–47.

Maigret, J., 1976. Contribution à l'étude des langoustes de la côte occidentale d'Afrique (Crustacés, Décapodes, Palinuridae). 1. Notes sur la biologie et l'ecologie des espéces sur les côtes du Sahara. *Bull. Inst. Fondam. Afr. Noire, A38:* 266–302.

Stephensen, K., 1923. Decapoda-Macrura excl. Sergestidae. *Rept. Danish Oceanogr. Exped. Mediterranean, 2 (D3):* 1–85.

Wolff, T., 1978. Maximum size of lobsters *(Homarus)* (Decapoda, Nephropidae). *Crustaceana, 34:* 1–14.

Zarenkov, N. A., and V. N. Semenov, 1972. Novyi vid roda Nephropides (Decapoda, Macrura) iz yugo-zapadnoi Atlantiki. *Zool. J. Moscow,* 51: 599–601.

Zariquiey Alvarez, R., 1968. Crustáceos Decápodos Ibéricos. *Invest. Pesqueras Barcelona, 32:* 1–510.

INFRAORDER ANOMURA H. Milne Edwards, 1832

Recent species	Approximately 1575.
Size range	Up to 185 cm (leg span).
Carapace	Variable in shape; not fused to epistome.
Eyes	Stalked, compound; generally well developed.
Antennules	Peduncle with 3 segments; flagella usually paired.
Antennae	Peduncles sometimes with 5, possibly 6, or fewer segments; exopod usually reduced to acicle; flagellum variable in length.

Mandibles	With or without palp; molar and incisor processes usually not distinct.
Maxillulae	With endopodal palp.
Maxillae	Biramous; usually with bilobed endites.
Maxillipeds	Flagella usually present; sometimes absent from 1st; often with crista dentata; usually not operculate.
Thoracic appendages	First usually, 2nd occasionally, chelate; 4th and 5th usually chelate or subchelate, one or both frequently reduced.
Abdominal appendages	Pleopods rarely well developed; often reduced or present only on 1 side; both sexes sometimes with copulatory structures. Uropods often reduced or modified, occasionally absent.
Telson	Occasionally reduced or absent; sometimes together with uropods forms tailfan.
Tagmata	Cephalothorax and abdomen.
Somites	Head with 5 + 3 thoracic (maxillipeds); thorax with 5; abdomen with 6, excluding telson, although segmentation frequently obscured.
Sexual characters	Gonopores on coxae of 3rd pereopods of female, on 5th of male; 1st and/or 2nd pleopods often modified as gonopods in both sexes; female sometimes with abdominal brood pouch; male sometimes with sexual tube(s).
Sexes	Presumably separate.
Larval development	Metamorphic; zoea → megalopa (glaucothöe).
Fossil record	L. Jurassic to Recent.
Feeding types	Various.
Habitat	Most frequently marine, but with few semiterrestrial and freshwater species.
Distribution	Worldwide.

Among the decapod crustaceans, the infraorder Anomura presents the greatest diversity in body form; there are five superfamilies. The shrimplike or crayfishlike elongate body is represented by the Thalassinoidea. The abdomen, although often not drastically reduced in length in the Coenobitoidea and Paguroidea, is most frequently membranous or only weakly calcified and an extrinsic protective covering often is adopted. In certain families the familiar crablike body form has been developed (i.e., Lomidae and Lithodidae). In these families, the abdomen is tucked under the cephalothorax; however, in contrast to that of brachyurans, the abdomen remains relatively soft and membranous or is covered by individual calcified plates. The superfamily Galatheoidea has representatives with crayfishlike body forms (Galatheidae) and crablike forms (Porcellanidae), while representatives of the Hippoidea exhibit variations between the two extremes. If a number of anomuran types are available, identify the characters that are possessed in common. As the paguroidean body form is perhaps the most atypical, a pagurid will be used as a basis for discussing anomuran characters (see Figures 48 and 49).

Begin your study with an overall examination of the general morphology of a typical hermit crab. Aside from the membranous abdomen, perhaps the most noticeable character and one that is not common in other crustaceans is the strong tendency toward asymmetry. Depending upon the species under study, the left or the right 1st pereopod (cheliped) usually is noticeably enlarged. A second indication of asymmetry is the presence of pleopods on only the left side of the abdomen; some primitive genera of coenobitoideans have paired pleopods. Usually, unless modified as gonopods, pleopods are absent from the 1st abdominal somite. Sexual modifications of the 1st pair may occur in both male and female; in the male the 2nd also may be modified. In some genera of pagurids the vas deferens is markedly extended from the gonopores to form 1 or a pair of sexual tubes. The asymmetrical uropods common to the majority of paguroideans are another example of the secondary asymmetry characteristic of the group.

In dorsal view, the eyes with their prominent peduncles usually are obvious. At the base of each ocular peduncle observe the ocular acicle; it may have several

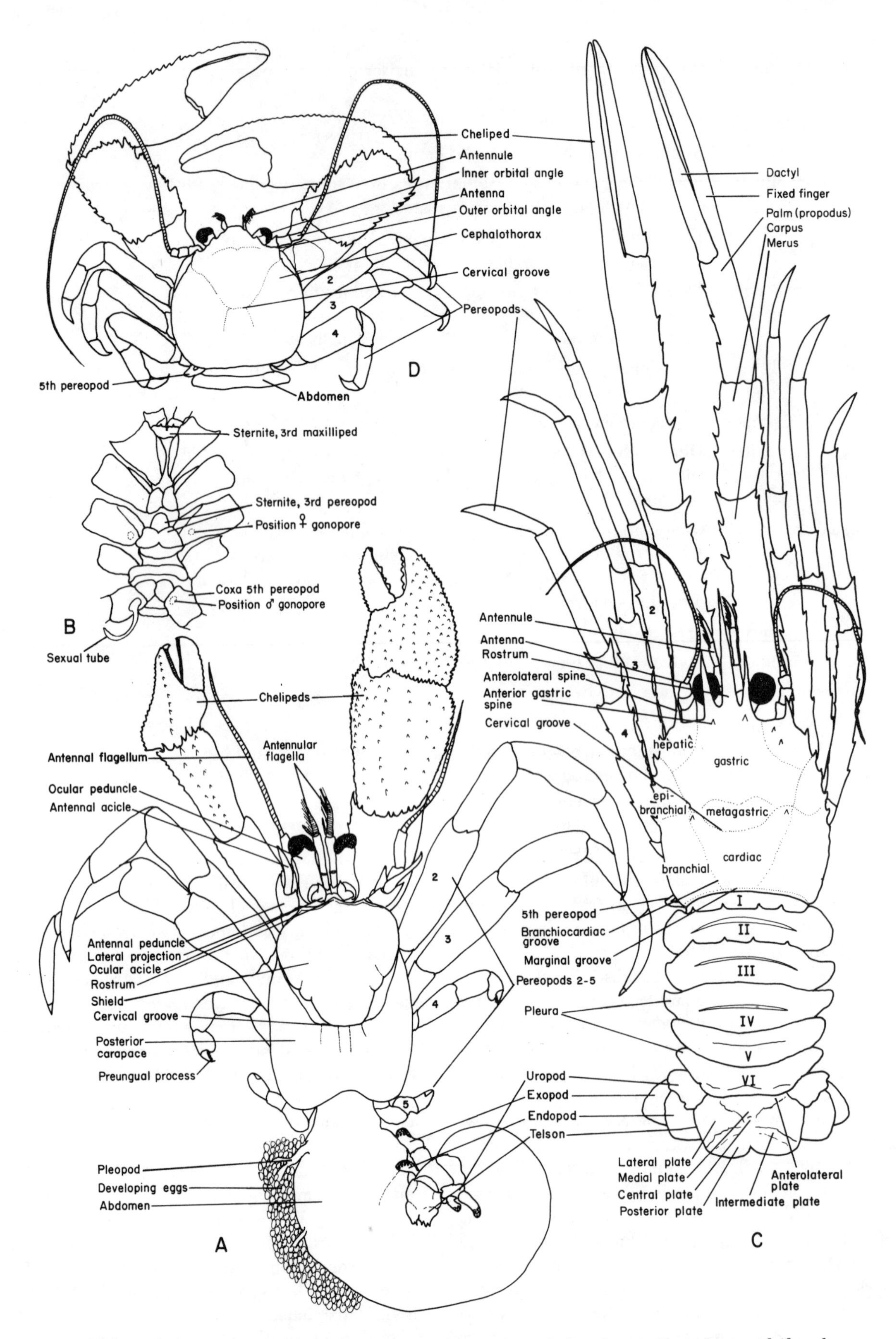

Figure 48 Decapoda, Anomura: A, B, Paguroidea; C, D, Galatheoidea. A. Typical pagurid (dorsal view); B. Pagurid thorax (ventral view); C. Typical galathoidean (dorsal view), showing regions of carapace; D. Typical porcellanid.

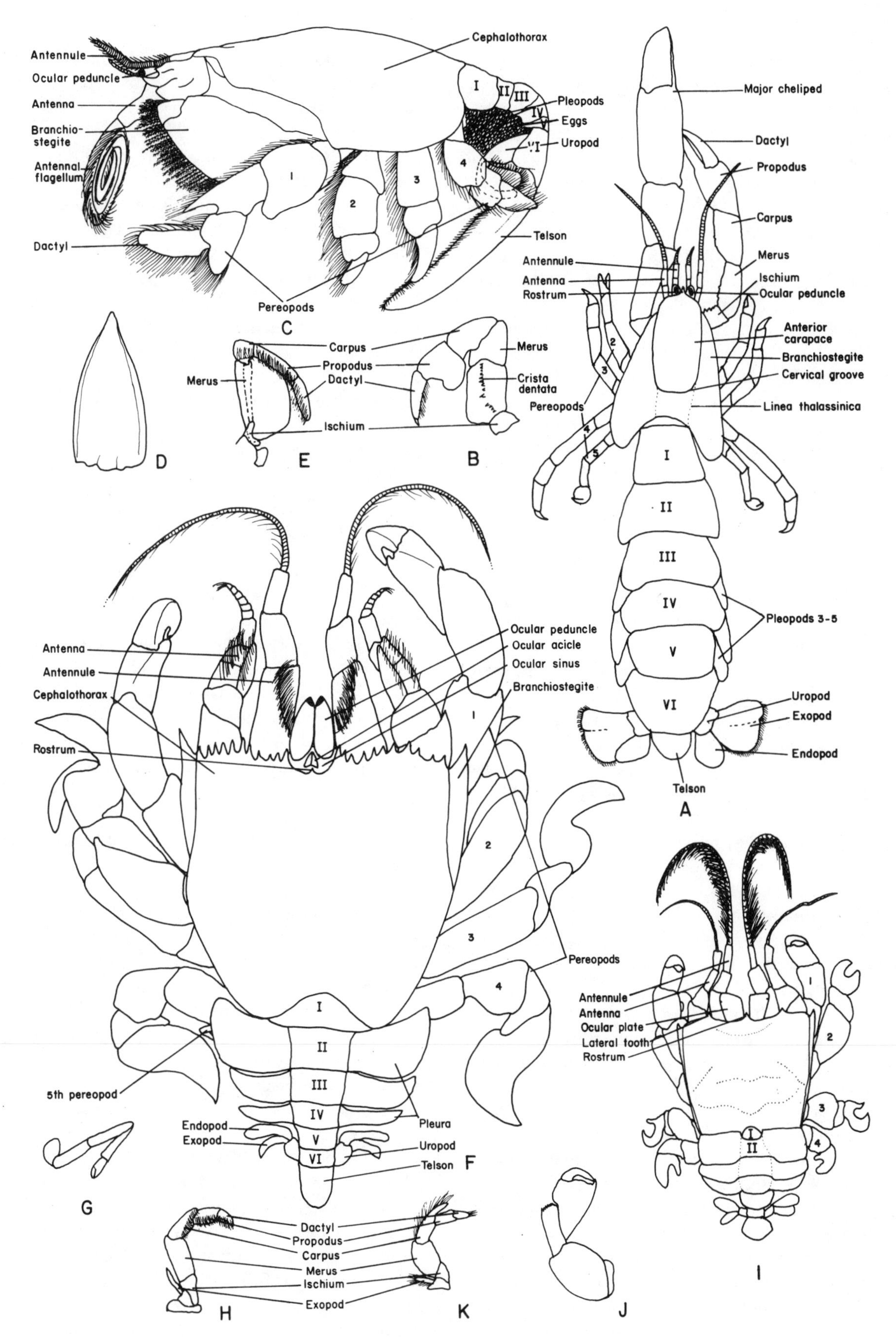

Figure 49 Decapoda, Anomura; A, B, Thalassinoidea; C—K, Hippoidea. A. Typical callianassid (dorsal view); B. 3rd maxilliped; C. Typical hippid (lateral view); D. Hippid telson; E. Hippid 3rd maxilliped; F. Typical albuneid (dorsal view); G. Albuneid 5th pereopod; H. Albuneid 3rd maxilliped; I. Typical *Lepidopa* (Albuneid); J. Left cheliped of *Lepidopa;* K. 3rd maxilliped of *Lepidopa.*

spines, a single spine, or less frequently be unarmed. The antennules are located ventrally between the ocular peduncles; the 1st segment has a prominent statocyst. The antennae are lateral to the ocular peduncles and appear to have, in addition to the typical 5 segments of the peduncle, a small supernumerary segment between the 3rd and 4th segments dorsolaterally. The antennal exopod is reduced to a slender acicle. The pagurid cephalothorax usually is calcified only in the anterior part, referred to as the shield. In some species the branchiostegites are calcified as well. The shield is separated from the posterior part of the carapace by the cervical groove. Usually 3 distinct pairs of lines or grooves can be distinguished on the posterior carapace. In the midline are a pair of elongate sutures, the sulcus cardiobranchalis; 2 short lines or grooves slightly laterad are referred to simply as sulcus "a"; a linea anomurica on each side of the carapace represents the 3rd pair. If the branchiostegite is lifted or removed, the gills are exposed. Typically the gill number in pagurids varies from 10 to 13 pairs, depending upon the taxa. What is the gill formula for your specimen(s)?

In addition to the asymmetrical chelipeds (except in the genera *Paguristes* and *Clibanarius*, and the family Pylochelidae), the 2nd and 3rd pereopods also may be asymmetrical and/or dissimilarly armed from right to left. The 4th pereopods are usually considerably shorter than the preceding pairs and usually are provided with a rasp on the external surface of each propodus; in some taxa they have a preungal process at the base of each dactyl (the function of this structure is unknown). The 5th pereopods also are reduced in size and most frequently are chelate. They too are equipped with rasps.

Although the pagurid abdomen is usually membranous, it often is possible to distinguish the outline of some of the abdominal tergites. The pagurid telson and uropods are distinctive. Notice that the uropods also are provided with rasps. These are used by the animal to help in holding the shell in place.

Before removing the mouthparts, observe the position of the 3rd maxillipeds relative to one another. This is a valuable diagnostic character; the two superfamilies are differentiated on the position of these maxillipeds. In the Coenobitoidea the 3rd maxillipeds are approximate at their bases; in the Paguroidea the bases are widely separated. The approximate position of the 3rd maxillipeds in the Thalassinoidea and the widely separated condition in the Galatheoidea suggest that the Coenobitoidea may be closely related to the former and the Paguroidea to the latter. If such relationships are true, the similarities in body form that exist among representatives of the latter two superfamilies would have to have resulted from convergent evolution.

If specimens of galatheoideans or thalassinoideans are available, list the similarities and differences that you observe between representatives of these taxa and hermit crabs. If specimens of lithodids and porcellanids are available, what characters do you find that relate the former closely with the pagurids and the latter with the galatheids? The superfamily Hippoidea is represented by a relatively few species, many of which are rather specialized. The common "mole crabs" *Emerita* spp. have the antennae extremely setose and adapted for filter feeding. Some of the Albuneidae have the antennules modified for a similar function.

The description of major organ systems also is based on the conditions found in pagurids (see Figure 50), although these are not necessarily always typical of all anomurans. Differences between pagurids and other anomurans will be pointed out. Remove the carapace very carefully to avoid damaging the heart; the pericardium attaches to the membrane underlying the carapace. As the cephalothoracic body cavity of typical pagurids is narrow, remove the gill chamber and inner branchial wall from one side. Locate the heart in the posterodorsal region of the cephalothorax. The pericardium extends from the cervical groove to the 8th thoracic somite, but usually will not be easily distinguished from neighboring tissues. Blood flows from the gills into the pericardium and into the heart via 3 pairs of ostia, 1 anterodorsal pair and 2 lateral pairs. Anteriorly the heart gives off 3 arteries, the unpaired, median ophthalmic artery and the paired lateral arteries. Follow the ophthalmic artery; shortly after leaving the heart a cor frontale is formed. After passing over the dorsal surface of the stomach the ophthalmic artery turns ventrally and divides into 2 branches that provide blood to the anterior cephalic area and supraesophageal ganglion. Returning to the heart trace the path of one of the lateral arteries anterolaterally. Enroute to the cephalic appendages, it gives off branches to the stomach and musculature. Ventrally from the heart locate the small hepatic arteries. In pagurids this pair of arteries no longer supplies blood to the hepatopancreas, which lies almost exclusively in the abdomen, but terminates instead on the stomach or midgut. From the posterior margin of the heart identify the large, posteriorly directed superior abdominal artery and the ventrally directed sternal artery. The path of the latter will be described, but should not be traced until after the stomach and midgut have been removed. Follow the superior abdominal artery posteriorly, cutting open the abdomen if necessary. At the level of the 1st abdominal somite the artery branches. Trace the large segmental artery ventrally; at the level of the 3rd somite it divides into sub- and supramuscular branches. The former passes ventrally along the ventral nerve cord and terminates in the 6th somite. The latter provides numerous branches to the hepatopancreas and gonads and terminates with branches to the telson and uropods. The superior abdominal artery supplies blood to the gonads and to the pleopods. Follow the sternal artery as it passes first ventrally into the 7th thoracic somite and then turns

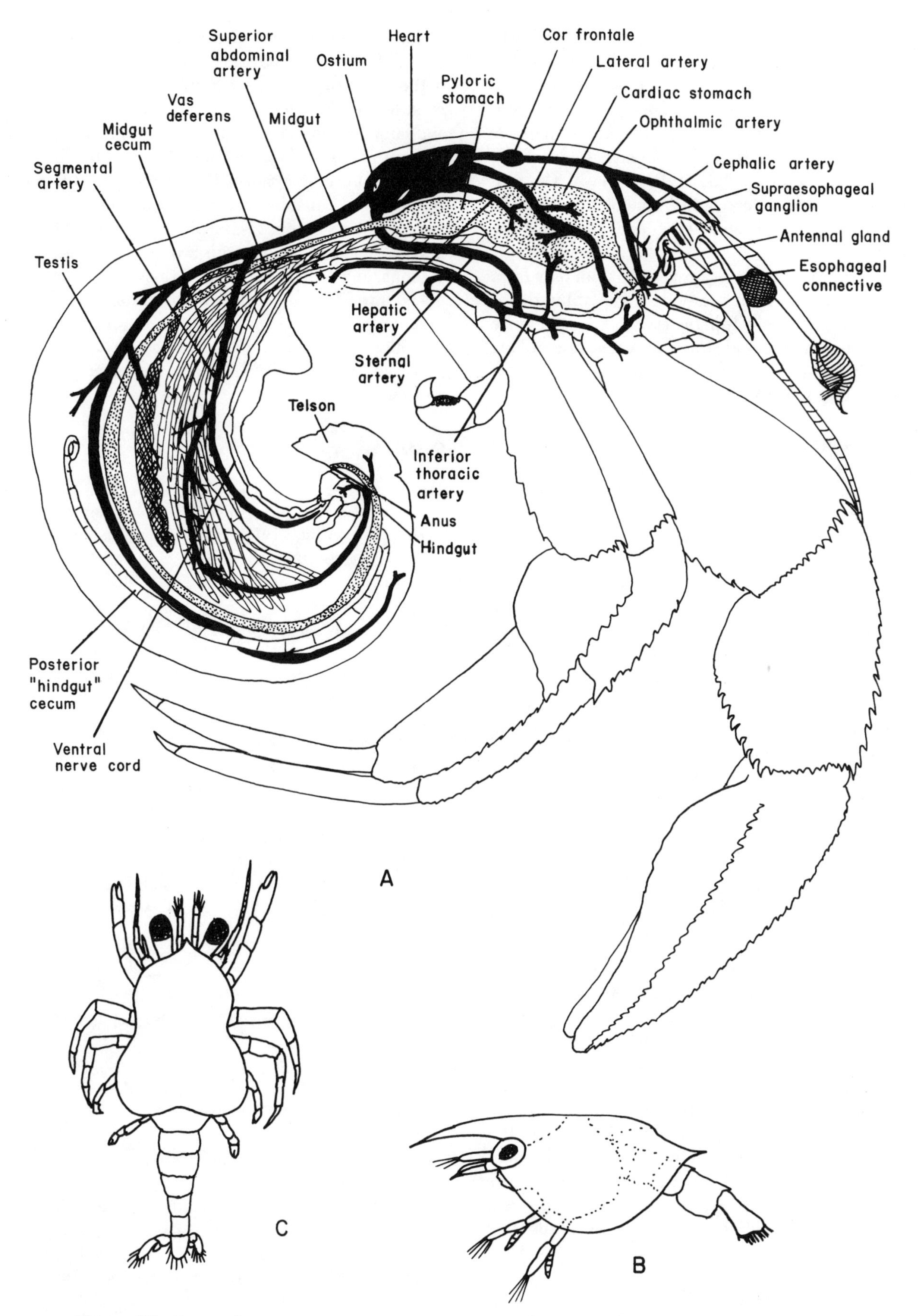

Figure 50 Decapoda, Anomura: A. Diagrammatic pagurid with musculature removed to show major organ systems; B. Zoea; C. Megalopa (Glaucothöe.)

horizontally. At the level of the 5th somite it turns ventrally again and pierces the central ganglionic mass between the nerves of the 2nd and 3rd pereopods. Beneath the nerve cord the vessel divides into anterior and posterior branches. The anterior branch, the inferior thoracic artery, provides blood to the chelipeds, mouthparts, renal gland, and ventral region of the stomach. The posterior branch supplies blood to the remaining pereopods, but in contrast to other anomurans, does not enter the abdomen.

The pagurid reproductive system differs not only from other anomurans but from decapods in general in that it is located almost exclusively in the abdomen. In the male the testes are paired, ovoid structures consisting of long convoluted tubules, frequently nearly imbedded in tubules of the hepatopancreas. Anteriorly from the testes, at about the level of the 3rd abdominal somite, trace one of the vas deferens anteriorly to its opening on the coxa of the 5th pereopod. The paired ovaries of the female lie in approximately the same position; the oviducts are distinct and lead from the abdomen, through the thorax, to their openings on the coxae of the 3rd pereopods.

The muscular esophagus leads into the large stomach. The cardiac portion easily is recognized by its baglike appearance and membranous dorsal surface. After examining the remainder of the digestive system, remove the stomach and examine the gastric mill of the cardiac stomach. The considerably smaller pyloric stomach is separated from the former by a cardiopyloric valve. At the junction of the pyloric stomach and midgut a pair of small, anteriorly directed ceca usually arise dorsally; ducts of the hepatopancreas enter the midgut at this level too. The midgut is an elongate thin-walled smooth tube extending almost the full length of the abdomen. Just anterior to its junction with the hindgut or rectum, a prominent, anteriorly directed cecum arises. It has been called the hindgut cecum, but this is a misnomer as it originates from the midgut. The hindgut terminates in a ventrally directed anus at the terminal end of the telson. Excretion in anomurans is via the antennal glands; in pagurids this usually is supplemented by an unpaired abdominal sac or bladder often hidden in the lobes of the hepatopancreas.

The prominent supraesophageal ganglion can be seen in the midline between the ocular peduncles and above the epistome. Trace the major nerves radiating from the ganglion; that is, the optic, antennal, antennular, and tegumental. Follow one of the esophageal connectives around the esophagus; the swelling is the paraesophageal ganglion. The connective terminates in a thoracic ganglionic mass overlaying the inferior thoracic artery. Three masses of fused ganglia, separated by constrictions, comprise this thoracic mass. The 3rd cluster of ganglia, which is pierced by the sternal artery, is composed of the ganglia of the 4th and 5th pereopods and 1st abdominal somite. In the abdomen the nerve cord is of the ladder type with 5 pairs of fused ganglia. As a result of the flexure of the abdomen, the nerve cord in pagurids is skewed to the left from the 2nd to the 4th somites. The abdominal flexure also has resulted in the atypical development of the abdominal musculature. The flexor muscles have become abnormally enlarged and those of the right side are considerably larger than those of the left.

Larval development is metamorphic; the first postembryonic stage is a zoea, with 2 or 3 pairs of maxillipeds present, and a carapace with the posterior margins produced into 2 lateral spines. The zoeal stage is followed by metamorphosis to a megalopa. In pagurids the zoea is basically symmetrical, but the megalopa (also referred to as a glaucothöe) may exhibit asymmetry in the uropods and chelipeds.

References

Alcock, A., 1905. *Catalogue of the Indian decapod Crustacea in the collection of the Indian Museum.* Part 2, Anomura, Fasc. 1. Pagurides. 197 pp. Calcutta: Indian Museum.

Balss, H., 1940–1957. Decapoda. In H. G. Bronn, *Klassen und Ordnungen des Tierreichs.* 5(1) 7: 321–480. Leipzig: Akad. Verl.

Benedict, J. E., 1892. Preliminary descriptions of 37 new species of hermit crabs of the genus *Eupagurus* in the U.S. National Museum. *Proc. U.S. Natl. Mus., 15:* 1–26.

———, 1903. Revision of the crustaceans of the genus *Lepidopa. Proc. U.S. Natl. Mus., 26:* 889–895.

———, 1904. A new genus and two new species of crustaceans of the family Albuneidae from the Pacific Ocean; with remarks on the probable use of the antennule in *Albunea* and *Lepidopa. Proc. U.S. Natl. Mus., 27:* 621–625.

Bliss, D. E., 1968. Transition from water to land in decapod crustaceans. *Am. Zool., 8:* 355–392.

Calman, W. T., 1909. Crustacea. In R. Lankester (ed.), *A treatise on zoology,* pt. 7, Appendiculata, fasc. 3, 346 pp. London: Ray Soc.

Chace, F. A., Jr., 1942. Report on the scientific results of the Atlantis expeditions to the West Indies, under the joint auspices of the University of Havana and Harvard University. The anomuran Crustacea. I. Galatheidae. *Torreia, Havana,* no. 11: 1–106.

Chace, F. A., Jr., and H. H. Hobbs, Jr., 1969. The freshwater and terrestrial decapod crustaceans of the West Indies with special reference to Dominica. *Bull. U.S. Natl. Mus.,* 292: 1–258.

Efford, I. E., 1966. Feeding in the sand crab *Emerita analoga. Crustaceana, 10:* 167–182.

———, 1967. Neoteny in *Emerita. Crustaceana, 13:* 81–93.

Forest, J., and M. de Saint Laurent, 1968. Campagne de la "Calypso" au large des côtes Atlantiques de l'Amérique de sud (1961–1962). 6. Crustacés Décapodés: Pagurides. *Ann. Inst. Océanogr., 45:* 47–169.

Gordan, J., 1956. A bibliography of pagurid crabs, exclusive of Alcock, 1905. *Bull. Am. Mus. Nat. Hist.,* 108: 257–352.

Haig, J., 1956. The Galatheidae (Crustacea Anomura) of the Allan Hancock Expedition with a review of the Porcellanidae of the western Atlantic. *Allan Hancock Pacific Exped.:* Rept. no. 8: 1–44.

———, 1960. The Porcellanidae (Crustacea Anomura) of the eastern Pacific. *Allan Hancock Pacific Exped.*, 24: 1–440.

Hay, W. P., and C. A. Shore, 1918. The decapod crustaceans of Beaufort, N.C., and the surrounding region. *Bull. Bur. Fish.*, *35:* 369–475.

Hazlett, B. A., 1966. Social behavior of the Paguridae and Diogenidae of Curaçao. *Stud. Fauna Curaçao Carib. Isl.*, *23:* 1–143.

Holthuis, L. B., 1959. The Crustacea Decapoda of Suriname (Dutch Guiana). *Zool. Verhandl.* no. 44: 1–296.

Kensley, B., 1977. The South African Museum's Meiring Naude cruises. Part 2. Crustacea, Decapoda, Anomura and Brachyura. *Ann. S. Afr. Mus.*, *72:* 161–188.

Lewinsohn, Ch., 1969. Die Anomuren des Roten Meeres (Crustacea Decapoda: Paguridea, Galatheidea, Hippidea). *Zool. Verhandl.*, no. 104: 1–213.

———, 1978. Bemerkungen zur Taxonomie von *Paguritta harmsi* (Gordon) (Crustacea Decapoda, Anomura) und Beschreibung einer neuen Art der gleichen Gattung aus Australien. *Zool. Meded.*, *53:* 243–252.

MacDonald, J. D., R. B. Pike, and D. I. Williamson, 1957. Larvae of the British species of *Diogenes*, *Pagurus*, *Anapagurus* and *Lithodes* (Crustacea, Decapoda). *Proc. Zool. Soc. London*, *128:* 209–257.

McLaughlin, P. A., 1974. The hermit crabs (Crustacea Decapoda, Paguridea) of northwestern North America. *Zool. Verhandl.*, no. 130: 1–396.

Makarov, V. V., 1938 (1962). *Fauna of USSR, Crustacea.* 10, Anomura. Zool. Inst. Akad. Nauk SSSR. 278 pp. (Translated by F. D. Por, for the National Science Foundation and Smithsonian Institution).

Man, J. G., de, 1927. A contribution to the knowledge of 21 species of the genus *Upogebia* Leach. *Capita Zool.*, *2:* 1–56.

Melin, G., 1939. Paguriden und Galatheiden von Prof. Sixten Bock's Expedition nach den Bonin-Inseln 1914. *K. Svenska. Vetensk. Akad. Handl. 18:*, 1–119.

Miyake, S., 1978. *The crustacean Anomura of Sagami Bay.* 200 pp. Tokyo: Biol. Lab., Imperial Household (ed.).

Pike, R. B., 1947. Galathea. *Liverpool Mar. Biol. Comm., Mem.* 34: 1–179.

Pilgrim, R. L. C., 1965. The morphology of *Lomis hirta* and a discussion of its systematic position and phylogeny. *Aust. J. Zool.*, *13:* 545–557.

Schmitt, W. L., 1921. The marine decapod Crustacea of California. *Univ. Calif. Publ. Zool.*, *23:* 1–470.

———, 1935. Mud shrimps of the Atlantic coast of North America. *Smiths. Misc. Coll.*, *93:* 1–21.

Snodgrass, R. E., 1952. The sand crab *Emerita talpoida. Smiths. Misc. Coll. 117:* 1–34.

Vinogradov, L. G., 1950. Opredelitel krevetok, rakov i karbov Dalnego Vostlka. *Izvest. Tikhookean. Nauchno-issledovat. Inst. Rybiogo Khozyaistva Okean. 33:* 179–358.

Williams, A. B., 1965. Marine decapod crustaceans of the Carolinas. *Fish. Bull.*, *65:* 1–298.

INFRAORDER BRACHYURA Latreille, 1803

Recent species	Approximately 4500.
Size range	Up to 365 cm (leg span).
Carapace	Progressively shortened and widened; fused to epistome.
Eyes	Stalked, compound; sometimes reduced.
Antennules	Peduncles with 3 segments; flagella often reduced or vestigial.
Antennae	Peduncles usually with 1 or 2 segments; usually without exopod; flagella short, sometimes absent.
Mandibles	With or without palp; molar and incisor processes usually not distinct.
Maxillulae	Usually with endopodal palp.
Maxillae	Biramous; usually with bilobed endites.
Maxillipeds	Flagella often reduced or absent; 3rd often with ischium and merus flattened, operculate.
Thoracic appendages	First chelate; 5th, or 4th and 5th, occasionally subchelate; 5th occasionally paddle-shaped.
Abdominal appendages	In males 1st and 2nd pairs developed as gonopods, 3rd–5th absent; in females 2nd–5th usually developed, rarely 1st pair also developed; uropods in both sexes usually absent; sometimes present in Dromioidea.
Telson	Usually reduced.

Tagmata	Cephalothorax and abdomen.
Somites	Head with 5 + 3 thoracic (maxillipeds); thorax with 5; abdomen with 6 or fewer (some frequently coalesced), excluding telson.
Sexual characters	Gonopores on coxae or sternite of 3rd pereopods in female, on 5th in male; 1st and 2nd pairs of pleopods usually modified as gonopods in male.
Sexes	Presumably separate.
Larval development	Metamorphic; zoea → megalopa.
Fossil record	L. Jurassic to Recent.
Feeding types	Various.
Habitat	Marine, freshwater, and semiterrestrial.
Distribution	Worldwide.

The infraorder Brachyura is divided into five major sections, the Dromiacea, Oxystomata, Oxyrhyncha, Cancridea, and Brachyrhyncha. Although considerable variation occurs, the basic brachyuran characters can be recognized. Examine the external morphology of all the available taxa, paying particular attention to the regions of the carapace, the structure and position of the antennules and antennae, and the ocular peduncles and orbits (see Figures 51 and 52). What differences from the structures in anomurans do you notice? In the oral region, observe the complete closure often obtained by the expanded and operculate segments of the 3rd maxillipeds.

In contrast to the anomurans, the gonopores of brachyurans are not readily discernible, as they are hidden by the tightly folded-under abdomen. The sex of the animal may be determined by the shape of the abdomen. In males the abdomen is acutely triangular; in females it is broadly triangular to semisubcircular. Care must be exercised in attempting to sex juveniles or young adults as the abdomen frequently is not appreciably expanded in immature females. If the abdomen is extended, the pairs of modified pleopods that form the gonopods of the males or the posterior 4 pairs of female pleopods can be observed.

After completing your general examination of the external morphology proceed to an examination of the major organ systems before studying the mouthparts, as the removal of these appendages often will cause serious damage to the cephalic region of the animal (see Figure 53). The description of the internal anatomy is based on specimens of the Portunidae; however, the information is sufficiently general to be applicable to most of the larger, unspecialized taxa. Begin by carefully removing the dorsal carapace. This can best be done by cutting, with a pair of sharp scissors, first the posterior margin and then about 1 to 5 mm from the dorsolateral margin on each side and on the anterior margin just behind the orbits. Starting at the posterior margin, carefully begin to separate the carapace from the underlying hypodermis. In the region of the cervical groove, a pair of prominent muscles, the anterior dorsal pyloric muscles, attached to the carapace will have to be severed. A few additional muscle strands probably also will have to be cut to remove the carapace. With the removal of the hypodermis, certain muscles, the stomach, heart, and the gills will be apparent. In the dorsal midline anteriorly to either side of the stomach identify the median posterior gastric muscles. Posteriorly in the midline observe the large heart; a pair of prominent ostia is present on the dorsal surface. Two additional pairs can be seen laterally if the heart is gently pushed to one side. Three arteries, the unpaired ophthalmic and paired antennal, leave the heart anteriorly. Follow the ophthalmic artery across the dorsal surface of the stomach; anteriorly it turns ventrally and branches to provide blood to the supraesophageal ganglion, eyes, and antennules. It will be difficult to trace the path of the antennal arteries until components of the digestive system have been examined and removed. Posterolaterally of the heart, behind the branchial chambers, observe the flattened pericardial sacs. From the posterior margin of the heart the unpaired superior abdominal artery turns ventrally and gives off a pair of branches that provides blood to the dorsal proximal muscles of the more posterior pairs of pereopods. It will be difficult to trace these and the ventral hepatic arteries until some of the other organs are removed.

Laterally the gills may be seen to converge to an apex abutting a portion of the endoskeleton. Overlying the gills observe the long epipodite of the 1st maxilliped. What is the gill formula of your specimen? Remove the

Figure 51 Decapoda, Brachyura: A. Typical raninoidean; B. Typical dromioidean; C. Typical caloppoidean; D. Typical majoidean; E—I, General structures from Portunoidea. E. 3rd maxilliped; F. 1st maxilliped; G. Maxilla; H. Maxillule; I. Male gonopod (1st pleopod).

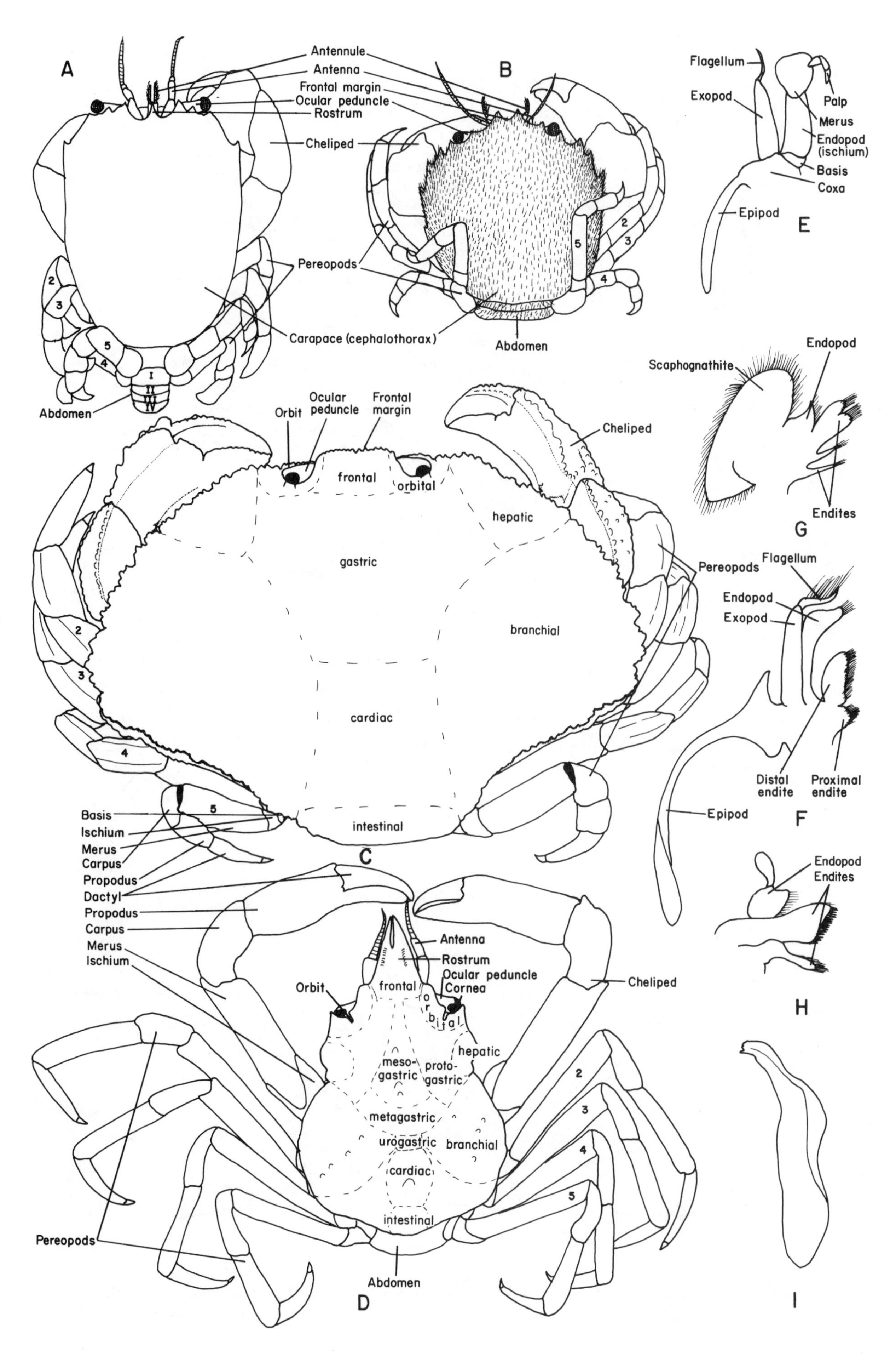
A
Antennule
Antenna
Frontal margin
Ocular peduncle
Rostrum
Cheliped
Pereopods
Carapace (cephalothorax)
Abdomen
B
Abdomen
E
Flagellum
Exopod
Palp
Merus
Endopod (ischium)
Basis
Coxa
Epipod
G
Scaphognathite
Endopod
Endites
C
Orbit
Ocular peduncle
Frontal margin
Cheliped
frontal
orbital
hepatic
gastric
branchial
cardiac
intestinal
Pereopods
Basis
Ischium
Merus
Carpus
Propodus
Dactyl
F
Flagellum
Endopod
Exopod
Distal endite
Proximal endite
Epipod
D
Propodus
Carpus
Merus
Ischium
Antenna
Rostrum
Ocular peduncle
Cornea
Orbit
frontal
orbital
hepatic
meso-gastric
proto-gastric
metagastric
urogastric
branchial
cardiac
intestinal
Cheliped
Pereopods
Abdomen
H
Endopod
Endites
I

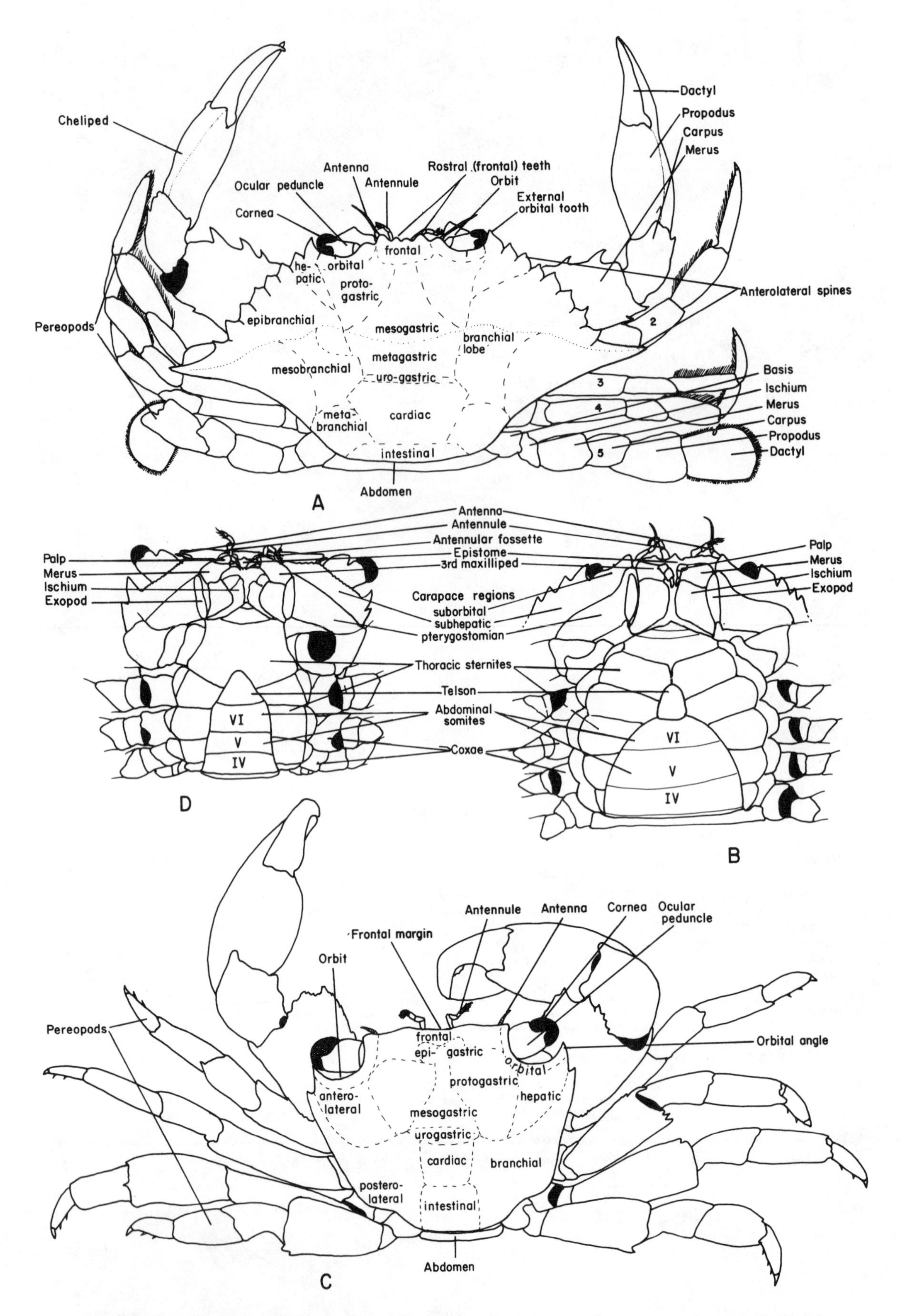

Figure 52 Decapoda, Brachyura: A, B, Portunoidea; C, D, Grapsoidea. A. Typical portunid (dorsal view) showing regions of carapace; B. Female portunid (ventral view, with pereopods removed); C. Typical grapsid (dorsal view) showing regions of carapace; D. Male grapsid (ventral view, with pereopods removed).

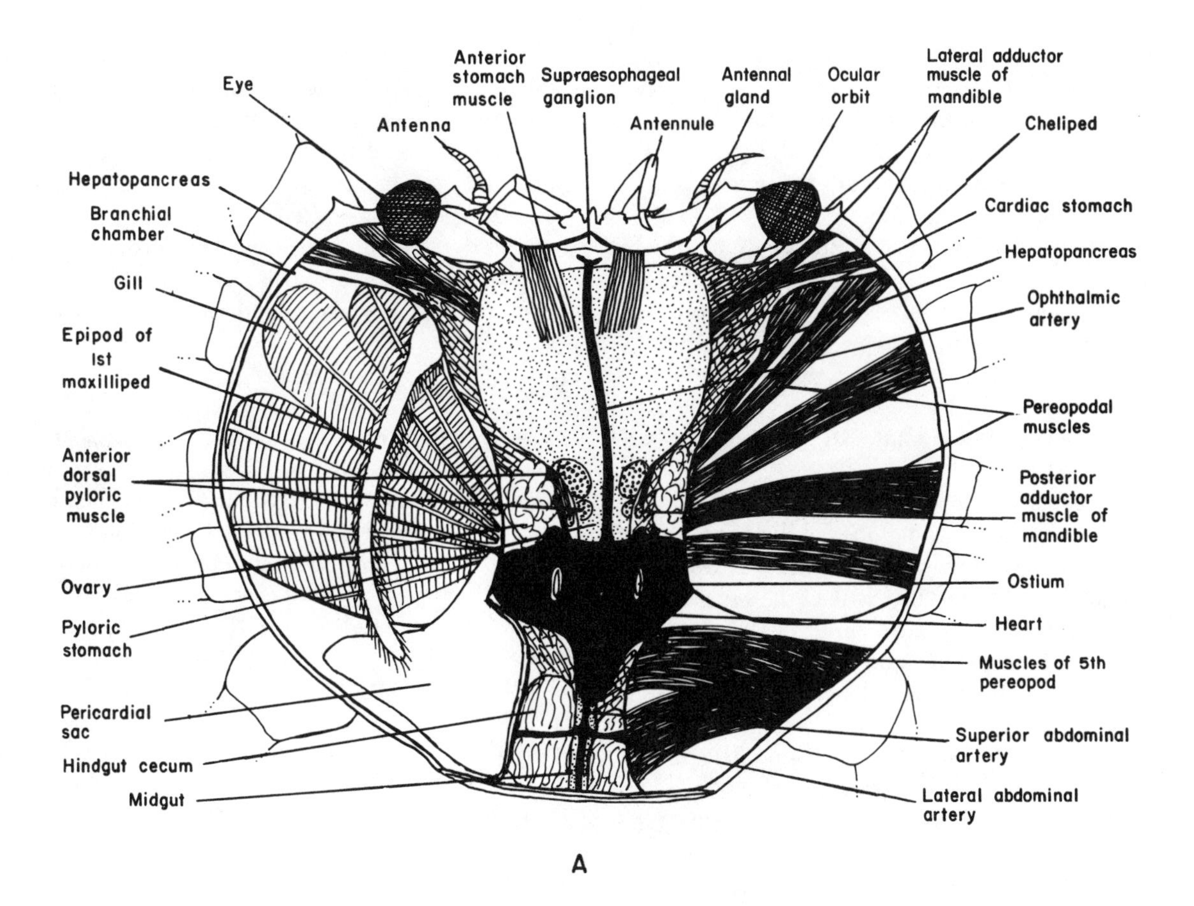

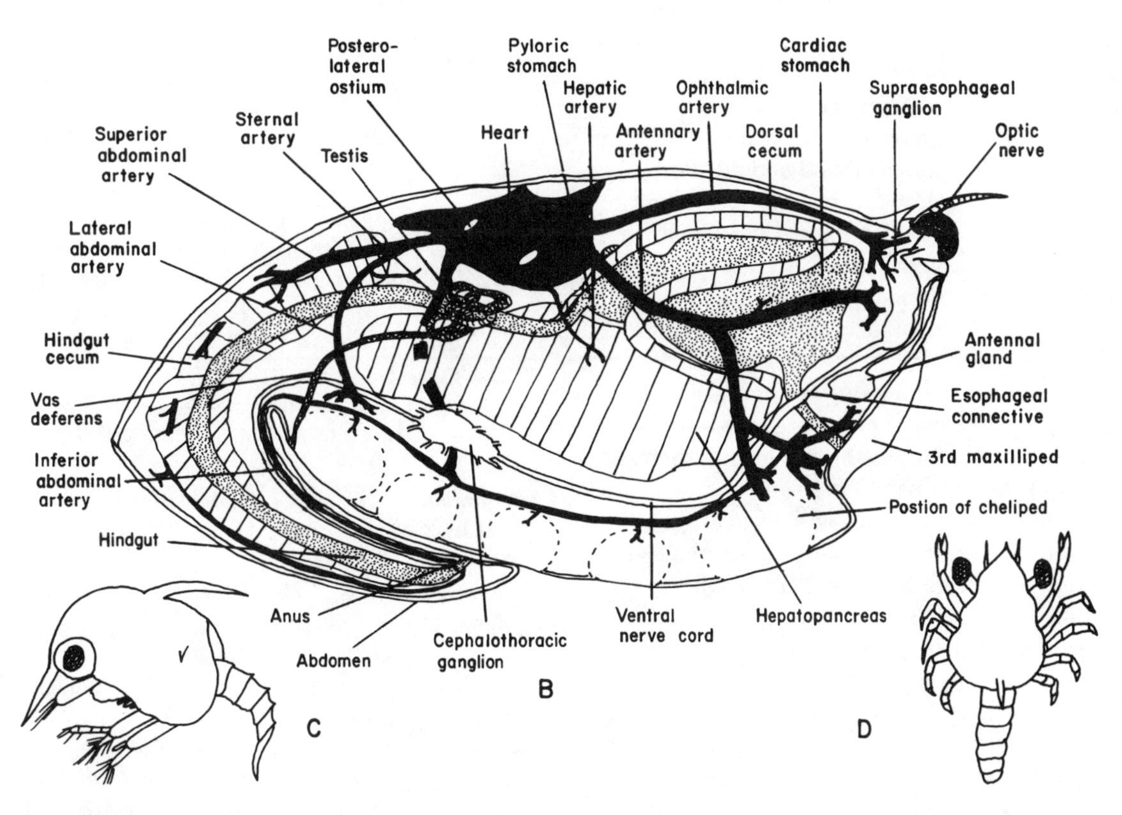

Figure 53 Decapoda, Brachyura: A. Diagrammatic portunid (dorsal view) with carapace removed to show gills and pericardial sac (left) and pereopodal musculature (right); B. Diagrammatic portunid (lateral view) with appendages, gills, and musculature removed to show major organ systems; C. Zoea; D. Megalopa.

gills and locate the smaller but similarly appearing epipodites of the 2nd and 3rd maxillipeds. Remove the stiff membranous floor of the branchial chamber to expose the muscles of the cheliped and first 3 pereopods; removal of the pericardial sac and underlying membrane will expose the muscles of the 5th pereopod.

Lateral to the stomach the mass of tissue filling much of the cephalothorax represents lobes of the hepatopancreas or 1st pair of ceca. If this is removed carefully, 2 mandibular muscles can be distinguished: the large lateral adductor muscle and the smaller external abductor muscle. The latter can be traced from the mandible to a point on the endoskeleton just anterior to the apex of the gills.

Ovaries or testes lie slightly posterior to the pyloric stomach on top of or among the lobes of the hepatopancreas. The difference in texture of the gonadal tissue will aid in distinguishing it from the tissue of the hepatopancreas. The ovaries actually are connected by a cross-bridge just behind the pyloric stomach. A short distance behind the bridge are the paired oviducts, but these usually are concealed in the ovaries. Near the median border of each duct is an oval-shaped seminal receptacle that extends ventrally almost to the sternal floor; ducts lead from the receptacles to the gonopores on the 6th somite. The testes similarly are connected by a bridge behind the pyloric stomach. The vas deferens leaving the testes are coiled, and in mature individuals can be separated into 4 distinct regions related to spermatophore production.

After tracing the midgut from the pyloric stomach to the abdomen, carefully remove the stomach by cutting it at its junctions with the esophagus and midgut. Posterior and lateral of the stomach a 2nd pair of thin, elongate ceca can be observed. A 3rd, unpaired cecum, arising from the hindgut, is located behind or posterior to the heart. Examine the gastric mill of the cardiac stomach. How does this grinding apparatus compare with those of the other decapods studied?

The general plan of the nervous system is similar to that of the crayfish but may not be observed as easily. The major difference between the nervous systems of the two is that in the brachyuran the ventral ganglia are fused into a single mass. The supraesophageal ganglion lies immediately behind the rostrum beneath the transverse apodeme between the eyes. Trace the major nerves radiating from the ganglion; these are the same as have been seen in other decapods. The esophageal connectives are considerably longer than in other decapods because the fused ventral ganglia are situated between the 4th and 5th thoracic somites. Now remove and examine the mouthparts. What structural similarities and differences do you find between the mouthparts of brachyurans and those of other decapods?

Larval development is metamorphic. The first post-embryonic stage is a zoea; however, since rudiments of the posterior thoracic appendages often are present, it may be considered a later substage than typically seen in anomuran zoeae. The brachyuran megalopal appendages resemble those of the adult; however, the abdomen is large and usually carried in an extended position; the 5 pairs of pleopods are used in swimming. In cases of abbreviated development, the megalopal stage may be suppressed. Metamorphosis apparently is suppressed in most, if not all, freshwater and terrestrial species.

References

Alcock, A., 1901. *Catalogue of the Indian decapod Crustacea in the collection of the Indian Museum.* Part 1. Brachyura, Fasc. 1. Introduction and Dromides or Dromiacea (Brachyura Primigenia). 80 pp. Calcutta: Indian Museum.

Balss, H., 1940–1957. Decapoda. In H. G. Bronn, *Klassen und Ordnungen des Tierreichs.* 5(1) 7: 321–480. Leipzig: Akad. Verl.

Bennett, E. W., 1964. The marine fauna of New Zealand: Crustacea Brachyura. *Mem. N.Z. Oceanogr. Inst.*, no. 22: 7–120.

Bliss, D. E., 1968. Transition from water to land in decapod crustaceans. *Am. Zool.*, *8:* 355–392.

Bliss, D. E., and L. H. Mantel, 1968. Adaptations of Crustacea to land. *Am. Zool.*, *8:* 673–685.

Broekhuysen, G. J., Jr., 1936. On growth and distribution of *Carcinides maenas* (L.). *Arch. Néerland. Zool.*, *2:* 257–400.

Bourne, G. C., 1922. The Raninidae; a study in carincology. *J. Linn. Soc. London (Zool.)*, *35:* 25–79.

Calman, W. T., 1909. Crustacea. In R. Lankester (ed.), *A treatise on zoology,* pt. 7, Appendiculata, fasc. 3, 346 pp. London: Ray Soc.

Chace, F. A., Jr., 1940. Reports on the scientific results of the "Atlantis" expeditions to the West Indies, under the joint auspices of the University of Havana and Harvard University. The brachyuran crabs. *Torreia Havana*, *4:* 3–67.

———, 1951. The oceanic crabs of the genera *Planes* and *Pachygrapsus. Proc. U.S. Natl. Mus.*, *101:* 65–103.

Chace, F. A., Jr., and H. H. Hobbs, Jr., 1969. The freshwater and terrestrial decapod crustaceans of the West Indies with special reference to Dominica. *Bull. U.S. Natl. Mus.*, 292: 1–258.

Christensen, A. M., and J. J. McDermott, 1958. Life history and biology of the oyster crab *Pinnotheres ostreum. Biol. Bull.*, *114:* 146–179.

Christiansen, M. E., 1969. Crustacea Decapoda Brachyura. *Mar. Invert. Scand.*, *2:* 1–143.

Crane, J., 1937. The Templeton Crocker Expedition. III. Brachygnathous crabs from the Gulf of California and the west coast of lower California. *Zoologica*, *22:* 47–78.

———, 1975. *Fiddler crabs of the world.* Ocypodidae: Genus *Uca.* 736 pp. Princeton, N.J.: Princeton University Press.

Cronin, L. E., W. A. van Engel, D. G. Cargo, and F. J. Wojcik, 1957. A partial bibliography of the genus *Callinectes. Spec. Sci. Rept., Virginia Fish. Lab.*, no. 8, 21 pp.

Felder, D. L., 1973. An annotated key to crabs and lobsters

(Decapoda, Reptantia) from the coastal waters of the northwestern Gulf of Mexico. *La. St. Univ. Publ.* no. LSU SG-73-02: 1–103.

Flipse, H. J., 1930. Die Decapoda Brachyura der Siboga-Expedition. Part 6. Oxyrrhyncha: Parthenopidae. *Siboga-Exped. Mongr.* 39c[2]: 1–96.

Garth, J. S., 1940. Some new species of brachyuran crabs from Mexico and the Central and South American mainland. *Allan Hancock Pacific Exped.*, *5:* 53–127.

———, 1946. Littoral brachyuran fauna of the Galapagos Archipelago. *Allan Hancock Pacific Exped.*, *5:* 341–601.

———, 1958. Brachyura of the Pacific coast of America, Oxyrhyncha. *Allan Hancock Pacific Exped.*, *21*, pts. 1, 2: 1–499, 500–854.

———, 1960. Distribution and affinities of the brachyuran Crustacea. In The biogeography of Baja California and adjacent seas (Symposium). *Syst. Zool.*, *9:* 105–123.

Garth, J. S., and J. Haig, 1971. Decapod Crustacea (Anomura and Brachyura) of the Peru-Chile Trench. *Anton Bruun Rept.*, no. 6: 1–20.

Glaessner, M. F., 1969. Decapoda. In R. C. Moore (ed.), *Treatise on invertebrate paleontology,* Part R Arthropoda. *2:* R400–R651. Lawrence, Kans.: Geol. Soc. America and Univ. Kansas.

Gordon, I., 1963. On the relationship of Dromiacea, Tymolinae and Raininidae to the Brachyura. In H. B. Whittington and W. D. I. Rolfe (eds.), *Phylogeny and evolution of Crustacea*, pp. 51–57. Spec. Publ. Mus. Comp. Zool. Harvard. Cambridge, Mass.: Harvard University Press.

Gore, R. H., and L. E. Scotto, 1979. Crabs of the family Parthenopidae (Crustacea Brachyura: Oxyrhyncha) with notes on specimens from the Indian River region of Florida. *Mem. Hourglass Cruises 3:* 1–98.

Griffin, D. J. G., 1976. Spider crabs of the family Majidae (Crustacea: Brachyura) from the Philippine Islands. *J. Nat. Hist.*, *10:* 179–222.

Griffin, D. J. G., and H. A. Tranter, 1974. Spider crabs of the family Majidae (Crustacea: Decapoda: Brachyura) from the Red Sea. *Israel J. Zool.*, *23:* 162–198.

Guinot, D., 1967. Recherches préliminaires sur les groupements naturels chez les crustacés decapodes brachyoures. I. Les affinites des genres *Aethra*, *Osachila*, *Hepatus*, *Hepatella* et *Actaeomorpha*. *Bull. Mus. Natl. Hist. Nat.* (2) *38:* 744–762, 828–845.

———, 1968a. Recherches préliminaires sur les groupements naturels chez les crustacés decapodes brachyoures. IV. Observations sur quelques genres de Xanthidae. *Bull. Mus. Natl. Hist. Nat. 39:* 695–727.

———, 1968b. Recherches préliminaires sur les groupements naturels chez les crustacés decapodes brachyoures. VI. Les Carpilinae. *Bull. Mus. Natl. Hist. Nat. 40:* 320–334.

———, 1969a. Recherches préliminaires sur les groupements naturels chez les crustacés decapodes brachyoures. VII. Les Goneplacidae. *Bull. Mus. Natl. Hist. Nat.*, *41:* 241–265.

———, 1969b. Recherches préliminaires sur les groupements naturels chez les crustacés decapodes brachyoures. VII. Les Goneplacidae (suite). *Bull. Mus. Natl. Hist. Nat.*, *41:* 507–528.

———, 1969c. Recherches préliminaires sur les groupements naturels chez les crustacés decapodes brachyoures. VII. Les Goneplacidae (suite et fin). *Bull. Mus. Natl. Hist. Nat.*, *41:* 688–724.

———, 1971. Recherches préliminaires sur les groupements naturels chez les crustacés decapodes brachyoures. VIII. Synthése et Bibliographie. *Bull. Mus. Natl. Hist. Nat.*, *42:* 1063–1090.

Hagen, H. O. von, 1975. Klassifikation und phylogenetische Einordnung der Lautäusserungen von Ocypodiden und Grapsiden (Crustacea, Brachyura). *Z. Zool. Syst. Evol.*, *13:* 300–316.

Holthuis, L. B., 1959. The Crustacea Decapoda of Suriname (Dutch Guiana). *Zool. Verhandl.*, no. 44: 1–296.

Kensley, B., 1977. The South African Museum's Meiring Naude cruises. Part 2. Crustacea, Decapoda, Anomura and Brachyura. *Ann. S. Afr. Mus.*, *72:* 161–188.

Lewinsohn, Ch., 1977. Die Dromiidae des Roten Meeres (Crustacea Decapoda, Brachyura). *Zool. Verhandl.*, no. 151: 1–41.

———, 1977. Die Ocypodidae des Roten Meeres (Crustacea Decapoda, Brachyura). *Zool. Verhandl.*, no. 152: 45–84.

Man, J. D., de, 1891. Decapoden des Indischen Archipels. *Zool. Ergebn.*, *2:* 265–527.

———, 1902. Ergebnisse einer zoologischen Forschungsreise in den Molukken und Borneo, im Auftrage der Senckenberg. naturforsch. Gesellschaft ausgeführt von Dr. Willy Kükenthal. *Abh. Senckenberg. Naturf. Ges.*, *25:* 467–929.

Monod, Th., 1956. Hippidea et Brachyura ouest-Africains. *Mem. Inst. Fr. Afr. Noire*, no. 45: 1–674.

Nation, J. D., 1975. The genus *Cancer* (Crustacea: Brachyura): systematics, biogeography and fossil record. *Nat. Hist. Mus. L.A. Count. Sci. Bull.*, no. 23: 1–104.

Ortmann, A., 1892. Die Decapoden-Krebse des Strassburger Museum. *Zool. Jahrb. Syst.*, *6:* 532–588.

———, 1893. Die Decapoden-Krebse des Strassburger Museum. Pts. 6–8 (Brachyura). *Zool. Jahrb. Syst.*, *7:* 23–88, 411–495, 683–772.

Pequegnat, W. E., 1970. Deep water brachyuran crabs. In W. E. Pequegnat, and F. A. Chace, Jr. (eds.), *Contributions on the biology of the Gulf of Mexico.* Texas A & M Univ. Oceanogr. Stud. *1:* 171–204. Houston, Texas: Gulf Publ. Co.

Pequegnat, W. E., L. H. Pequegnat, R. W. Firth, Jr., B. M. James, and T. W. Roberts, 1971. *Gulf of Mexico deep-sea fauna. Decapoda and Euphausiacea.* 12 pp. Serial Atlas of the Marine environment, Folio 20. Am. Geogr. Soc.

Rathbun, M. J., 1893. Scientific results of explorations by the U.S. Fish Commission steamer "Albatross." XXIV. Descriptions of new genera and species of crabs from the west coast of North America and the Sandwich Islands. *Proc. U.S. Natl. Mus.*, *16:* 223–260.

———, 1905. Fauna of New England. 5. Lists of the Crustacea. *Occ. Pap. Boston Soc. Nat. Hist.*

———, 1925. The spider crabs of America. *U.S. Natl. Mus. Bull.* 129: 1–613.

———, 1930. The cancroid crabs of America of the families Euryalidae, Portunidae, Atelecyclidae, Cancridae and Xanthidae. *U.S. Natl. Mus. Bull.* 152: 1–609.

Rathbun, M. J., 1937. The oxystomatous and allied crabs of America. *U.S. Natl. Mus. Bull.* 166: 1–278.

Sakai, T., 1965. *The crabs of Sagami Bay.* 330 pp. Tokyo: Maruzen Co. Ltd.

———, 1975. *Crabs of Japan and the adjacent seas,* vols. 1–3. 773 pp., 461 pp., 251 pls. Tokyo: Kodansha Ltd.

Schmitt, W. L., 1921. The marine decapod Crustacea of California. *U. Calif. Publ. Zool. 23:* 1–470.

Serène, R., 1973. A key for the separation of the Indo-Pacific species of *Uca* Leach 1914. *Spl. Publ. Mar. Biol. Assoc. India,* pp. 349–361.

Serène, R., and P. Lohavanijaya, 1973. The Brachyura (Crustacea: Decapoda) collected by the Naga Expedition, including a review of the Homolidae. *Naga Rept., 4:* 1–146.

Stephensen, K., 1945. The Brachyura of the Iranian Gulf. *Danish Sci. Invest. Iran,* Pt. 4: 57–237.

Stimpson, W., 1907. Report on the Crustacea collected by the North Pacific Exploring Expedition 1853–1856. *Smiths. Misc. Coll., 49:* 1–240.

Tesch, J. J., 1918a. The Decapoda Brachyura of the Siboga-Expedition. I. Hymenosomidae, Retroplumidae, Ocypodidae, Grapsidae, and Gecarcinidae. *Siboga-Exped., Mongr. 39c:* 1–148.

———, 1918b. The Decapoda Brachyura of the Siboga Expedition. II. Goneplacidae and Pinnotheridae. *Siboga-Exped., Mongr. 39c;* 149–295.

Türkay, M., 1974. Die Gecarcinidae Asiens und Ozeaniens (Crustacea: Decapoda). *Senckenberg. Biol.* 55: 223–259.

Warner, G. F., 1967. Life history of the mangrove tree crab *Aratus pisoni. J. Zool. London, 153:* 321–335.

———, 1977. *The biology of crabs.* 202 pp. New York: Van Nostrand Reinhold Co.

Williams, A. B., 1965. Marine decapod crustaceans of the Carolinas. *Fish. Bull., 65:* 1–298.

———, 1976. Distinction between a Gulf of Mexico and a Carolinian Atlantic species of the swimming crab *Ovalipes* (Decapoda: Portunidae). *Proc. Biol. Soc. Wash., 89:* 205–214.

GENERAL CRUSTACEAN REFERENCES

Anderson, D. T., 1973. *Embryology and phylogeny in annelids and arthropods.* xiv + 495 pp. New York: Pergamon Press.

Balss, H., 1957. Decapoda. In H. G. Bronn, *Klassen und Ordnungen des Tierreichs.* 2nd ed. 5(1) (7): 1–1770. Leipzig: Acad. Verlags.

Bliss, D. E., L. G. Abele, H. L. Atwood, L. H. Mantel, A. J. Provenzano, Jr., D. C. Sandeman, F. J. Vernberg, and W. B. Vernberg (eds.), in press. *The biology of Crustacea: a multi-volume treatise.* New York: Academic Press.

Bliss, D. E., and L. H. Mantel (eds.), 1968. Terrestrial adaptations in Crustacea. *Amer. Zool., 8:* 307–685.

Bowman, T. E., 1971. The case of the nonubiquitous telson and the fraudulent furca. *Crustaceana, 21:* 165–175.

Calman, W. T., 1909. Crustacea. In R. Lankester (ed.), *A treatise on zoology.* Pt. 7, Appendiculata, fasc. 3. vii + 346 pp. London: Adam & Charles Black (Reprinted by A. Asher & Co., Amsterdam, 1964.)

———, 1911. *The life of Crustacea.* 289 pp. New York: Macmillan.

Cisne, J. L., 1974. Trilobites and the origin of the arthropods. *Science, 186:* 13–18.

Crustaceana. 1960– International Journal of crustacean research. Leiden: E. J. Brill.

Giesbrecht, W., 1913. Crustacea. In A. Lang, and Hescheler (eds.) *Handbuch der Morphologie der Wirbellosen Tiere.* v. 4.

Glaessner, M. F., 1960. The fossil decapod Crustacea of New Zealand and the evolution of the order Decapoda. *N.Z. Geol. Surv., Paleont. Bull.* 31: 1–79.

Green, J., 1961. *A biology of Crustacea.* 160 pp. London: Witherby.

Gurney, R., 1942. *Larvae of decapod Crustacea.* viii + 306 pp. London: Ray Society.

Hansen, H. J., 1925. *On the comparative morphology of the appendages in the Arthropoda. A. Crustacea.* 176 pp. Copenhagen: Gyldendalske.

———, 1930. *On the comparative morphology of the appendages in the Arthropoda.* 376 pp. Copenhagen: Gyldendalske.

Hazlett, B. A., 1975. Ethological analyses of reproductive behavior in marine Crustacea. *Pubbl. Staz. Zool. Napoli, 39* suppl. 1: 677–695.

Hessler, R. R., and W. A. Newman, 1975. A trilobitomorph origin for the Crustacea. *Fossils and Strata,* no. 4, pp. 437–459.

Kaestner, A., 1970. *Invertebrate zoology, 3. Crustacea.* xi + 523 pp. New York: John Wiley and Sons [1969] (translated from the German by H. W. and L. R. Levi).

Knowles, F. G. W., and D. B. Carlisle, 1956. Endocrine controls in the Crustacea. *Biol. Rev., 31:* 396–473.

Kükenthal, W., and T. Krumbach (eds.), 1926–27. Crustacea. *Handbuch der Zoologie.* vol. 3. Berlin: Der Gruyter.

Lockwood, A. P. M., 1967. *Aspects of the physiology of Crustacea.* x + 328 pp. San Francisco: W. H. Freeman and Company.

Makarov, R. R., 1978. Kaudalnaya tagma vysshikh rakoobraznykh (Crustacea: Malacostraca), yee biologicheskaya spetsifika i proiskhozhdenie. *Zh. Obshchei Biol. 39:* 927–939.

Manton, S. M., 1960. Concerning head development in the arthropods. *Biol. Rev., 35:* 265–282.

———, 1973. Arthropod phylogeny. A modern synthesis. *J. Zool.,* London, *171:* 111–130.

———, 1978. *The Arthropoda. Habits, functional morphology and evolution.* 550 pp. New York: Oxford University Press.

Milne Edwards, H., 1834–1840. *Histoire naturelle des Crustacés,* vols. 1–3. Paris: Atlas.

Moore, R. C., (ed.), 1969. *Treatise on invertebrate paleontology.* Pt. R. Arthropoda, 4, vols. 1, 2. xxxvi + 651 pp. Lawrence, Kans.: Geological Society of America and University of Kansas.

Richards, A. G., 1951. *The integument of arthropods.* 411 pp. Minneapolis, Minn.: University of Minnesota Press.

Schminke, H. K., 1976. The ubiquitous telson and the deceptive furca. *Crustaceana, 30:* 292–300.

Schmitt, W. L., 1965. *Crustaceans.* 204 pp. Ann Arbor, Mich.: University of Michigan Press.

Siewing, R., 1957. Morphologie der Malacostraca. *Zool. Jahrb. Abt. Anat., 75:* 39–176.

Snodgrass, R. E., 1952. *A textbook of arthropod anatomy.* 363 pp. Ithaca, N.Y.: Comstock Press.

———, 1956. Crustacean metamorphosis. *Smiths. Misc. Coll., 131:* 1–78.

———, 1958. Evolution of arthropod mechanisms. *Smiths. Misc. Coll., 138:* 1–77.

Stebbing, T. R. R., 1893. *A history of Crustacea, Recent Malacostraca.* xvii + 466 pp. London: Kegan Paul, Trench, Trubner and Co.

Tiegs, O. W., and S. M. Manton, 1958. The evolution of the Arthropoda. *Biol. Rev., 33:* 255–337.

Waterman, T. H. (ed.), 1960–61. *The physiology of Crustacea,* vols. 1, 2. xvii + 670, xiv + 681 pp. New York: Academic Press.

Whittington, H. B., and W. Rolfe (eds.), 1963. *Phylogeny and evolution of Crustacea.* 192 pp. Spec. Publ. Mus. Comp. Zool. Harvard. Cambridge, Mass.: Harvard University Press.

Young, D. (ed.), 1973. *Developmental neurobiology of arthropods.* vii + 268 pp. London: Cambridge University Press.

GLOSSARY OF MORPHOLOGICAL TERMS

Abdomen. Trunk somites (tagma) between thorax and telson; somites with or without limbs; syn., *pleon.*

Abdominal process. Fingerlike projection(s) on dorsal surface of abdomen (Cladocera).

Abdominal somite. Any single division of body between thorax and telson; syn., *pleomere, pleonite.*

Abreptor. Postabdomen, bent forward from junction with body, terminating in 2 claws (Cladocera).

Acanthopod. Appendage (cirrus) with short row of strong sharp spines distally at each articulation of greater curvature and few or no spines along lesser curvature (Cirripedia).

Acron. Anteriormost part of body carrying eyes; not considered true cephalic somite; syn., *ophthalmic somite, presegmental region.*

Adductor muscle. (1) Muscle attached to carapace for pulling it to body or connecting halves of bivalve carapace (e.g., Conchostraca, Ostracoda, Leptostraca); or (2) Muscle attached to scutum for closing aperture (thoracic Cirripedia).

Adductor muscle scar. Impression of adductor muscle on inner surface of each valve (Ostracoda); see *muscle scar.*

Adductor pit. Depression on inner surface of scutum between adductor ridge and occludent margin for attachment of adductor muscle (thoracic Cirripedia).

Adductor ridge. Linear elevation on inner surface bounding adductor pit on tergal side in sessile barnacles.

Aesthetasc. Sensory seta covered by delicate cuticle, often projecting from antenna or antennule; syn., *olfactory hair, ethetasc, esthete..*

Afferent channel. Opening through which water passes to gills.

***Ala* (pl. *alae*).** One of pair of posteriorly directed cephalic-shield extensions (Branchiura and some parasitic Copepoda); triangular lateral part of compartmental plate, with or without radii, overlapped by adjacent compartmental plate (balanomorph Cirripedia).

Aliform apophyses. Incurved anterior and posterior extremities of growth lines (Conchostraca).

Ambulatory leg. See *pereopod.*

Anal spines. Single row of spines on either side of postabdomen (Cladocera) or prominent telsonal spines (Conchostraca).

Annulus ventralis. Seminal receptacle of female crayfish.

Antenna. One of pair of appendages of second cephalic somite; syn., *2nd antenna.*

Antennal carina. See *decapod carapace, carina a.*

Antennal gland. One of pair of complex excretory glands in many malacostracans with duct opening on antenna; syn., *green gland.*

Antennal groove. See *decapod carapace, groove a.*

Antennal muscle scar. Impression of antennal muscle on inner surface of valve, situated in front of adductor muscle scar, generally above (in some podocopans behind) mandibular muscle scar (Ostracoda).

Antennal region. See *decapod carapace, region a.*

Antennal scale. See *scaphocerite.*

Antennal spine. See *decapod carapace, spine a.*

Antennular fossette. Depression, pit or socket containing basal portion of the antennule.

Antennular scale. See *stylocerite.*

Antennule. One of pair of appendages of first cephalic somite; syn., *1st antenna.*

Anterolateral region. See *decapod carapace, region b.*

Aperture. Posteroventral opening into mantle cavity (Cirripedia); cf. *orifice.*

Apex. Upper angle of scutum or tergum (thoracic Cirripedia).

Apodeme. Infold of exoskeleton for attachment of muscles.

Appendix interna. Median process of pleopodal endopod uniting members of each pair; syn., *stylamblys.*

Appendix masculina. Complex median process of 2nd pleopodal endopod of male Caridea and some other Eucarida used in copulation or spermatophore transfer.

Areola. Area between branchiocardiac grooves and posterior to cervical groove on dorsal surface of carapace (Astacidea and Austroastacidea).

Arthrobranch. Gill attached to articular membrane between appendage and body (Decapoda); syn., *arthrobranchiata.*

Arthrobranchiata. See *arthrobranch.*

Arthrophragm. See *endophragm.*

Article. Subdivision of antennal or antennular flagella or appendage (cf. *segment*).

Articular furrow. Groove near tergal margin of scutum or scutal margin of tergum forming part of articulation between the two valves (balanomorph Cirripedia).

Articular ridge. Linear elevation on scutum or tergum bordering articular furrow and with it forming articulation between the two valves (balanomorph Cirripedia).

Atrium oris. Preoral cavity, bounded ventrally by posteriorly directed labrum, dorsally by ventral surface of cephalon just behind mouth, and laterally by paragnaths and mandibles.

Attractor epimeralis muscle. Prominent muscle in many decapods, inserting along line of branchiocardiac groove.

Basal margin. Lower edge of scutum or tergum or other plate (thoracic Cirripedia).

Basicarinal angle. Intersection of basal and carinal margins of tergum (thoracic Cirripedia).

Basicerite. Second segment of antennal peduncle, bearing scaphocerite (Caridea).

Basioccludent angle. Intersection of basal and occludent margins of scutum (thoracic Cirripedia).

Basiophthalmite. Proximal segment of eyestalk, articulating with distal segment (podophthalmite) bearing corneal surface of eye.

Basipod(ite). See *basis.*

***Basis* (pl. *bases*).** Segment of protopod adjoining coxa and carrying exopod and endopod distally; also basal calcareous or membranous plate furnishing anchorage to substrate in sessile cirripeds.

Basiscutal angle. Intersection of basal and scutal margins of tergum (thoracic Cirripedia).

Basitergal angle. Intersection of basal and tergal margins of scutum (thoracic Cirripedia).

Biformes. Carapaces reflecting sexual dimorphism (e.g., Conchostraca), marked by differing valve proportions for each sex of the same species.

Biramous. Having two branches. Crustacean appendage with two rami; also antennule or antenna with two flagellar elements.

Biserial branch. Primary branch of dendrobranchiate gill subdivided into two rows or series.

Bopyridum. Postlarva of epicaridean isopod that is attached to permanent host.

Brain. See *supraesophageal ganglion.*

***Branchia* (pl. *branchiae*).** Thin-walled, fingerlike or leaflike structure extending outward from appendage or secondarily from side of body, functioning in respiration; syn., *gill.*

Branchial carina. See *decapod carapace, carina b.*

Branchial chamber. Area between body and carapace enclosing branchiae; syn. *gill chamber.*

Branchial glands. Masses of connective-tissue cells without ducts surrounding venous channels in branchiae.

Branchial region. See *decapod carapace, region c.*

Branchiocardiac carina. See *decapod carapace, carina c.*

Branchiocardiac groove. See *decapod carapace, groove b.*

Branchiostegal area. Part of carapace extending laterally and ventrally over branchiae.

Branchiostegal spine. See *decapod carapace, spine b.*

Branchiostegite. Expanded dorsal and lateral branchial region of carapace.

Buccal cavity. Area of cephalon containing mouthparts; bounded by epistome anteriorly and free margins of carapace laterally (Malacostraca).

Buccal frame. Structural region of cephalon enclosing mouthparts (Brachyura).

Buccal groove. See *decapod carapace, groove c.*

Bullate. Inflated, blisterlike.

***Calceolus* (pl. *calceoli*).** Complex sensory filaments on antennules (Amphipoda).

Calyptopis stage. Third larval stage, characterized by differentiation of abdomen and appearance of compound eyes (Euphausiacea); see also *zoea.*

Capitulum. Portion of carapace enclosing body, commonly protected by calcareous plates (lepadomorph cirripeds); or anterior prominence in complex tooth and socket hingement (Ostracoda).

Carapace. Cuticular, usually calcified, structure arising from posterior margin of cephalon, extending anteriorly and posteriorly, often covering head and thorax (cf. Eucarida); also fold of integument extending from maxillary segment, forming bivalved shell of cyprid larvae and of ascothoracicans and mantle of other Cirripedia; mantle usually with calcified plates in thoracicans.

Carapace angles. Intersection of straight dorsal margin by anterior rib (α) and posterior rib (β) (Conchostraca).

Carapace carina. Narrow ridge on surface of carapace (cf. *decapod carapace* and *stomatopod integumental ornamentation).*

Carapace costae. Closely spaced radial ridges, grading from fine to coarse, not crossing umbo (Conchostraca); syn., *radial lirae.*

Carapace costellae. Fine radial ridges extending from ventral margin across umbo (Conchostraca).

Carapace groove. Furrow on surface of carapace (cf. *decapod carapace* and *stomatopod integumental ornamentation).*

Carapace growth line. Peripheral margin of successive membranes added to shell during each molt (Conchostraca).

Carapace lirae. Linear concentric ridges parallel to and between growth lines (Conchostraca).

Carapace region. Differentiated portion of carapace surface (cf. *decapod carapace).*

Carapace spine. Sharp projection from carapace (cf. *decapod carapace).*

Carapace tooth. Generally blunt projection of carapace, often broader than spine (cf. *decapod carapace* and *stomatopod integumental ornamentation).*

Cardiac notch or incision. Indentation on posterior margin of carapace.

Cardiac region. See *decapod carapace, region d.*

Cardiac tooth. See *decapod carapace, teeth a.*

Cardo. Basal segment of maxillule articulating with cephalon.

Caridean lobe. External rounded projection on basal part of exopod of 1st maxilliped (Caridea).

Caridoid facies. Basic group of characters distinguishing eumalacostracan crustaceans: enclosure of thorax by carapace, movable stalked eyes, biramous antennules, antennae with scaphocerites, thoracopods with natatory exopods, ventrally flexed abdomen, and tailfan.

Carina. Any keellike structure; any well-defined projecting ridge on outer surface of carapace (podocopan Ostracoda); unpaired posterodorsal plate of thoracic Cirripedia (in lepadomorphs 1 of up to 4 unpaired plates of capitulum; in verrucomorphs compartmental plate between rostrum and fixed tergum; in balanomorphs compartmental plate, with alae on each side, opposite rostrum).

Carinal latus. See *latus.*

Carinal margin. Edge of any plate adjacent to carina; occluding with carinal margin of opposed tergum (thoracic Cirripedia).

Carinate. Valve bearing rib(s) (Conchostraca).

Carinolateral. One of pair of compartmental plates typically overlapping carina on each side, with radius on carinal side and ala on lateral side, sometimes absent; homologous with lepadomorph carinal latus (balanomorph Cirripedia).

Carpocerite. Distal (5th) segment of antennal peduncle.

Carpopod(ite). See *carpus.*

Carpus. Antepenultimate segment of thoracopod or pereopod; syn., *carpopod.*

Caudal appendage. One of terminal, multiarticulate or simple, uniramous paired appendages homologous with caudal furca (Cirripedia).

Caudal fan. Powerful swimming structure formed of laterally expanded uropods and telson; syn. *tailfan.*

Caudal furca. Paired caudal rami of terminal abdominal segment or telson; syn., *furca.*

Caudal process. Posterior projection of valve border generally above midheight, or posteroventral, directed upward (Ostracoda).

Caudal ramus. One of paired appendages constituting caudal furca, usually rodlike or bladelike, sometimes filamentous and multiarticular; syn., *caudal filament, caudal style, cercus, cercopod, furcal ramus, stylet.*

Caudal siphon. Posteroventral opening in valve border; sometimes produced as tubular structure (Ostracoda).

Caudal style. See *caudal ramus.*

Cement gland. Special concentration of cells in dermal cover that secretes calcareous substances of valves (Cirripedia).

Cephalic constriction. Constriction delimiting anterior antennulary part of head from posterior part (Mystacocarida).

Cephalic flexure. Forward, or sometimes upward, deflection of anterior sterna (Decapoda).

Cephalic shield. Chitinous or more or less calcified covering of the head region, formed of fused tergites of cephalic somites, commonly having pleura.

Cephalic somite. One unit of cephalon bearing distinctive pair of appendages; syn., *cephalomere.*

Cephalomere. See *cephalic somite.*

Cephalon. Anteriormost tagma, bearing eyes, mouth, two pairs of antennae, and three pairs of mouthparts; syn., *head.*

Cephalosome. Head region with one or two fused thoracic somites (Copepoda).

Cephalothorax. Anterior part of body composed of fused cephalic and thoracic somites; latter with appendages modified as mouthparts, sometimes also with relatively unmodified appendages.

Cercopod. See *caudal ramus.*

***Cercus* (pl. *cerci*).** See *caudal ramus.*

Cervical furrow. See *decapod carapace, groove d.*

Cervical groove. See *decapod carapace* or *stomatopod integumental ornamentation.*

Cervical notch or incision. Strong indentation of carapace at level of cervical groove.

Cervical sinus. Rounded to angular indentation anteriorly along dorsal edge of carapace (Cladocera).

Cervical suture. See *cervical groove.*

***Chela* (pl. *chelae*).** Distal part of appendage; pincerlike, with opposed movable and immovable fingers; occasionally both fingers movable.

Chelate. Pincerlike.

Cheliped. Any thoracopod bearing chela.

Cincinnulus. See *retinaculum.*

***Cirrus* (pl. *cirri*).** Thoracic, usually biramous, multiarticulate appendage generally flattened laterally and curled anteriorly, with food-gathering function; anterior and posterior margins designated lesser and greater curvature respectively (Cirripedia, except Ascothoracica).

Clasper. Appendage, including antenna, modified for holding female during copulation; or an organ for fixation in parasites.

Clypeus. Part of cephalon carrying labrum; plate anteromedially on head formed by fusion of basal segments of antennae (Anostraca).

Colleteric gland. Single or paired gland in female or hermaphrodite producing viscid material for binding eggs together (Rhizocephala).

Comb collar. Retractable membrane supporting row of uniform setae, at superior angle of aperture (Acrothoracica).

Compartmental plate. One of several rigidly articulated plates forming wall in sessile cirripeds.

Compound eyes. Paired array of contiguous ommatidia having common optic nerve trunks.

Compound rostrum. Compartmental plate formed by fusion of rostrolaterals with rostrum or of rostrolaterals, with rostrum missing (balanomorph Cirripedia).

Copepodid. Postnaupliar developmental stage (Copepoda).

Cor frontale. Special pulsating structure or accessory heart formed from enlargement of blood vessel; contraction caused by outer tangential muscles or internal muscles derived from muscles having other functions.

Cormopod(ite). See *thoracopod.*

Cornea. Transparent cuticle covering ommatidia of compound eye.

Corpus mandibulae. See *mandible body.*

Coxa. Segment of appendage adjoining sternite, except in forms having precoxa; syn., *coxopodite.*

Coxal endite. Lobe produced from inner margin of coxa.

Coxal exite. Lobe produced from outer margin of coxa; syn., *coxepipod.*

Coxal plate. Lateral expansion of coxa broadly joined to lateral margin of tergite.

Coxepipod(ite). See *coxal exite.*

Crista dentata. Toothed crest on ischium of 3rd maxilliped (Decapoda).

Cryptoniscus. Planktonic larval stage of epicaridean isopod with pereopods modified as holdfasts. Stage at which larva seeks permanent host.

Ctenopod. Appendage (cirrus) usually with long paired setae on segments of lesser curvature and few setae distally on each articulation of greater curvature.

Cycladiformes. Carapace with dorsal margin of valve forming obtuse angle with posterior margin (Conchostraca).

Cyclops stage. Post-metanaupliar stage (Copepoda).

Cyprid. Bivalve larval stage (Cirripedia).

Cyrtopia. Formerly considered 5th larval stage with antennae no longer used in locomotion; currently included in furcilia stage (Euphausiacea).

Dactyl(us). Ultimate segment of thoracopod; syn., *dactylopodite.*

Dactylopodite. See *dactyl.*

Decapod carapace. Weakly or strongly calcified integuments covering cephalothorax; variously subdivided marked, and armed; named parts:

Regions:

(a) antennal. Anterior marginal part bordering orbital region laterally, adjoining hepatic, pterygostomial, and occasionally also frontal regions.

(b) anterolateral. Lateral part bordering subhepatic or hepatic regions.

(c) branchial. Lateral part posterior to ptergostomial region, overlying branchiae; epibranchial, mesobranchial, and metabranchial lobes or areas sometimes distinguished.

(d) cardiac. Median part posterior to cervical groove between urogastric and intestinal regions.

(e) frontal. Anteromedian part including rostrum and region behind it.

(f) gastric. Median part anterior to cervical groove and posterior to frontal region; sometimes epigastric, mesogastric, metagastric, protogastric, and urogastric (genital) areas distinguished.

(g) hepatic. Part adjoining antennal, cardiac, and pterygostomial regions.

(h) intestinal. Short transverse part posterior to cardiac region; sometimes referred to as posterior cardiac lobe.

(i) jugal. See pterygostomial region.

(j) orbital. Part posterior to eyes bordered by frontal and antennal regions.

(k) pterygostomial. Anterolateral part on ventral surface located on opposite sides of buccal cavity; syn., *jugal region, pterygostome.*

(l) subhepatic. Part on ventral surface below hepatic region, bounded by pterygostomial and suborbital regions.

(m) suborbital. Part on anteroventral surface beneath orbit.

Carinae:

(a) antennal. Extending posteriorly from antennal spine.

(b) branchial. Extending posteriorly from orbit over branchial region.

(c) branchiocardiac. Dividing branchial from cardiac region.

(d) gastroorbital. Extending posteriorly from supraorbital spine; syn., *supraorbital.*

(e) lateral. On lateral margin of carapace.

(f) orbital. On margin of orbit.

(g) posterior. Transverse ridge anterior to marginal groove.

(h) postorbital. Slightly posterior and parallel to orbital margin.

(i) postrostral. Posterior to rostrum along dorsal midline.

(j) rostral. Continuous with lateral margin of rostrum.

(k) subhepatic. Extending posteriorly from branchiostegal spine.

(l) submedian. On either side of, and parallel to, postrostral carina, sometimes joining rostral carina.

(m) supraorbital. See *decapod carapace, carina d.*

Grooves:

(a) antennal. Extending posteriorly from vicinity of antennal spine.

(b) branchiocardiac. Oblique groove approximately in middle of posterior half on each side of carapace, separating branchial and cardiac regions and reaching dorsomedian part well posterior to cervical or postcervical grooves; sometimes longitudinal, connecting cervical and postcervical grooves or extending posteriorly from submedian point on postcervical groove.

(c) buccal. Transverse groove crossing mandibular elevation behind antennal spine, connecting gastroorbital and antennal grooves (Nephropidae).

(d) cervical. Transverse groove medially between gastric and cardiac regions, curving toward antennal spine; syn., *cervical suture, cervical furrow.*

(e) dorsomedian. Longitudinal groove extending from tip of rostrum to posterior carapace margin dorsomedially (Nephropidae).

(f) gastroorbital. Short, longitudinal groove branching from cervical groove at level of orbit and directed toward it.

(g) hepatic. Short, longitudinal groove connecting cervical with postcervical and branchiocardiac grooves, more or less continuous with antennal groove.

(h) inferior. Extending from junction of hepatic and cervical grooves toward lateral margin, more or less continuous with cervical groove.

(i) intercervical. Oblique groove connecting postcervical and cervical grooves (Nephropidae).

(j) intestinal. Short, transverse groove in median part of posterior carapace, interrupted by intestinal tubercle (Nephropidae).

(k) marginal. Close to, and parallel with, posterior margin.

(l) parabranchial. Groove below, behind and almost parallel with branchiocardiac and postcervical grooves, joining latter in lower part (Nephropidae).

(m) postcephalic. One of three transverse furrows on carapace of many fossils.

(n) postcervical. Posterior to, and parallel with, cervical groove, bisecting cardiac region.

(o) sellar. Short transverse groove dorsally anterior to cervical groove (Nephropidae).

(p) submedian. Longitudinal groove in submedian dorsal part, contiguous with postrostral carina.

(q) urogastric. Short transverse groove in median or submedian region posterior to postcervical groove, sometimes joining upper part of postcervical groove (Nephropidae).

Spines:

(a) antennal. On anterior margin slightly below orbit.

(b) branchiostegal. On or close to anterior margin medially between antennal and pterygostomial spines.
(c) hepatic. Below and posterior to lower branch of cervical groove.
(d) infraorbital. On lower angle of orbit.
(e) postorbital. At moderate distance posterior to middle of orbit.
(f) postrostral. Immediately posterior to rostrum.
(g) pterygostomial. On anterolateral angle.
(h) suborbital. Slightly below and posterior to middle of orbit.
(i) supraorbital. At moderate distance obliquely behind and above orbit; sometimes on postorbital carina.

Teeth:
(a) cardiac. On midline of carapace just posterior to cervical groove.
(b) gastric. On midline of carapace immediately anterior to cervical groove.
(c) lateral. On lateral margin of carapace; anterolateral, mediolateral, and posterolateral teeth distinguished.
(d) orbital. On orbital margin.
(e) posterior. On midline of carapace between posterior margin and marginal groove.
(f) pregastric. On midline of carapace between gastric tooth and rostrum.
(g) rostral. On rostrum, either single or multiple; upper, lower, and lateral teeth distinguished.

Deflected front. Broadly downturned front margin of carapace in some decapods.

Dendrobranch(ia). Type of gill with lamellae divided into arborescent bundles.

Depressor muscle. Muscle inserted at basicarinal angle of tergum (balanomorph Cirripedia).

Depressor muscle crests. Elevated denticles or ridges on inner surface of tergum near basicarinal angle for attachment of depressor muscles (balanomorph Cirripedia).

Dermal gland. Cell or concentration of cells in epidermis traversed by canals communicating with surface through fine ducts.

Deuterocerebrum. See *mesocerebrum.*

Diacresis. Transverse groove on posterior part of exopod (also rarely of endopod) of uropod; sometimes dividing exopod into 2 movable parts.

Distal. Part of structure farthest from midline of body or base of attachment; opposed to proximal.

Dorsal organ. Thickened glandular area of hypoderm on dorsal surface, usually in posterior part of cephalon, sometimes in anterior part; not homologous in all taxa.

Dorsal plate. Spindle-shaped division of carapace intercalated with median suture in some decapods.

Dorsomedian groove. See *decapod carapace, groove e.*

Dorsoventralis posterior. Prominent muscle connecting head apodemes with inner surface of carapace posterior to cervical groove in many decapods.

Doublure. Reflected margin of carapace (Stomatopoda).

Duplicature. That part of border of shell with calcareous peripheral part of inner lamina in contact with, or separated by vestibule from, outer lamina, generally narrow, sometimes extensive (Ostracoda).

Ecdysis. Act of molting the integument.

Efferent channels. Passageways through which water moves away from gills and out of branchial region.

Endite. Inwardly (medially) directed lobe of precoxa, coxa, basis, or ischium.

Endognath. Endopod of maxilliped.

Endophragm. Septum formed by cephalic and thoracic apodemes; syn., *arthrophragm.*

Endophragmal skeleton. Complex internal structure formed by fusion of apodemes, providing framework for muscle attachment.

Endopleurite. Lateral apodeme of endoskeleton (Decapoda).

Endopod(ite). Inner ramus of biramous appendage.

Endosternite. Mesodermal tendonous plate below anterior part of alimentary canal (Notostraca); also firm calcareous plate between nerve cord and alimentary canal anteriorly in thorax of some decapods.

Endostome. Palatelike part of buccal frame in some brachyuran decapods; syn., *palate.*

***Ephippium* (pl. *ephippia*).** Semielliptical modification of cuticle in dorsal region of carapace valves forming encasement for eggs, capable of withstanding dessication after being shed (Cladocera).

Epibranchial lobe. See *decapod carapace, region c.*

Epicaridum. First larval stage of epicaridean isopod; syn., *micronicus.*

Epigastric area. See *decapod carapace, region f.*

Epimeral fold. Steep fold of endopleurites connected with branchiostegite to form branchial chamber in some decapods.

Epimere. Each lateral part of integument of somite; syn., *epimeron, pleurepimere, pleurite, pleuron* (pl. *pleura*), *pleura* (pl. *pleurae*), *pleural lobe, tergal fold.*

Epimeron. See *epimere.*

Epipod(ite). Laterally directed exite of protopod, usually branchial in function.

Episternum. Posterolateral projection of various sterna (Decapoda).

Epistome. Plate of varying shape between labrum and bases of antennae; also sternum of antennal somite.

Esthestasc. See *aesthetasc.*

Esthete. See *aesthetasc.*

Exhalant passage. Anterior to gill chamber leading to large anterior opening, with scaphognathite for driving water outward.

Exite. Laterally directed lobe arising from external margin of protopodal segment.

Exognath. Exopod of maxilliped.

Exopod(ite). Outer ramus of biramous appendage.

Exoskeleton. Chitinous or calcified outer integument of crustaceans.

Eyestalk. See *ocular peduncle.*

Falcate. Sickle-shaped or hooked.

Filamentary appendage. Membranous process developed on body in some cirripeds commonly at base of cirrus (Lepadomorpha, Ascothoracica).

Filter chamber. Space beneath thorax enclosed by ventral body wall and moving thoracopods used for food-gathering.

First antenna. See *antennule.*

First maxilla. See *maxillule.*

Fixed finger. Immovable distal part of propodus of chela; syn., *pollex.*

Flabellum. Thin distal exite (Branchiopoda); or epipodite of thoracopod.

***Flagellum* (pl. *flagella*).** Multiarticulate distal portion of antennule, antenna, or exopod.

Flange. Ridge along valve margin formed by projection of outer lamella as narrow brim (Ostracoda).

Foregut. See *stomodeum.*

Fornix. Ridge in lateral part of cephalon above insertion of antennal muscles (Cladocera).

Free edge. Line of contact between closed valves except along hinge line marking distal limit of contact margin, sometimes inside free margin (Ostracoda).

Free margin. All parts of margin except hingement (Ostracoda).

Frena. Tegumentary folds holding eggs (Cirripedia).

Frontal appendages. Paired filaments arising from bases of antennae but independent of them (Anostraca).

Frontal band. Glandular organ of adhesion in frontal region of various parasitic copepods used for attachment to host.

Frontal eye. See *frontal organ.*

Frontal organ. Sensory cells on anterior surface of cephalon; syn., *haft organ* or *frontal eye* in non-malacostracans; not homologous among taxa.

Frontal plate. Modified rostrum with downward projecting process united with epistome (brachyuran Decapoda).

Frontal region. See *decapod carapace, region e.*

Frontolateral horn. One pair of tubular frontolateral extensions of cuticle of cirriped nauplius (except Ascothoracica).

Furca. See *caudal furca.*

Furcal ramus. See *caudal ramus.*

Furcilia stage. Last larval stage marked by movable compound eyes projecting beyond margin of carapace; antenna not used for locomotion in later substages (Euphausiacea).

Galea. Outer distal hoodlike lobe of second segment of maxillule.

Gastric groove. See *stomatopod integumental ornamentation.*

Gastric mill. Apparatus in cardiac stomach (stomodeum) with framework of movably articulated ossicles developed as thickened and calcified parts of stomodeal lining used to break up food, mostly highly specialized in decapods; also chitinous triturating apparatus in foregut of some acrothoracicans (Cirripedia).

Gastric region. See *decapod carapace, region f.*

Gastric tooth. See *decapod carapace, teeth b.*

Gastrolith. Discoid calcareous nodule, common in stomodeum (Decapoda).

Gastroorbital carina. See *decapod carapace, carina d.*

Gastroorbital groove. See *decapod carapace, groove f.*

Geniculate. Bent; having upper part of filament forming more or less obtuse angle with lower, e.g., antennule.

Genital region. See *decapod carapace, region f (urogastric).*

Gill. See *branchia.*

Gill chamber. See *branchial chamber.*

Glaucothöe. Stage in larval development of hermit crabs, equivalent to megalopa.

Gnathal lobe. Masticatory endite of mandible; syn., *masticatory process.*

Gnathobases. Paired endites used to manipulate or move food.

Gnathopod. Prehensile maxilliped; also 1st 2 prehensile pereopods of amphipods, either chelate or subchelate.

Gnathothorax. Cephalothorax with all appendages of fused thoracic somites modified as maxillipeds.

Gonad. Reproductive organ of either sex, communicating with pair of efferent ducts.

Gonapophysis. Median process from base of 1st or 2nd pleopods of male (Syncarida).

Gonochoristic. Sexes separate; producing distinct males and females.

Gonopod. Pleopod modified for reproductive purposes.

Gonopore. Outlet for genital products; syn., *sexual pore.*

Green gland. See *antennal gland.*

Haft organ. See *frontal organ.*

Head apodeme. Fused endopleurite and endosternite forming place for muscle attachment at anterior end of skeleton (Astacidea).

Hemocoel. Lacunar system extending throughout much of body, filled with blood.

Hemocyanin. Copper-containing respiratory pigment in blood (Malacostraca).

Hemoglobin. Oxygen-carrying protein colored substance of red plasma in blood of some crustaceans.

***Hepatic caecum* (pl. *caeca*).** See *hepatic cecum.*

***Hepatic cecum* (pl. *ceca*).** Pouchlike diverticulum generally connected with mesenteron, with liver function; also see hepatopancreas.

Hepatic groove. See *decapod carapace, groove g.*

Hepatic region. See *decapod carapace, region g.*

Hepatic spine. See *decapod carapace, spine c.*

Hepatopancreas. Digestive gland consisting of tubules ramifying through cephalothorax, with both liver and pancreas functions.

Hermaphrodite. Organism with both male and female reproductive organs.

Heterochelate. Chelae of left and right chelipeds differing in shape and size.

Heteromorph. Adult female, inferred by carapace structure, in dimorphic genera (Ostracoda).

Hindgut. See *proctodeum.*

Hinge line. Middorsal line of junction of two valves of carapace, permitting movement between them.

Hingement. Collective term for structures comprising articulation of valves (Ostracoda).

Hinge nodes. Localized thickened parts of right valve hinge (Phyllocarida).

Hinge selvage. Structure of hinge area corresponding to, and sometimes continuous with, selvage of contact margin (Ostracoda).

Hypobranchial space. Area of gill chamber below gills.

Hypopharynx. See *metastoma.*

Hypostoma. See *metastoma.*

Hypostome. See *metastoma.*

Incisor process. Biting portion of gnathal lobe of mandible; syn., *pars incisiva.*

Inferior groove. See *decapod carapace, groove h.*

Inframedian latus. See *latus.*

Infraorbital spine. See *decapod carapace, spine d.*

Inner lamina. Inner shell layer of compartmental plates separated from outer lamina by longitudinal tubes (balanomorph Cirripedia).

Interantennular septum. Plate separating antennular cavities in some malacostracans; syn., *proepistome.*

Intercervical groove. See *decapod carapace, groove i.*

Interlaminate figure. Simple or arborescent lines extending between epicuticle of outer lamina through longitudinal septa into inner lamina in sections parallel to base in some balanomorph Cirripedia.

Intermediate carina. See *stomatopod integumental ornamentation.*

Intermediate denticle. See *stomatopod integumental ornamentation.*

Intermediate tooth. See *stomatopod integumental ornamentation.*

Intestinal groove. See *decapod carapace, groove j.*

Intestinal region. See *decapod carapace, region h.*

***Intraparies* (pl. *intraparietes*).** Secondary lateral margin of carina in some lepadomorph cirripeds.

Ischiocerite. Third segment of antennal peduncle.

Ischiopod(ite). See *ischium.*

Ischium. Third segment of pereopod or 1st segment of endopod articulating with basis; syn., *ischiopod(ite).*

Jugal region. See *decapod carapace, region i.*

Kentrogon. Undifferentiated cells following cyprid (Rhizocephala).

Labium. See *metastoma.*

Labrum. Unpaired outgrowth arising just in front of mouth and often more or less covering it; syn., *upper lip.*

Lacinia. Inner distal spiny lobe of second segment of maxillule.

Lacinia mobilis. Small, generally toothed, process articulated with incisor process of mandible (Peracarida).

Lappet. Ventrally projecting subdivisions of pleura (Mysidacea).

Lasiopod. Appendage (cirrus) with setae in transverse row at each articulation (Cirripedia, limited usage).

Latera. See *latus.*

Lateral. One of pair of compartmental plates typically located between carino- and rostrolaterals, with radius on carinal side and ala on rostral side, sometimes between carina and rostrolateral or compound rostrum; homologous with lepadomorph median latus (balanomorph Cirripedia).

Lateral bar. Pair of external chitinous thickenings extending from apertural thickenings medially down each side of mantle sac (Acrothoracica).

Lateral carina. See *decapod carapace, carina e,* or *stomatopod integumental ornamentation.*

Lateral denticle. See *stomatopod integumental ornamentation.*

Lateral depressor pit. Small depression near basitergal angle of scutum for attachment of lateral depressor muscle (balanomorph Cirripedia).

Lateral cups. Paired element of nauplius eye of nonmalacostracan crustaceans.

Lateral gastrocardiac markings. Insertions of attractor epimeralis muscle in brachyuran decapods lacking branchiocardiac groove.

Lateral tooth. See *decapod carapace, teeth c,* or *stomatopod integumental ornamentation.*

***Latus* (pl. *latera*).** Any of lepadomorph capitular plates except paired scuta and terga and unpaired rostrum, carina, subrostrum, and subcarina. Smaller plates in basal whorls below paired latera referred to as lower latera; sometimes all or some absent. Types of paired plates:

(a) carinal. Plate on each side of carina (cf. *carinolateral*).

(b) inframedian. Plate below upper latus.

(c) median. Plate between rostral and carinal latera in forms with paired latera in one whorl (cf. *lateral*).

(d) rostral. Plate on each side of rostrum or below scutum (cf. *rostrolateral*).

(e) upper. Plate in upper whorl between scutum and tergum or carina.

Limadiiformes. Carapace with recurvature of posterior margin near dorsal line (Conchorstraca).

***Linea* (pl. *lineae*).** Linear marking on carapace.

Linea anomurica. Longitudinal groove or uncalcified line on carapace of many anomuran decapods.

Linea branchiostegalis. Longitudinal groove or uncalcified line extending posteriorly from anterior margin of carapace, slightly above branchiostegal spine, to or slightly beyond hepatic spine in some caridean decapods.

Linea dromica. Longitudinal groove or uncalcified line on dromioidean decapods comparable with linea thalassinica; syn., *linea dromiidica.*

Linea dromiidica. See *linea dromica.*

Linea homolica. Longitudinal groove or uncalcified line on homoloidean decapods comparable with or equivalent to linea thalassinica.

Linea lateralis. Longitudinal groove or uncalcified line extending posteriorly from frontal margin of carapace below orbit sometimes to posterior extremity of carapace in some penaeoideans.

Linea thalassinica. Longitudinal groove or uncalcified line on dorsal part of carapace extending from anterior margin below antennal spine to posterior margin in most thalassinoideans.

List. Ridge on proximal side of selvage on contact margin (Ostracoda).

Longitudinal canal. See *longitudinal tube.*

***Longitudinal septum* (pl. *septa*).** Partition disposed normal to inner and outer laminae of compartmental plate in some balanomorph cirripeds, resulting in longitudinal tubes; syn., *parietal tubes.*

Longitudinal tube. Canal formed in compartmental plate of some balanomorph cirripeds between longitudinal septa and inner and outer lamina; syn., *longitudinal canal; parietal tube, parietal pore.*

Lower lip. See *metastoma.*

Lunule. Attachment disc at base of antennule in some parasitic copepods.

Male-cell receptacle. Pocket or pair of pockets within mantle cavity of female receiving cells of male cyprid, later differentiated into "testes" (rhizocephalan Cirripedia).

Manca. Young of some Peracarida lacking last thoracopod at time of release from marsupium.

Mancoid stage. Postlarval stage with rudimentary 4th pleopod (Phyllocarida).

Mandible. One of 3rd pair of cephalic appendages, used to masticate food.

Mandible body. Inflated base of mandible for attachment of mandibular muscles; syn., *corpus mandibulae.*

Mandibular foramen. Relatively large opening in mandibular body for passage of transverse adductor muscle.

Mandibular palp. Distal articulated part of mandible used in feeding or cleaning; also nonsegmented part of mandible in Acrothoracica; attached to labrum in thoracic Cirripedia.

Mantle. Membranous covering of body, often strengthened by calcareous plates in thoracic Cirripedia (cf. *carapace).*

Mantle cavity. Space occupied by body, with aperture (Cirripedia).

Manus. Broad proximal part of propodal cheliped; syn., *palm.*

Marginal carina. See *stomatopod integumental ornamentation.*

Marginal groove. See *decapod carapace, groove k.*

Marsupium. Brood pouch.

Masticatory process. See *gnathal lobe.*

Mastigobranch. Slender respiratory process at base of epipod; syn., *mastigobranchia.*

***Mastigobranchia* (pl. *mastigobranchiae*).** See *mastigobranch.*

Mastigopus stage. Larval stage in development of some decapods equivalent to adult of *Leucifer.*

***Maxilla* (pl. *maxillae*).** Paired appendage of 5th cephalic somite, used in feeding, often also in respiration; syn., *2nd maxilla.*

Maxillary gland. Excretory organ in maxillary somite with duct opening on maxilla; syn., *shell gland.*

Maxillipeds. Paired appendages modified for feeding on 1st, up to 3rd, thoracic somites, usually fused to cephalon.

Maxillule. One of pair of 4th cephalic appendages, usually serving as mouthpart; syn., *1st maxilla.*

Median carina. See *stomatopod integumental ornamentation.*

Median dorsal plate. Elongate plate separating carapace valves posterodorsally in some phyllocarids.

Median eye. See *nauplius eye.*

Median latus. See *latus.*

Megalopa. First postlarval stage in development of many Eucarida.

Mereopod(ite). See *merus.*

Merus. Fourth segment (distally from body), articulating proximally with ischium and distally with carpus; syn., *meropod(ite).*

Mesenteron. Midportion of alimentary tract, of endodermal origin, with surface commonly increased by pouchlike extensions serving as digestive glands; syn., *midgut.*

Mesobranchial lobe or area. See *decapod carapace, region c.*

Mesocerebrum. Ganglion of antennular somite; syn., *deuterocerebrum.*

Mesogastric lobe or area. See *decapod carapace, region f.*

Mesosome. Collective term for all free thoracic somites behind head (not common usage).

Mesosternum. Median plate of sternum in some brachyuran decapods.

Metabranchial lobe or area. See *decapod carapace, region c.*

Metacerebrum. Ganglion of antennal somite; syn., *tritocerebrum.*

Metagastric lobe or area. See *decapod carapace, region f.*

Metanauplius. Postnaupliar larva with generally same body and body shape as nauplius, but with additional appendages.

Metasoma. See *metasome.*

Metasome. Part of prosome, consisting of free thoracic somites anterior to major articulation (Copepoda); or first three abdominal somites (Amphipoda, not general usage); syn., *metasoma.*

Metastoma. Lower lip posterior to mandibles, usually cleft into pair of lobes (paragnaths); syn., *hypostoma, hypostome, hypopharynx, labium.*

Metazoea. Last zoeal substage of larval brachyurans.

Metapon. Entire preoral area, including part of mandibular somite (Decapoda).

Micronicus. See *epicaridum.*

Midgut. See *mesenteron.*

Molar process. Grinding portion of gnathal lobe of mandible; syn., *pars molaris.*

Mouth cirri. First pair of cirri considerably modified (acrothoracican Cirripedia).

Movable finger. Dactyl of chela.

***Mucro* (pl. *mucrones*).** Spine on inferoposteal angle of carapace in some Cladocera.

Muscle scar. Mark on interior of valve or carapace indicating position of muscle attachment, generally distinguishable by localized difference in surface texture, elevation, depression, or delimiting narrow groove.

Mysis stage. Larval stage in penaeoidean development equivalent to zoeal stage of Nephropoidea; cf. *zoea;* syn., *schizopod larva.*

***Nauplius* (pl. *nauplii*).** Early larval stage with only 3 pairs of appendages, antennules, antennae, and mandibles.

***Nauplius eye*.** Unpaired median eye, common in nauplii and many adult crustaceans; structure simple, consisting of 1 to few light-sensitive cells; syn., *median eye*.

***Neck organ*.** See *nuchal organ*.

***Nephropore*.** Elevated opening of antennal gland on ventral surface of coxa of antenna.

***Nisto*.** Postlarval stage of some palinurans; syn., *pseudibacus*, *puerulus*.

***Notum*.** Posterior part of dorsal carapace in shrimplike decapods (limited usage).

***Nuchal organ*.** Sense organ on upper side of cephalon (Branchiopoda); syn., *neck organ*.

***Occipital notch*.** Angulated indentation at rear of cephalon (Conchostraca).

***Occludent margin*.** Margin of scutum and tergum bordering orifice (thoracic Cirripedia).

***Occludent teeth*.** Small projections on occludent scutal margin, formed by extensions of external growth ridges, interdigitating with teeth on margin of opposed scutum (balanomorph Cirripedia).

***Ocellus*.** Simple single eye or eye spot; common in some branchiopods and copepods (cf. *nauplius eye*).

***Ocular bulla*.** Knob on inner surface of carapace connecting lower and upper orbital margins with basal segment of antenna, serving to protect eye.

***Ocular papilla*.** Anterior projection on eyestalk (Mysidacea and one Anaspidacea).

***Ocular peduncle*.** Peduncle movably articulated with cephalon, with compound eye at distal end, sometimes with two or three segments, sometimes retractable; syn., *eyestalk*.

***Olfactory hair*.** See *aesthetasc*.

***Ommatidium* (pl. *ommatidia*).** Cylindrical or prismoidal visual constituent of compound eye covered by transparent cuticle (cornea).

***Oostegite*.** Inner, medially directed lamella arising from coxa of pereopod in female, part of midventral marsupium (Peracarida).

***Opercular membrane*.** Thin, flexible membrane attaching opercular valves to sheath, periodically shed during ecdysis, in balanomorph Cirripedia; represented by membranous hinge in verrucomorph Cirripedia.

***Opercular valve*.** One of 2 or 4 movable plates occluding aperture in sessile barnacles.

***Operculum*.** Scuta and terga (balanomorph) or movable scutum and tergum (verrucomorph) and associated membrane, forming apparatus occluding aperture.

***Ophthalmic somite*.** See *acron*.

***Optic lobe*.** Ganglion of brain or nervous system for innervation of eye.

***Orbit*.** Circular to rectangular opening in anterior face of carapace supporting ocular peduncle in some decapods.

***Orbital carina*.** See *decapod carapace, carina f*.

***Orbital hiatus*.** Gap or slit in orbital margin of carapace.

***Orbital region*.** See *decapod carapace, region j*.

***Orbital tooth*.** See *decapod carapace, teeth d*.

***Orifice*.** Opening in sessile barnacle wall occupied by operculum (cf. *aperture*).

***Ostium* (pl. *ostia*).** Valved opening in heart for return of blood.

***Oviduct*.** Passageway from ovary to genital aperture.

***Ovigerous frena* (pl. *frenae*).** Fleshy ridge or flap on inner surface of mantle adhering to and holding egg masses in place in some lepadomorph Cirripedia.

***Ovigerous lamella*.** Adherent egg masses forming one or more lamellae within mantle cavity, in some lepadomorph Cirripedia, held in place by ovigerous frena.

***Palate*.** See *endostome*.

***Palm*.** See *manus*.

***Palp*.** Usually one ramus (endopod), sometimes both, and basis, reduced distally to 1 to 3 segments, associated with mouthparts; also nonsegmented setose structure attached to mandible (Acrothoracica) and to lateral margin of labrum in thoracic Cirripedia.

***Palp foramen*.** Small opening in mandibular body communicating with mandibular palp.

***Parabranchial groove*.** See *decapod carapace, groove l*.

***Paracopulatory organ*.** Specialized endopod of pleopod in some isopods; used in copulation.

***Paragnath(s)*.** See *metastoma*.

***Paries* (pl. *parietes*).** Median triangular part of each compartmental plate in sessile barnacles.

***Parietal pore*.** See *longitudinal tube*.

***Parietal septum*.** See *longitudinal septum*.

***Parietal tube*.** See *longitudinal tube*.

***Pars ampullaris*.** Bottle-shaped diverticulum at entrance of ceca into pyloric chamber of stomach (Hoplocarida, Anaspidacea).

***Pars incisiva*.** See *incisor process*.

***Pars molaris*.** See *molar process*.

***Parva stage*.** First postlarval stage in development of caridean decapods.

***Pedigerous*.** Bearing footlike appendages.

***Peduncle*.** Stalk in lepadomorph cirripeds, supporting capitulum, attached to substrate by opposite end, commonly armed with calcareous scales.

***Penicilla* (pl. *penicillae*).** Dentate setae on mandible (Stygocaridacea).

***Penicillus* (pl. *penicilli*).** Tufts of fine hair resembling small brush.

***Penis* (pl. *penes*).** Male copulatory organ; probosciform in hermaphroditic thoracic Cirripedia and greatly distensible.

***Peraeon*.** See *pereon*.

***Peraeonite*.** See *pereonite*.

***Peraeopod*.** See *pereopod*.

***Pereion*.** See *pereon*.

***Pereionite*.** See *pereonite*.

***Pereiopod*.** See *pereopod*.

***Pereon*.** Anterior portion of trunk with thoracopods, exclusive of maxillipedal somites and appendages; syn., *peraeon, pereion.*

Pereonite. Somite of pereon; syn., *peraeonite, pereionite.*

Pereopod. Thoracic appendage used in locomotion; syn., *peraeopod, pereiopod, ambulatory leg, walking leg.*

Pericardium. Blood sinus surrounding heart and communicating with it by paired ostia.

Peritrophic membrane. Chitinous sheath secreted around feces.

Petasma. Abdominal structure developed from male pleopods, used in copulation.

P-4 structure (type A). Structure on lateral face of propodus of 4th pereopod in some pagurids; presumably sensory in function.

Photophore. Luminous organ.

Phyllobranch. Gill with leaflike filaments; syn., *phyllobranchia.*

Phyllobranchia. See *phyllobranch.*

***Phyllopodium* (pl. *phyllopodia*).** Leaflike thoracic appendages.

Phyllosoma. Larval stage in development of palinuroidean decapods.

Pleomere. See *abdominal somite.*

Pleon. See *abdomen;* also first 3 abdominal somites of amphipods.

Pleonite. See *abdominal somite.*

Pleopod. Paired appendage of any of first 5 abdominal somites (rarely 6) in Malacostraca, adapted for swimming; syn., *swimmeret.*

Pleotelson. Structure formed by fusion of one or more abdominal somites with telson.

***Pleura* (pl. *pleurae*).** See *epimere.*

Pleural lobe. See *epimere.*

Pleural suture. Line of separation of carapace in molting.

Pleurepimere. See *epimere.*

Pleurite. See *epimere.*

Pleurobranch. Gill attached directly to body wall (Decapoda); syn., *pleurobranchia.*

Pleurobranchia. See *pleurobranch.*

***Pleuron* (pl. *pleura*).** See *epimere.*

Pleuropod. See *precoxa.*

Podobranch. Gill arising from coxa of thoracopod; syn., *podobranchia.*

Podobranchia. See *podobranch.*

Podomere. See *segment.*

Podophthalmite. One of 2 segments of eyestalk (when segmented), bearing cornea.

Pollex. See *fixed finger.*

Pore canal. Minute tubular passageway extending through shell (Ostracoda).

Postabdomen. Terminal part of body (Cladocera).

Postcephalic groove. See *decapod carapace, groove m.*

Postcervical groove. See *decapod carapace groove n.*

Postcervical notch or incision. Strong indentation of carapace at level of postcervical groove.

Posterior cardiac lobe. See *decapod carapace, region h.*

Posterior carina. See *decapod carapace, region h.*

Posterior gastric pit. One of two small depressions dorsally near midline on exterior of carapace marking point of insertion of stomach muscle (Decapoda).

Posterior tooth. See *decapod carapace, teeth e.*

Postlarval stage. Developmental stage reached after completion of megalopal or equivalent metamorphosis; marked by initial appearance of adult characters.

Postorbital carina. See *decapod carapace, carina h.*

Postorbital spine. See *decapod carapace, spine e.*

Postrostral carina. See *decapod carapace, carina i.*

Postrostral spine. See *decapod carapace, spine f.*

Postsegmental region. See *telson.*

Precoxa. Segment of protopod proximal to coxa, rarely present; syn., *pleuropod.*

Pre-epipod(ite). Laterally directed lobe of coxa or from coxal position.

Pregastric tooth. See *decapod carapace, teeth f.*

Preischium. Segment of endopod between protopod and ischium, rarely present.

Prelateral lobe. Proximalmost lateromarginal lobe of telson (Stomatopoda).

Preoral sting. Retractile piercing mechanism with poison gland (Branchiura).

Presegmental region. See *acron.*

Preungal process. Structure at base of dactyl of 4th pereopod in many pagurids; function unknown but presumed sensory.

Prezoea. Just-hatched postnaupliar larva still covered by embryonic cuticle.

Primordial valve. One of 5 chitinous plates (scuta, terga, and carina) in cyprid larvae of lepadomorphs and verrucomorphs, site of calcification during metamorphosis, sometimes visible at umbones of these plates in adults.

Proctodaeum. See *proctodeum.*

Proctodeum. Posterior part of alimentary canal lined with cuticle of ectodermal origin; syn., *hindgut, proctodaeum.*

Proepistome. See *interantennular septum.*

Propodus. Penultimate segment of pereopod (thoracopod).

Prosartema. Scale implanted on inner margin of basal segment of antennular peduncle (Penaeoidea).

Prosoma. See *prosome.*

Prosome. Anterior region of body, commonly limited posteriorly by major articulation. In cirripeds, large saclike body in position of "head" in front of, and rostral to, thoracic appendages, supporting trophi and usually 1st cirri; syn., *prosoma.*

Protandry. Hermaphroditic condition with male elements maturing and being released before maturation of female elements.

Protocephalon. See *acron.*

Protocerebrum. Ganglion of preantennulary region.

Protogastric lobe or area. See *decapod carapace, region f.*

Protogyny. Hermaphroditic condition with female elements maturing and being released before maturation of male elements.

Protopod(ite). Proximal part of appendage, consisting of coxa and basis or less frequently of precoxa, coxa, and basis, sometimes fused; syn., *sympod.*

Protozoea. First two or three postnaupliar substages in development of Penaeiodea and Euphausiacea (cf. *calyptopis* in latter).

Proventriculus. Elaborated anterior part of alimentary canal in some crustaceans.

Proximal. Part of structure nearest midline of body or base of attachment; opposed to distal.

Pseudepipod(ite). Lateral lobe arising from distal part of basis or proximal part of exopod.

Pseudibacus. Postlarval stage of some palinurans; syn., *nisto, puerulus.*

Pseudorostrum. Anterior part of carapace formed by pair of forward projecting plates (Cumacea).

Pseudotrachea. Respiratory structure developed in pleopods of some terrestrial isopods for air-breathing.

Pterygostome. See *pterygostomial region.*

Pterygostomial region. See *decapod carapace, region k.*

Pterygostomial spine. See *decapod carapace, spine g.*

Puerulus. See *pseudibacus.*

Punctum **(pl. *puncta*).** Small pitlike depression(s) in valve surface (Ostracoda).

Pustula **(pl. *pustulae*).** Small protuberance(s) on valve surface with pore at summit (Ostracoda).

Radial lirae. See *carapace costae.*

Radius **(pl. *radii*).** Lateral part of compartmental plate when marked off from paries by change in direction of growth lines; overlaps ala of adjoining compartmental plate.

Ramus. Branch of appendage or other structure (e.g., flagellum).

Raptorial claw. Toothed dactyl, generally strong, curved backward on propodus, modified for quick motion in catching prey.

Rasp. One or several rows of chitinous plates or scales on surface of pereopodal or uropodal segments.

Receptaculum seminalis. See *seminal receptacle.*

Retinaculum **(pl. *retinacula*).** Small hook at tip of appendix interna, one of many serving to interlock right and left pleopods; syn., *cincinnulus.*

Rostral angle. Intersection of basal and occludent margins of scutum in lepadomorph Cirripedia.

Rostral carina. See *decapod carapace, carina j.*

Rostral incisure. Gap below rostrum in anterior margin of valve for protrusion of antenna (Ostracoda); syn., *rostral notch.*

Rostral latus. See *latus.*

Rostral notch. See *rostral incisure.*

Rostral plate. Anteriorly projecting, unpaired, movably articulated, median extension of carapace (Phyllocarida); antennulary portion of cephalon (Mystacocarida).

Rostral tooth. See *decapod carapace, teeth g.*

Rostrolateral. One of pair of compartmental plates typically overlapping rostrum on each side, with radii on both sides, sometimes fused with rostrum, laterally or to each other; homologous with lepadomorph rostral latus (balanomorph Cirripedia).

Rostrum. (1) Anteriorly projecting, unpaired, usually rigid median extension of carapace between eyes or ocular peduncles; (2) anterior beaklike projection of valve margins overhanging incisure or notch (Ostracoda); (3) unpaired anteroventral plate of thoracic Cirripedia; in lepadomorphs, 1 of up to 4 unpaired plates of capitulum; in verrucomorphs, compartmental plate between carina and fixed scutum; in balanomorphs, compartmental plate overlapping adjacent plates, sometimes compound, either fused with rostrolaterals or missing and formed of fused rostrolaterals; simple or compound with or without radii on each side.

Saw bristles. Heavy setae in row on gnathal lobe of mandible between molar and incisor processes.

Scale. Small calcareous plates on peduncle of lepadomorph Cirripedia; see also *scaphocerite.*

Scaphocerite. Exopod of antenna; syn., *scale, squama.*

Scaphognathite. Exopod of maxilla, often used to produce respiratory current in gill chamber.

Schizopod larva. See *mysis stage.*

Scutal margin. Edge of tergum articulating with scutum or edge of any other adjacent to scutum (thoracic Cirripedia).

Scutum **(pl. *scuta*).** Paired plate or valve of thoracic Cirripedia; in lepadomorphs, 1 on each side of occludent margin of capitulum. In verrucomorphs, of two types: fixed scutum, 1 of 4 compartmental plates; movable scutum, 1 of 2 opercular plates. In balanomorphs, 1 of 4 opercular plates.

Second antenna. See *antenna.*

Second maxilla. See *maxilla.*

Segment. Individual component of crustacean appendage connected by movable articulation with adjoining segments; syn., *podomere* (cf. *article*).

Seller groove. See *decapod carapace, groove o.*

Selvage. Middle (principal) ridge of contact margin sealing valves closed (Ostracoda).

Seminal receptacle. Diverticulum of oviduct or external pouch for storing spermatozoa delivered by male; syn., *receptaculum seminalis, spermatheca.*

Seminal vesicle. Sac in male, independent of testes, for storage of spermatozoa; syn., *vesicula seminalis.*

Serration. Irregular saw-toothed outline, e.g., on conchostracan valves.

Seta **(pl. *setae*).** Hairlike process of cuticle with which it articulates or through which it protrudes.

Sexual pore. See *gonopore.*

Sheath. Cylindrical structure inside shell, consisting of alae and thickened upper part of parietes, greatly strengthening orifice and providing for attachment of opercular membrane (balanomorph Cirripedia).

Shell fold. Part of carapace behind cephalon.

Shell gland. See *maxillary gland.*

Shield. Anterior part of cephalothorax in hermit crabs.

Simple velum. Velate structure having simple flangelike form or forming ridge (Ostracoda).

Sinus gland. Storage release site for molt-inhibiting and other hormones produced by x-organ and other sites of central nervous system.

Skeletal duplicature. Outer chitinous body cover shed during ecdysis (Conchorstraca, Notostraca, Cladocera).

Somite. Division of body, including exoskeleton, usually with pair of appendages.

Spermatheca. See *seminal receptacle.*

Spermatophore. Packet or capsule of spermatozoa.

Spur. Pendent projection from basal margin of tergum (sessile Cirripedia); also velate structure modified as flattened spinelike projection in some dimorphic genera (Ostracoda).

Spur fasciole. Nearly level, slightly depression on outer surface of tergum extending to apex in line with spur; usually delimited on 1 or both sides by narrow groove (balanomorph Cirripedia).

Spur furrow. Groove on outer surface of tergum extending to apex in line with spur, with sides sometimes folded in (sessile Cirripedia).

Squama. See *scaphocerite.*

Statocyst. Diminutive organ providing sense of balance.

Stenopodium. Slender, elongate appendage, composed of rod-like segments.

Sternum canal. Internal skeletal structure of some crabs formed by meeting of sternal apodemes of opposite sides above nerve cord.

Sternal plastron. See *sternum.*

Sternal process. Projection from midsection of sternite (Mysidacea).

Sternite. Sclerotized ventral surface of body somite.

***Sternum* (pl. *sterna*).** Structure comprised of ventral portions of somites taken together; syn., *sternal plastron.*

Stipe. Stemlike basal part of appendage with exopod sometimes squamate.

Stomatopod integumental ornamentation. Armature and ornamentation of carapace, thoracic somites, abdomen, and telson; named types include:

Carinae:

(a) intermediate. Between submedian and lateral carinae on each side of carapace.

(b) lateral. Between intermediate and marginal carinae on each side of carapace and abdomen; extending posteriorly from near anterolateral angle nearly to margin on carapace.

(c) marginal. Carinae of lateral margins on abdomen; extending along each lateral margin of carapace, often curving upward posteriorly to loop around lateral carinae.

(d) median. Middorsal carina, sometimes bifurcate anteriorly on either side of cervical groove.

(e) submedian. Slightly laterad of midline and median carina in abdominal region, not present on carapace.

Denticles:

(a) intermediate. Row of small projections between intermediate and submedian teeth on lateroterminal margin of telson.

(b) lateral. Small projection(s) at medial base of each lateral tooth on telson.

(c) submedian. Small projection(s) just laterad of midline on terminal margin of telson (medial to submedian teeth).

Grooves:

(a) cervical. Distinct transverse depression or groove in posterior half of carapace.

(b) gastric. Longitudinal pair of grooves, extending from base of rostrum to posterior margin of carapace.

Teeth:

(a) intermediate. Strong spinelike or blunt projection at distolateral angle of telson, between submedian and lateral teeth.

(b) lateral. Strong spinelike or blunt projection on lateral margin of telson distally, between intermediate tooth and prelateral lobe.

(c) submedian. Strong spinelike or blunt projection just laterad of midline on terminal margin of telson.

Stomodaeum. See *stomodeum.*

Stomodeum. Anterior part of alimentary tract, ectodermal in origin and lined with cuticle; cast with molt; syn., *foregut.*

Stridulating organ. Structure producing sound by 2 parts of exoskeleton being rubbed together; surface with ridge or tubercles or cross-ridges surface and opposing surface with single transverse ridge or tubercle.

Stylamblys. See *appendix interna.*

Style. See *telson.*

Stylet. See *caudal ramus.*

Stylocerite. Rounded or spiniform process on outer part of proximal segment of antennular peduncle; syn., *antennular scale.*

Subcarina. Unpaired plate below carina (lepadomorph Cirripedia).

Subchela. Distal end of appendage developed as prehensile structure by folding back of dactyl against propodus or some broadened part of it; also may result from propodus folded back against carpus; e.g., gnathopod.

Subchelate. Provided with subchela.

Subesophageal ganglion. Nerve plexus below esophagus.

Subhepatic carina. See *decapod carapace, carina k.*

Subhepatic region. See *decapod carapace, region l.*

Submedian carina. See *decapod carapace, carina l* or *stomatopod integumental ornamentation.*

Submedian denticle. See *stomatopod integumental ornamentation.*

Submedian groove. See *decapod carapace, groove p.*

Submedian tooth. See *stomatopod integumental ornamentation.*

Suborbital region. Narrow region bordering lower margin of orbit, sometimes defined, sometimes indistinguishable (Brachyura); see *decapod carapace, region m.*

Suborbital spine. See *decapod carapace, spine h.*

Subrostrum. Unpaired plate below rostrum (lepadomorph Cirripedia).

Suctorial structures. Mouthparts of ectoparasites modified for piercing bodywall of host and for sucking out body fluids.

Sulcus. Groove or furrow.

Supra-anal plate. Tongue-shaped, spatulate, or rounded plate produced posteriorly on dorsal side of telson (Notostraca).

Supraesophageal ganglion. Nerve plexus above esophagus, equivalent to brain.

Supraorbital carina. See *decapod carapace, carina m.*

Sutural edge. Margin of compartmental plate along suture (Cirripedia).

Suture. Line or seam at juncture of 2 compartmental plates (Cirripedia); also weakly calcified areas of integument separating at ecdysis.

Swimmeret. See *pleopod.*

Sympod(ite). See *protopod.*

Tagma (**pl.** ***tagmata).*** Major division of body (e.g., head, thorax, abdomen), each composed of varying number of somites.

Tailfan. See *caudal fan.*

Telopod. Part of appendage distal to coxa (limited usage).

Telson. Terminal portion of body (not considered to be true somite), usually bearing anus, sometimes with caudal furca; syn., *postsegmental region, style.*

Tergal fold. See *epimere.*

Tergal margin. Edge of scutum adjacent to tergum, or edge of any plate bordering tergum (Cirripedia).

Tergite. Sclerotized dorsal surface of single body somite.

Tergum **(pl.** ***terga*****).** Paired plate or valve of thoracic Cirripedia; in lepadomorphs, 1 on each side of occludent margin at apex of capitulum. In verrucomorphs, of two types: fixed tergum, 1 of 4 compartment plates, movable tergum, 1 of 2 opercular plates. In balanomorphs, 1 of 4 opercular plates.

Terminal cirri. Cirri, except first pair, located at posterior end of thorax (Ascothoracica).

Terminal claw spines. Toothlike projections of varying size at concave end of postabdomen (Cladocera).

Thelycum. External pocket on ventral side of thorax in penaeid female serving as seminal receptacle.

Thoracic appendage. Any appendage attached to somite of thorax (cf. *thoracopod).*

Thoracomere. Somite of thorax.

Thoracopod(ite). Appendage of any thoracic somite; see thoracic appendage, phyllopod, maxilliped, pereopod, syn., *cormopod.*

Thorax. Tagma between cephalon and abdomen, i.e., anterior part of trunk.

Transverse septum. Thin wall normal to longitudinal septum and parallel to basis, dividing longitudinal tubes into series of cells (balanomorph Cirripedia).

Trichobranch(ia). Gill with filamentous structure of hairlike projections from axis.

Tritocerebrum. See *metacerebrum.*

Trophi. All mouthparts of cirripeds.

Trunk. Postcephalic portion of body.

Umbo. Apical portion of either valve of bivalved crustaceans; also location on plate from which successive growth increments extend in lepadomorph Cirripedia.

Umbonal spine. Hollow, minute to large, curved, looped, or nodular spinose projection, sometimes involving entire umbo (Conchostraca).

Upcurved growth lines. Upwardly bent growth lines covering tear in shell margin at site of injury (Conchostraca).

Upper latus. See *latus.*

Upper lip. See *labrum.*

Urogastric groove. See *decapod carapace, groove q.*

Urogastric lobe or area. See *decapod carapace, region f.*

Uropod(ite). Appendage of 6th abdominal somite of Malacostraca, generally fanlike, sometimes reduced or modified; also last 3 pairs of abdominal appendages in amphipods.

Urosoma. See *urosome.*

Urosome. Part of body behind major articulation (Copepoda); also last 3 abdominal somites bearing modified appendages (Amphipoda).

Valve. Lateral part of divided carapace, commonly joined to opposite part by hingement along dorsal midline; also any one of opercular elements in sessile Cirripedia.

Vas deferens. Duct in male for passage of sperm from testis to penis.

Ventral comb. Row of setae or bristles on posteroventral margin of last abdominal somite or telson (Cephalocarida).

Ventral cup. Element of nauplius eye in nonmalacostracans.

Ventral frontal organ. Paired sensory structure associated with the nauplius eye of nonmalacostracan crustaceans.

Ventral nerve chain. See *ventral nerve cord.*

Ventral nerve cord. Ganglia or connectives in somites joined by single or double tract of nerve fibers running longitudinally beneath alimentary canal.

Vertex. Top part of head or cephalon (Cladocera).

Vesicula seminalis. See *seminal vesicle.*

Vestibule. Space between duplicature and outer lamella (Ostracoda).

Walking leg. See *pereopod.*

X-organ. Site of secretion of molt-inhibiting and other hormones. Located in eyestalks of most decapods or cephalon of sessile-eyed crustaceans. Also referred to as paired frontal organ in Anostraca.

Y-organ. Site of secretion of molting hormone in decapods.

Zoea. Larval stage characterized by natatory exopods on some or all thoracic appendages, pleopods absent or rudimentary; syn., *mysis, phyllosome, protozoea, schizopod larva.*

INDEX

Pages on which illustrations appear are indicated in italic.

Acron, 64
Acrothoracica, 3, 44–46, *45*
Ala(e), 38, 41, 52
Albuneidae, 148
Allanaspides, 71, 73
Alpheidae, 134, 136
Alpheoidea, 126
Amphionidacea, 3, 123–124, *125*
Amphionides, 124
  A. reynaudii, 123, 124
Amphipoda, 3, 109–117, *111*, *113*
Anaspidacea, 3, 70–74, *72*, *73*
Anaspides, 71, 73
  A. tasmaniae, 71, 73
Anaspididae, 71
Annulus ventralis, 138
Anomura, 126, 144–151, *146*, *147*, *149*
Anostraca, 3, 11–15, *13*, *14*
Antennal glands, 14, 36, 83, 114, 122, 131, 136, 141, 150
Anthuridea, 96, 100, 104
Appendix interna, 60, 66, 124, 136
Appendix masculina, 104, 129, 133, 136, 138
Apus. See *Triops*
Argulus, 41
Ascothoracica, 3, 54–56, *57*
Ascothorax ophioctenis, 55
Asellota, 98
Astacidea, 126, 137–142, *139*, *140*
Astacoidea, 126, 138
Austroastacidae, 143
Austroastacidea, 126, 142–143
Austroastacus, 142
  A. hemicirratulus, 142

Balanomorpha, 52–54, *53*
Balanus, 52
Basic characters
  Branchiopoda, 8
  Crustacea, 2
  Decapoda, 126
  Eucarida, 117
  Eumalacostraca, 70
  Malacostraca, 59
  Peracarida, 79
  Syncarida, 70
  Thoracica, 46–47
Bathynellacea, 75–78, *76*
Bathynellidae, 77
Bathynomus, 104, 105, 106
  B. giganteus, 100
Bellioidea, 127
Bentheuphausia, 117, 118, 120
Bivalve carapace, 16, 20, 24, 47, 55, 60
Bodotriidae, 90
Body flexure, 34
Body rings, 8, 9
Bopyridae, 107
Boreomysis, 83
Brachyrhyncha, 126, 152
Brachyura, 126, 151–158, *153*, *154*, *155*
Branchinecta, 12
  B. gigas, 12
Branchiopoda, 3, 8
Branchiura, 3, 41–43, *42*
Bresilioidea, 126
Brood pouch, 20, 60, 124, 145. *See also* Oostegites

Caenestheriella, 16
Calanoida, 3, 33, 34, 36–37
Calanus, 36, 37
Calappoidea, 126
Calmanostraca, 3
Cancridea, 126, 152
Caprellidea, 3, 110, *111*, 114–115
Caridea, 126, 133–137, *135*
Caudal appendage, 48
Caudal furca, 24, 28, 30, 55, 60, 75. *See also* Caudal rami
Caudal rami, 6, 9, 12, 18, 20, 34, 38, 43, 78
Cement gland, 15, 46, 51
Cephalic shield, 6, 34
Cephalocarida, 3, 4–8, *5*, *7*
Cephalosome, 34
Chiltoniella, 4, 6
  C. elongata, 4
Chonopeltis, 41
Chthamalus, 52
Circulatory system
  Amphipoda, 112–113
  Anaspidacea, 71
  Anostraca, 12
  Astacidea, 138, 141
  Brachyura, 152
  Caridea, 136
  Cirripedia, 51
  Cladocera, 20
  Copepoda, 34, 36
  Dendrobranchiata, 129–131
  Euphausiacea, 122
  Isopoda, 104–105
  Leptostraca, 62
  Mysidacea, 82–83
  Ostracoda, 24, 26
  Paguroidea, 148, 150
  Stomatopoda, 67
Cirri, 46, 47, 48, 49
Cirripedia, 3, 44–58, *49*, *50*, *53*, *57*
Cladocera, 3, 19–22, *21*
Classification
  Superclass Crustacea, 3
  Order Decapoda, 126–127
Clibanarius, 148
Clypeus, 12
Coenobitoidea, 126, 145, 148
Complemental male, 47, 54
Conchostraca, 3, 16–19, *17*
Copepoda, 3, 32–40, *35*, *36*, *39*
Cor frontale, 104, 122, 136, 138, 148
Coronida, 69
Corystoidea, 127
Coxal plate, 104, 112
Crangonidea, 126, 136
Cumacea, 3, 88–91, *89*
Cyclomorphosis, 19
Cyclopoida, 3, 33, 34, 36–37
Cymothoidae, 106
Cypridinidae, 24, 26–27
Cyzicidae, 16

Daphnia, 20–22
Decapoda, 3, 126–157
Dendrobranchiata, 126, 127–132, *128*, *129*, 138
Derocheilocaris, 28
Digestive system
  Amphipoda, 113–114
  Anaspidacea, 71, 73
  Anostraca, 14
  Astacidea, 141
  Brachyura, 156
  Cirripedia, 51
  Cladocera, 20
  Copepoda, 36
  Dendrobranchiata, 131
  Euphausiacea, 122
  Isopoda, 105–106
  Leptostraca, 62
  Mysidacea, 83
  Ostracoda, 26
  Paguroidea, 150
  Stomatopoda, 67
Dikonophora, 92, 94
Diplostraca, 3
Dipteropeltis, 41
Dolops, 41
Dorippoidea, 127
Dorsal organ, 9
Dorsal shield, 8, 9, 18
Dromiacea, 126
Dromioidea, 126, 151, 152
Duplicature, 24

Emerita, 148
Engaeus, 142
Ephippium (ephippia), 22
Epicaridea, 106
Eryonoidea, 126

Eucarida, 3, 117
Eucopia, 81
Eucopiidae, 81
Eumalacostraca, 3, 70
Euphausia, 120
Euphausiacea, 3, 117–123, *119, 121*
Euphausiidae, 118
External morphology
  Acrothoracica, 45–46
  Amphionidacea, 124
  Amphipoda, 110, 112, 115
  Anaspidacea, 71
  Anomura, 145, 148
  Anostraca, 12
  Ascothoracica, 55
  Astacidea, 138
  Austroastacidea, 143
  Bathynellacea, 77–78
  Brachyura, 152
  Branchiura, 41, 43
  Caridea, 134, 136
  Cephalocarida, 4, 6
  Cladocera, 20
  Conchostraca, 16, 18
  Copepoda, 34, 38
  Cumacea, 90–91
  Dendrobranchiata, 127, 129
  Euphausiacea, 118
  Isopoda, 100, 104, 107
  Leptostraca, 60, 62
  Mysidacea, 81–82
  Mystacocarida, 28, 30
  Notostraca, 9
  Ostracoda, 24
  Rhizocephala, 56
  Spelaeogriphacea, 88
  Stenopodidea, 133
  Stomatopoda, 64, 66–67
  Stygocaridacea, 75
  Tanaidacea, 92, 94
  Thermosbaenacea, 85
  Thoracica, 48, 50, 52

Filamentary appendages, 48
Flabellifera, 96–97, 106
Food basket, 118
Food groove, 9, 12
4-celled sense organ, 71
Frontal organ, 15, 18, 20, 24
Frontal plate. *See* Clypeus

Galatheidae, 145
Galatheoidea, 126, 145, 148
Gammaridea, 109–110, 112
Gastric mill, 131, 141, 150, 156
Gills (or branchial apparatus), 66, 81, 88, 90, 104, 115, 118, *126*, 129, 133, 138, 152, 156
Glypheoidea, 126
Gnathiidea, 99–100, 104, 106
Gnathophausia, 81
Gnathopod, 101, 112, 115
Gonopophyses, 70
Grapsidoidea, 127

Halobaena, 84
Hapalocarcinoidea, 127
Harpacticoida, 3, 33, 34, 36–37
Hippoidea, 126, 145, 148
Hippolyte, 136
Homoloidea, 127
Hoplocarida, 3, 63
Hutchinsoniella, 4, 6
  H. macrocanthus, 6
Hyperiidea, 110, 112

Idotea, 105
Ingolfiellidea, 110, 114
Isopoda, 3, 95–109, *101, 102, 103, 105*

Kalliapseudidae, 94
Kentrogon, 58
Koonuga, 71
Koonugidae, 71

Lacinia mobilis, 79, 81, 85, 88, 91, 94, 104, 112, 115
Larval development
  Acrothoracica, 45
  Amphipoda, 115–116
  Anaspidacea, 73
  Anomura, 150
  Anostraca, 15
  Astacidea, 141
  Bathynellacea, 78
  Brachyura, 156
  Branchiura, 43
  Caridea, 136
  Cephalocarida, 6
  Cladocera, 22
  Conchostraca, 18
  Copepoda, 37
  Cumacea, 91
  Dendrobranchiata, 131
  Euphausiacea, 122–123
  Isopoda, 106, 107
  Leptostraca, 60
  Mysidacea, 83
  Mystacocarida, 30
  Notostraca, 9
  Ostracoda, 27
  Rhizocephala, 56, 58
  Stenopodidea, 133
  Stomatopoda, 69
  Stygocaridacea, 75
  Tanaidacea, 94
  Thermosbaenacea, 85
  Thoracica, 47
Lepadidae, 48
Lepadomorpha, 47–51, *49, 50*
Lepas, 48, 50
Lepidurus, 8, 9
Leptobathynellidae, 77
Leptochelia, 92
Leptodora, 19, 20, 22
Leptodoridae, 20
Leptostraca, 3, 59–63, *61, 62*
Leuconidae, 90
Lifting spines, 85. *See also* Setal row
Lightiella, 4, 6
  L. floridana, 4
  L. incisa, 4, 6
  L. serendipita, 6
Ligia, 105
Limnadiidae, 18
Limnobaena, 84
Lithodidae, 145
Lomidae, 145
Lophogaster, 81
Lophogastrida, 79, 81
Loxoconcha, 24
Luminescent organs, 118
Lunule, 38
Lynceidae, 16
Lysiosquilla, 69

Majoidea, 127
Malacostraca, 3, 59
Maxillary gland, 9, 14, 36, 66, 73, 83, 106
Maxillary hooks, 38
Maxilliped somite, 30
Meganyctiphanes, 120
Metasome, 34
Micraspides, 71
Mimilambroidea, 127
Misophrioida, 3, 34
Monodella, 84, 85
Monodellidae, 85
Monokonophora, 92, 94
Monstrilloida, 3, 37
Myodocopa, 3, 23–27
Mysida, 79, 81
Mysidacea, 3, 79–84, *80*
Mysis, 83
Mystacocarida, 3, 28–31, *29*

Nannastacidae, 90
Nebalia, 59, 60, 62
  N. typhlops, 59
Nebaliacea, 60
Nebaliella, 59, 60, 62
Nebaliidae, 60
Nebaliopsis, 59, 60, 62
  N. typicus, 60
Nematobrachion, 120
Nematoscelis, 120
Nephropidae, 126, 138
Nephropoidea, 126, 138
Nervous system
  Amphipoda, 114
  Anaspidacea, 73
  Anostraca, 15
  Astacidea, 141
  Brachyura, 156
  Cirripedia, 51
  Cladocera, 20, 22
  Copepoda, 36
  Dendrobranchiata, 131
  Euphausiacea, 122
  Isopoda, 106
  Leptostraca, 62–63
  Mysidacea, 83
  Ostracoda, 26
  Paguroidea, 150
  Stomatopoda, 67–68
Notodelphyoida, 3, 37
Notostraca, 3, 8–11, *10*
Nuchal organ, 12. *See also* Dorsal organ
Nyctiphanes, 120

Ocypodoidea, 127
Oniscoidea, 100, 104
Oostegites, 79, 81, 83, 85, 88, 91, 94, 104, 106, 107, 112, 115
Oplophoridea, 126
Oral pyramid, 55
Ostracoda, 3, 23–27, *25, 26*
Ovisac, 9, 12, 120
Oxyrhyncha, 127, 152
Oxystomata, 127, 152

Pagurapseudes, 92, 94
Paguristes, 148
Paguroidea, 126, 145, 148
Palaemonoidea, 126
Palinura, 126, *139*, 143–144
Palinuridae, 144
Palinuridea, 144
Palinuroidea, 126, 144
Pandalidae, 136
Pandaloidea, 126
Pandalus, 136
Parabathynellidae, 77
Paranaspides, 71
Paranebalia, 59, 60, 62
Parastacoidea, 126, 138
Parastygocaris, 74, 75
Pars ampullaris, 67, 71
Parthenopoidea, 127
Pasiphaeoidea, 126
Peltogastrella, 56
Penaeidae, 126, 129, 136
Penaeoidea, 126, 127, 134
Penaeus, 127, 131
Penicillae, 75
Penis (penes), 12, 27, 48, 51, 52, 66, 88, 112, 113, 114, 115
Peracarida, 3, 79
Petasma, 66, 75, 129
Phreatoicidea, 98–99, 100, 104
Phyllocarida, 3, 59
Physetocaridoidea, 126
Pinnotheroidea, 127
Pleocyemata, 126, 132, 138
Pleotelson, 66, 77, 85, 90, 94, 104, 106
Pleuron (pleura), 6, 81, 88, 129, 134, 138
Podocopa, 3, 23–27
Poecilostomatoida, 3, 38
Pollicipes, 48
Polyartemia, 11, 12
Polyartemiella, 12
Polyartemiidae, 12
Polyphemidae, 20
Porcellanidae, 145
Portunidae, 152
Portunoidea, 127
Potamoidea, 127
Precoxa, 24, 66
Preischium, 71, 81
Preoral sting, 43
Preparatory stage, 94
Presoma, 44
Preungual process, 148
Procaridoidea, 126
Psalidopodoidea, 126
Pseudoeuphausia, 120
Pseudorostrum, 90
Pylochelidae, 148

Raninoidea, 127
Raptorial claw, 66
Reproductive system
  Amphipoda, 113
  Anaspidacea, 73
  Anostraca, 15
  Astacidea, 141
  Brachyura, 156
  Cirripedia, 51
  Cladocera, 22
  Copepoda, 36–37
  Dendrobranchiata, 131
  Euphausiacea, 122
  Isopoda, 105
  Leptostraca, 62
  Mysidacea, 83
  Ostracoda, 27
  Paguroidea, 150
  Stomatopoda, 67
Rhizocephala, 3, 56–58, *57*
Rostral plate, 60, 64, 75

Sandersiella, 4, 5
Sarsostraca, 3
Scalpellidae, 48
Scyllaridae, 144
Sergestidae, 126
Sergestoidea, 126, 127, 134
Setal row, 91, 104, 112, 129
Sexual tube, 145
Siphonostomatoida, 3, 38
Spelaeogriphacea, 3, 87–88, *89*
Squillidae, 64
Statocyst, 71, 75, 82, 96, 129, 148
Stenopodidae, 126
Stenopodidea, 126, 132–133, *126*, 134
Stenopus, 133
  S. hispidus, 133
Stomatopoda, 3, 63–69, *65, 68*
Stygocaridacea, 3, 74–75, *76*
Stygocaris, 74, 75
Stylocheiron, 118, 120
Stylodactyloidea, 126
Supernumerary segment, 148
Supra anal plate, 9, 30
Syncarida, 3, 70

Tanaidacea, 3, 91–95, *93*
Tanais, 94
Telsonic comb, 6, 28
Tessarabrachion, 120
Thalassinoidea, 126, 145, 148
Thaumastochelidae, 126
Thelycum, 129, 138
Thermosbaena, 84, 85
Thermosbaenacea, 3, 84–87, *86*
Thermosbaenidae, 85
Thoracica, 3, 46–54
Thysanoessa, 120
Thysanopoda, 118, 120
Toothed furrow, 30
Triops, 8, 9
Trophi, 48, 52

Urosome, 34, 112, 115

Valvifera, 95–96, 104
Verruca, 51
Verrucomorpha, 51, *53*

Xanthoidea, 127
X-organ, 15